Basic Statistical Concepts

THIRD EDITION

Albert E. Bartz
Concordia College · Moorhead, Minnesota

MACMILLAN PUBLISHING COMPANY
New York

To Sally,
who has always been significantly different

PRINTED IN THE UNITED STATES OF AMERICA

MACMILLAN PUBLISHING COMPANY
866 Third Avenue, New York, New York 10022

Collier Macmillan Canada, Inc.

Library of Congress Cataloging-in-Publication Data

Bartz, Albert E.
Basic statistical concepts.

Bibliography: p.
Includes index.
1. Mathematical statistics. I. Title.
QA276.12.B37 1988 519.5 87-14184
ISBN 0-02-306445-5

Printing: 7 8 Year: 2 3 4 5 6 7

Foreword

Eleven years ago, I chose to adopt the first edition of this textbook for my statistics course. What I theorized would be a good choice has turned out to be an excellent decision. Let me tell you why.

First of all, the range of coverage of topics is appropriate for an introductory quarter or semester statistics course. Whether Bartz is writing about descriptive statistics, correlation, or nonparametric tests, his presentation is clear and complete.

Second, to use some computer jargon, this textbook is "user friendly." Through his choice of words and examples, Bartz gives you a feeling of personal involvement with statistics. Readers of his book have reported that they feel as if they are engaged in a conversation with the author. Such a relationship can do much to lessen math anxiety.

Third, the textbook de-emphasizes the mathematical complexities that you may have come to expect in the study of statistics. This is accomplished without losing the book's focus on promoting your understanding of the application of statistical concepts and procedures. In case you have a fear of mathematics, the approach taken in this book lessens your desire to avoid the study of statistics.

Finally, Bartz's book is more than a presentation of statistical methods and procedures; this textbook *teaches* you statistics. Al Bartz becomes your personal instructor.

Gary Narum
Associate Professor of Education
North Dakota State University

Preface

The fact that a textbook has stayed around long enough to be a third edition should serve as a reminder to the author to proceed with caution in planning omissions, additions, and alterations in text material. Although some necessary changes have been made in this edition to faithfully represent the current state of the discipline, the three central emphases of the first edition remain:

1. *A student-centered approach.* A deliberate effort has been made to develop techniques that will assist the student in grasping statistical concepts. A series of Notes set off in colored boxes call the student's attention to helpful hints that will aid him or her in understanding concepts, avoiding errors, developing computational skills, or discovering the historical significance of statistical events. By isolating selected subject matter in this way, the student is able to associate these smaller bits of information with the larger whole, and more easily remember the connection between the two. Another learning aid is the set of Sample Problems that appears at the end of each chapter. Each one shows collected data, decision steps, calculations, and conclusions that can be drawn from the results. These techniques—in addition to the step-by-step progression from simpler to more complex material, the attempt to combine an intuitive approach with the usual mathematical development, and the Study Questions and Exercises for each chapter—should provide the learner with ample exposure to the conceptualizations and calculations necessary for proficiency in statistical techniques.
2. *Selection and presentation of topics.* A 10-year-old girl, returning a book on penguins to the public library, remarked to the librarian, "This book told me more about penguins than I really want to know." Many textbooks have included much more material than can reasonably be covered in a typical one-semester or two-quarter

course in introductory statistics. The topics chosen for this book are concepts that form the foundation for measurement in education and the behavioral sciences. However, each instructor may have individual preferences, and a different order of topics may be used or some of the later topics may be omitted or abbreviated at the instructor's discretion. Most of the chapters here are organized to include an introduction to the concept with several familiar illustrations, progressing to formulas that define or demonstrate the concept, and concluding with computational formulas, worked examples, and applications and limitations of the concept.

3. *Simplified language.* The idea for this textbook began in 1958 with a small manual, *Elementary Statistical Methods for Educational Measurement.* It was intended as a supplement for textbooks in tests and measurements and attempted to lead the student down statistical pathways in a language and style that caused as little trauma as possible. As subsequent editions appeared, more and more material was added and an increasing number of users suggested the addition of inferential techniques employing the same simplified language. Although it is one thing to discuss percentiles and an entirely different thing to describe a two-way ANOVA, I have attempted to present both kinds of material in an uncomplicated, narrative style, with a profusion of lucid examples. A considerable portion of the material may seem redundant to the mathematically sophisticated student, but, in teaching the introductory statistics course for 25 years, I have found that a certain amount of redundancy is not only desirable but essential for realizing optimum gain as the student progresses through the material. There is a difference between a "watered-down" approach and one that attempts to be rigorous yet readable. I have attempted to present the traditionally difficult and more complex material in the same manner as the easier concepts.

One change that will be immediately obvious to users of previous editions is the inclusion of the Mathematics Refresher (Appendix 2) for students who feel the need for a review of basic mathematics and wish to upgrade their arithmetic and algebra skills. I thank Julie Legler of Moorhead State University for this sensitive and thorough contribution. Another substantial change from previous editions is the inclusion of a large number of exercises at the end of each chapter, many of which are drawn from recent research projects in education, psychology, and the health sciences.

I thank Dr. N. Jo Campbell of Oklahoma State University, Dr. Joseph L. Daly of Colorado State University, and Dr. Frederick R. Korf of the U.S. International University for their critical reviews of the third-edition manuscript. Their suggestions were very helpful; however, I must bear the responsibility for the finished product. A note of gratitude is also due all the users of the second edition who filled out questionnaires and answered queries on suggestions for improvement. I am grateful to Dr. Jayne Rose of Concordia College for her critique of the material on test theory in Chapter 14. I especially thank Dr. Gary Narum of North Dakota State University and Dr. George Seifert of Bowling Green State University for their many helpful comments during the revision process. And I am grateful to my wife, Sol, for providing manuscript typing, ideas for research applications in the field of special education, and many encouraging words. Finally I thank the many students who have consented to serve as guinea pigs while different teaching methods and techniques were evaluated. Their willing participation in efforts to improve the teaching and learning of statistical methods is greatly appreciated.

Albert E. Bartz

Contents

CHAPTER 1

Some Thoughts on Measurement

"There are three kinds of lies:
lies, damned lies, and statistics."

—Disraeli

This comment by one of England's great prime ministers does not exactly inspire confidence in statistics or statisticians. It, of course, implies that part of a statistician's duties is to bend the data in some form or fashion to prove whatever point needs proving. Even if you were not acquainted with Disraeli's pessimistic viewpoint, you have at least heard statements such as "You can prove anything with statistics" or "Figures don't lie, but liars can figure."

It is not surprising, then, that the average person has a jaundiced view of statistics and statisticians. In Figure 1.1 we note that even Charlie Brown is not immune from the attitudes of the lay public toward statisticians.

The Need for Statistics

It is hoped that a first course in statistics will clear up some of the misconceptions about statistics mentioned in the opening paragraphs.

FIGURE 1.1. **Even the comics reflect the public's attitude about statistics.**

There are three general objectives that you should hope to achieve through this introductory course. It is expected that you will (1) be able to read the professional literature in your field of study, (2) appreciate the fact that statistics is a necessary tool for research, and (3) develop an increased ability to identify situations where statistics are used in an inappropriate or misleading manner. Let us examine these three objectives in detail.

Professional Literature

As your course of study progresses, you are in the process of becoming a professional in your chosen field, and you will be expected to read the professional literature in books and journals. Since statistics are used to communicate results of surveys, experiments, and tests, you must be able to understand what an author is saying by means of graphs, averages, correlation coefficients, *t* tests, and the like. You can no longer "skip the hard places," as you might have done back in grade school, but will have to be able to read—and understand what you are reading—to be a professional. And since statistics permits the most concise and exact way of describing data, you can expect that the majority of your professional reading will use a statistical analysis of some sort.

Research Tool

Statistics is a necessary tool for all research, where *research* is broadly defined. Whether you are a classroom teacher, guidance counselor, social caseworker, physiological psychologist, or personnel manager, a rudimentary knowledge of statistics is a prerequisite for analyzing data. As education and the behavioral sciences move increasingly toward a quantitative approach, the results of research studies of all kinds become the basic diet for everyone in the field. The days of armchair speculation are numbered, and more and more hard-nosed research results will be demanded as evidence for a particular point of view.

Misleading Statistics

Although we are all rather suspicious of claims using such statistics as "Three out of four doctors recommend . . ." or ". . . dissolves 50% more stomach acid than . . .," there are times when data manipulation is not so obvious. One candidate for ease of misinterpretation is the graphical method; several examples are shown in Figures 1.2 and 1.3.

Figure 1.2 shows the familiar *pictogram*, where the height of the figures represents the amount of the quantity being graphed—in this case, the number of members of several church denominations in one

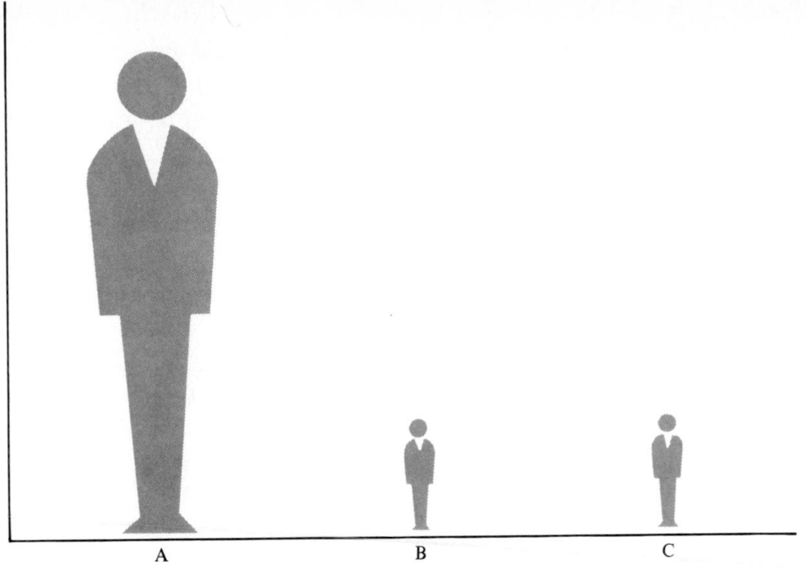

FIGURE 1.2. **Pictogram for church membership among three religious denominations.**

small midwestern state. On first examination of the graph you would probably conclude that denomination A has anywhere from 6 to 10 times as many members as either B or C. The graph is misleading because you are being influenced not by the *height* of the figures alone but by the proportionate *space* they occupy. Actually, the height of the figure representing denomination A is correct. The figure is just over 4 times taller (representing 78,426 members) than either B (18,192 members) or C (17,513 members).

Two graphs representing unemployment figures for the last 4 months of a year are shown in Figure 1.3. Both graphs are constructed from the same basic data, but they differ with respect to the units of measurement on the vertical axis. Someone who wished to demonstrate that there is no cause for alarm might represent the data by Figure 1.3a. Since the units in this graph are large, an increase of 200,000 unemployed causes little change from September to December. On the other hand, someone who wished to exaggerate the increase in unemployment would demonstrate that point by means of Figure 1.3b, where the smaller units magnify the change of 200,000.

The inappropriate use of graphical methods demonstrated above is only one example of how statistics can be misleading. As you develop expertise with the statistical concepts to be presented in the following chapters, you will become familiar with the appropriate applications

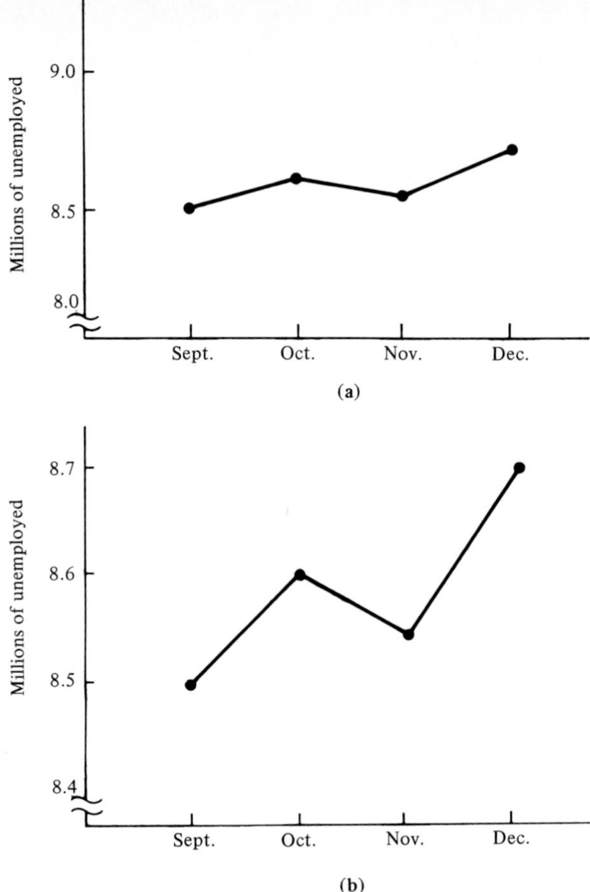

FIGURE 1.3. **Unemployment figures graphed with different units.**

for each statistic and will be sensitized to possible uses and abuses in the presentation of data.

The Importance of Measurement

The growth in importance of the physical sciences in the last few decades has been in part due to increased precision of measurement. The different teams of astronauts could not have explored the surface of the moon were it not for the ability of scientists to measure time, space, and matter with extreme precision. A visit to any facility for research in biology, medicine, or any other scientific discipline would show a large amount of floor and counter space devoted to precision-measuring devices of one kind or another.

Although we do not claim the same degree of precision as the natural sciences, measurement is of supreme importance in education and the behavioral sciences as well. The use of statistical methods enables us to make *quantitative* statements about data that could not be formulated by any other means. The football coach who says he has a "big" team is making a less precise statement than one who states that the average weight of his defensive line is 250 pounds. To say that John is older than Joe is not as informative as saying that John is about a year older than Joe. And saying that John is 11 months and 3 days older than Joe would give us information of even greater precision. Similarly, we make quantitative statements about Greg's IQ test score, Sheri's reading readiness score, Ryan's anxiety level, and Sue's college grade-point average.

So far we have been using the term *statistics* in several different ways, and at this point it might be helpful to sort out exactly what we mean. An acquaintance of mine informs his classes that "we calculate statistics from statistics by statistics." A good way to begin this course would be to clear up this apparent confusion.

1. "We calculate statistics . . ." What he means here are the results of a computation: the graphs and percentiles and correlation coefficients. These are the final figures that tell us what a particular batch of data means.
2. ". . . from statistics . . ." The term here refers to the data, the collection of numbers that constitute the raw material we work with. They may be test scores, reaction times, frequency of divorces, or number of juvenile delinquents, or what have you.
3. ". . . by statistics." The last use of the term is the process or method that we use to get our averages, percentiles, or the like. Formulas have been derived and methods have been developed so that we follow a certain procedure and perform the necessary calculations to obtain the desired result.

The emphasis in this book will be on the first and third uses of the term. As the concepts are introduced in the following chapters, we will study the *method* or *process* that is appropriate for the desired results and learn how the *end product* tells what we need to know about our data. Viewed in this way, the task of statistics *is to reduce large masses of data to some meaningful values.* Whether we are working with 20 algebra test scores, 200 freshmen entrance exam scores, or 200,000 automobile driver records, our objective is the same: to come up with some meaningful values that tell us something about our set of data.

The second use of the term, the statistics as *raw data* or *scores,* will not be discussed at any great length in the chapters to follow. Since this book is written to cover broadly the concepts applied to education and the behavioral sciences, it must be left to each individual discipline to define the kinds of data that characterize its own peculiar domain. It must be emphasized that the statistical process yielding a statistical

FIGURE 1.4. **Statisticians and computer programmers can suffer from the same malady. © 1985 Newspaper Enterprise Association, Inc.**

result will never be any better than the raw material on which it is based. Figure 1.4 illustrates the computer programmer's lament, "Garbage in, garbage out." It is a cliché to be remembered. An average or a correlation coefficient or any other statistic will never give respectability to a group of raw data that is essentially inaccurate.

From time to time we will take a look at some general considerations that may tend to bias the raw data and thus influence the accuracy of the final results. We will devote several of the remaining sections in this chapter to this problem, as well as noting the limitations of each statistic in the following chapters.

The Descriptive and Inferential Approaches

The field of statistics has traditionally been divided into two broad categories—descriptive and inferential (sampling) statistics. After you have completed this first course, you will see that these two categories are not mutually exclusive and that there is a great deal of overlap in what may be labeled "descriptive" and what may be labeled "inferential." However, it is useful at this point to distinguish these two concepts.

Descriptive Statistics

We noted earlier that the task of statistics in general was to reduce large masses of data to some meaningful values. In terms of descriptive statistics this would mean that these meaningful values *describe* the results of a particular sample of behavior. We might use statistics to describe the distribution of scores on a ninth-grade algebra exam, the distribution of weights of the entire football squad at a local high school, or the incidence of Asian flu among all college students at a particular university. Your grade-point average, minimum checking account balance (sometimes a little too descriptive), and rank in high school graduating class are also descriptive statistics. The purpose of a descriptive statistic, as you can see, is to tell us something about a particular group of observations.

Inferential Statistics

In the study of inferential statistics we find our attention shifting from describing a limited group of observations to making *inferences* about the population. We will define *sample* and *population* precisely and in more detail in later chapters, but for now let us say that a sample is a smaller group of observations drawn from the larger group of observations, the population. For example, if we are conducting a poll on voting preferences, we might choose 100 voters out of a certain neighborhood. In this case the 100 voters would constitute our sample, while *all* possible voters in that neighborhood would make up the population. The task, then, of inferential statistics is to draw inferences or make predictions concerning the population on the basis of data from a smaller sample. In the case of the poll of voters, we are not particularly interested in the 100 voters but rather in making a prediction of how *all* the eligible voters in that neighborhood would vote. Similarly, if a research project is conducted by an automobile manufacturer to determine which of two designs of rear turning signals is easier to see, a sample of 500 drivers might be tested on their reaction time to the two types of lights. The engineers are not concerned only with the sample of 500 drivers but with how these results can be used to predict how *all* drivers will react.

Whether a given statistic is descriptive or inferential depends on the *purpose* for which it is intended. If a group of observations (heights, IQ scores, birth rates) is used merely to describe an event, the statistics calculated from these observations would be *descriptive*. If, on the other hand, a sample is selected (hence the term *sampling statistics*) with the intent of predicting what the larger population is like, the statistics would be *inferential*.

This book was designed to present both the descriptive and inferential approaches. Descriptive statistics, for example, would be found in measuring classroom performance (achievement tests), stock market trends, baseball records, and vital statistics. But as much a part of our life as these statistics are, I guess we would have to say that we are most often interested in the inferential approach.

We are rarely content as students of behavior merely to *measure* some characteristic but would rather use a single observation or a small number of observations to make statements about the world in general. We may measure Cathy's IQ, Jim's grade-point average, and Tony's sociability, and, using these as descriptors, put them to good use in advisement or counseling. But quite often we go beyond a mere description and contemplate such things as how Cathy's low socioeconomic level has affected her IQ test performance, how Jim's grade-point average is related to his father's achievement motivation, and what effect Tony's being an only child has on his introversion–extroversion score. It is the inferential approach that can help us answer questions such as these, and the majority of the chapters to follow will describe techniques designed to measure these variables.

Measurement and Scales

Ever since our early childhood years we have been bombarded by numbers. We probably began by learning to count our fingers, graduated to toes, and continued into elementary school, sharpening our skills in the marvelous numerical system. We've reached a point now where we feel relatively comfortable with numbers because we have been adding and subtracting and multiplying and dividing for years.

Occasionally, the numbers that we worked with in arithmetic classes represented something concrete like ducks or money or people, but more often than not we worked with just simple rows or columns of digits. This bothered us not at all and we performed arithmetic operations on acres of numbers with the sole objective of getting the right answer.

We would now like to apply our quantitative skills to observations of behavior, but before we perform arithmetic manipulations on numbers that represent something (e.g., test scores) we have to be aware of the properties that these numbers have. We cannot blindly manipulate a set of numbers without knowing something about their properties and what the numbers represent. Somewhere back in your arithmetic training a teacher no doubt said you couldn't add apples and oranges together. This probably didn't make much of an impact on you at the time (other than the fact that you have successfully avoided adding apples and oranges), but the statement could not be more true than in the field of measurement. We must be sensitive to precisely what the numbers represent.

Measurement is formally defined as the *assignment of numbers to objects or events according to certain prescribed rules.* Once this has been done, the numbers have certain properties of which we must be fully aware as we perform the arithmetic operations. We treat these numbers differently and with varying degrees of precision, depending upon what they represent. The number 9 representing the hardness of a rock sample is different from a glove size of 9 or nine items correct on a short geography quiz.

A very convenient way of looking at the properties that a given number may have is to examine four kinds of scales that describe different levels of precision in treating numbers. These are called, in increasing order of precision, *nominal, ordinal, interval,* and *ratio* scales. Each type of scale represents a way of assigning numbers to objects or events, and the description of each type that follows shows the rules that were applied.

Nominal Scale

The simplest and most elementary type of measurement uses the nominal scale. Numbers are assigned for the sole purpose of differentiating one object from another. Steve has a book locker labeled 80, Joe wears basketball jersey number 22, and Dave is taking his motorbike for a trip on U.S. Highway 40. The nominal scale has only the property of differentiating one object or event from another.

In fact, it would not be necessary to use *numbers* to differentiate the categories. We could just as easily use letters of the alphabet, nouns, or proper names. When we categorize groups as male/female or freshman/sophomore/junior/senior, for example, we are categorizing according to a nominal scale.

When numbers are used in a nominal scale, we would not dream of adding them together or trying to calculate an average, because this scale does not have the necessary properties to enable us to do so. People would start avoiding you if you went around stating that the average basketball jersey number at the NCAA tournament was 37.63 or that the sum of the U.S. Interstate Highway numbers traveled on your vacation last summer was 208! In a similar vein book locker number 80 is not twice as large as book locker 40, and Joe's basketball jersey number 22 does not mean he is only half as good as his teammate wearing jersey number 44.

Such examples may seem to be belaboring the obvious, but these extreme situations are used to make you fully aware of the single property of the nominal scale—that of differentiating one object or event from another. Crude as it may be, this scale is still a form of measurement and will be useful in certain statistical techniques.

Ordinal Scale

As the term implies, measurements in an ordinal scale have the property of *order*. Not only can we differentiate one object from another as in the nominal scale, but we can specify the *direction of the difference*. We can now make statements using "more than" or "less than," since our measuring system has the property of order, and the objects or events can be placed on a continuum in terms of some measurable characteristic.

The *ranking* of objects or events would be a good example of an ordinal scale. A third-grade teacher, for example, might rank Ann, Beth, and Cindy as 1, 2, and 3, respectively, on the trait of sociability. As a result of the teacher's ranking, we know that Ann is more sociable than Beth and Beth is more sociable than Cindy. This is a definite improvement over the nominal scale, where no indication of the *direction* of the difference could be made. Under a nominal scale, all we could say is that Ann, Beth, and Cindy are different in sociability.

But note that our scale still does not permit us to say *how much* of a difference exists between two or more objects or events. Ann may be just slightly more sociable than Beth, while Cindy may be considerably less sociable than either. Yet the ranks of 1, 2, and 3 do not communicate this information. In this example, there would be hardly any difference between ranks 1 and 2, while there would be a large difference between ranks 2 and 3.

As another example, consider an art critic ranking three paintings in terms of their realism. He could conceivably give the most realistic a rank of 1, the next most realistic a rank of 2, and so on. According to the critic, the paintings may have been very reliably put into their proper order in terms of realism, but there is no way to assess the *amount* of realism or to say that one is twice as realistic as another. This is an example of an ordinal scale.

Interval Scale

The most important characteristic of an interval scale is equality of units. This means that there are equal distances between observation points on the scale. Not only can we specify the direction of the difference, as we did in the ordinal scale, but we can indicate the amount of the difference as well. A very common example of an interval scale is our familiar Fahrenheit temperature scale. There are equal intervals between the points on the scale, and the difference between 30° and 34° is the same as the difference between 72° and 76°. We can also say that an increase from 50° to 70° is twice as much as an increase from 30° to 40°. With the interval scale we can now assess the *amount* of the difference between two or more objects or events, something we could not do with the ordinal scale.

Many of the measurements that characterize education and the behavioral sciences are of the interval type. For example, most test scores (in terms of number of items correct) are treated as if they were based on interval measures.

Ratio Scale

The ratio scale has all the characteristics of the interval scale plus an *absolute zero*. With an absolute zero point we can make statements involving ratios of two observations, such as "twice as long" or "half as fast."

In the Fahrenheit temperature scale discussed above, we *cannot* say, for example, that a temperature of 80° is twice as warm as one of 40°. This is due to the fact that the Fahrenheit temperature scale has only an arbitrary zero, not an absolute zero. A ratio scale must have a meaningful absolute zero. If Carla can jog 2 miles in 12 minutes, but it takes Lisa 24 minutes, we know that Carla can run twice as fast. But in intelligence testing, for example, we cannot say that a person with an IQ of 100 is twice as intelligent as someone with an IQ of 50. This is because zero intelligence cannot be defined, and thus there is no absolute zero point. Most physical scales such as time, length, and weight are ratio scales, but very few behavioral measures are of this type.

Comparing the Four Scales

Table 1.1 is a summary of the characteristics of the four types of scales, and, as you can see, each successive scale possesses the characteristics of the preceding scales.

It is easy to get the impression that the ideal measurement would be in a ratio scale and that measurements in the other three scales are something to be avoided if at all possible. Although this may be the theoretical goal or ideal of measurement purists, we are faced with the fact that most behavioral measures miss this ideal by a significant amount, and practical considerations require our making the most of the precision that we do have. As was stated earlier, many of the measurements that we deal with in education and the behavioral sciences are treated as interval measures. However, a good part of our data is of the ordinal type, and even the nominal scale serves us very well. If a teacher ranks his students on "social participation," or a poll taker asks her respondents to indicate their preferences of political candidates' views on some national issue, or a football coach judges his backfield candidates on their proficiency, all are making use of ordinal measures. And any study involving categorization according to some attribute (men/women, freshman/senior, alcoholic/social drinker/teetotaler) would involve the nominal scale.

TABLE 1.1. **Characteristics of Measurement Scales**

Scale	Properties
Nominal	Indicates a difference
Ordinal	Indicates a difference Indicates the direction of the difference (e.g., more than or less than)
Interval	Indicates a difference Indicates the direction of the difference Indicates the amount of the difference (in equal intervals)
Ratio	Indicates a difference Indicates the direction of the difference Indicates the amount of the difference Indicates an absolute zero

As the various statistical methods are covered in the chapters to follow, you will notice very soon that a lot of information can be obtained from measurements that do not meet the theoretical ideal of a ratio scale. The important thing is that you must be aware of what sort of scale a given body of measurements represents and use the statistical procedures that are appropriate for that scale. After you have done this, the statistics that you do obtain can provide a wealth of meaning about a given set of measurements, whether they are nominal, ordinal, interval, or ratio in nature.

Some Statistical Shorthand

Statistics, like many other disciplines, has developed its own set of symbols or notation to help summarize or condense words and statements in an efficient manner. As the different statistical concepts are presented in the chapters to follow, each abbreviation or symbol will be defined and will continue to be used throughout the book. One set of symbols is so common that it might be a good idea to spend a little time at this point covering them in some detail, because they will be appearing in almost every chapter. These symbols are collectively called *summation notation*, and they are simply a way of expressing some simple mathematical operations in a very efficient way. Those of you with a mathematics background that included summation notation may skip this section.

Variables

Table 1.2 shows a group of test scores made by a number of students. The test score is the *variable* (so called because the scores *vary* from student to student) and is denoted by the letter X. Each test score in the column is one of the values of the variable X.

TABLE 1.2. **Summation Notation**

Student Number	X (Test Score)
1	72
2	31
3	83
4	42
5	57
6	91
7	42
8	31
⋮	⋮
i	X_i
⋮	⋮
N	X_N

Adding all values: $(X_1 + X_2 + X_3 + \cdots + X_N) = \sum_{i=1}^{N} X_i$

Adding the first five values: $(X_1 + X_2 + X_3 + X_4 + X_5) = \sum_{i=1}^{5} X_i$

Adding the third through the sixth: $(X_3 + X_4 + X_5 + X_6) = \sum_{i=3}^{6} X_i$

In order to identify *which* value of X we are considering, we give a tag or label to each value by attaching a *subscript* to X that corresponds to its position in the column (Table 1.2). X_1 is the value of the test score in the *first* row and student No. 1 received a score of 72, so we say that $X_1 = 72$. Similarly, X_3 is the score of student No. 3, so $X_3 = 83$. In order to be able to express the *general* case of any student's score, we use the subscript i. Thus X_i (read "X sub i") means *any* value of the variable, and it corresponds to a score made by *any* student who has the general number designated by i. So X_i could be the score made by the 5th student, the 23rd student, or the 99th student. If you have trouble visualizing the ith student, just picture the values of X as belonging to particular people. Then your values of X could be X_{Joe}, X_{Pete}, and so on, while X_i would be X_{Someone}!

The last row in a set of numbers always contains the final observation. This row is denoted as N, and the corresponding score is X_N. So the score of the Nth student in Table 1.2 is X_N. If there were a total of 50 students, N would be 50, and the score of the 50th student would be X_{50}.

Since N is always the last row containing an observation, the symbol N is also used to denote the number of observations. If there is a group of 75 scores, we say that $N = 75$, or if 29 rats are run in a maze-learning experiment, $N = 29$. It is a convenient way of expressing the *size* of the group.

The letter X is usually used to express a variable quantity (test scores, IQs, heights, etc.), but theoretically we could use A, B, C, J, K, or any uppercase letter to stand for the variable. Traditionally, the letters X, Y, and Z have been used in statistics to express a variable, with the letter X being the most common. In a later chapter we will have occasion to use both X and Y, when we need to identify two variables that are being studied simultaneously—such as height and weight or grade-point average and socioeconomic level.

Summation Notation

Many statistical techniques require the *addition* of columns or rows of numbers, so it is worthwhile to examine a shorthand way of expressing addition. This shorthand method is called *summation notation*.

The Greek letter sigma (Σ) is used to signify the addition of a group of quantities, and it means "sum of." Thus ΣX means "sum of the X values." If the X values are 5, 17, 22, and 25, $\Sigma X = 69$. It is just a shorthand way of saying. "The sum of the X values is 69."

It is often convenient to express quantities in *general*, so that the terms apply to any group of numbers. It is for this reason that it is necessary to define summation notation a little more precisely. Let us consider the case where we have a *single* column (or row) of numbers such as is shown in Table 1.2.

The symbol for adding any row or column of X values is $\sum_{i=1}^{N} X_i$.

This rather imposing little set of symbols simply says, "Add up the values of X, starting with the first and stopping with the one called N." Note that the subscript of X is i. The $i = 1$ under the sigma symbol is the *starting* point for the addition. It is, in effect, telling you to start adding with the first value of X (when $i = 1$), go on to the next, and continue adding each value to the total. The *stopping* point is the value at the top of the sigma symbol. Since N is the stopping point, you are to keep adding the column until you get to the last value, or N.

In other words, when you see $\sum_{i=1}^{N} X_i$, it is saying, "Take the value X_1, then X_2, and add the two. Then take X_3, and add it to the previous two. Then take X_4, and add it to the total, and so on, until you come to the value labeled X_N. You add X_N to the total and then you stop." You have then completed the operation that $\sum_{i=1}^{N} X_i$ calls for.

Suppose that you want the sum of the first five values. You would then want to begin with $i = 1$ as usual and end with the fifth value, so your summation sign would be $\sum_{i=1}^{5} X_i$. If this were applied to Table 1.2, you would take $X_1 = 72$ and add $X_2 = 31$, $X_3 = 83$, $X_4 = 42$, and stop with $X_5 = 57$. So the value of $\sum_{i=1}^{5} X_i$ would be $72 + 31 + 83 + 42 + 57$, or $\sum_{i=1}^{5} X_i = 285$. In a similar fashion $\sum_{i=3}^{6} X_i$ would start with $X_3 = 83$ and would sum all values through $X_6 = 91$, so $\sum_{i=3}^{6} X_i = 273$.

In an introductory textbook such as this, most of the uses of the operation of addition will involve adding *all* the numbers in a *single* column, so there is little need to specify the starting and stopping points in a summation operation. If you are going to add *all* the numbers, the starting point will always be $i = 1$ and the stopping point will always be N. For that reason it is customary to omit the limits from the summation symbol and use only Σ, since what is to be added is understood. For example, when ΣX is called for, it simply means that you are to add all the X values. Since ΣX is easier to write and less confusing to read than $\sum_{i=1}^{N} X_i$, we will use the abbreviated form throughout the text except for one occasion in Chapter 11. However, this one occasion will be very important, so it was essential to present the elements of summation at this point.

Summation Rules

There will be times when we will have to manipulate several formulas during the course of the remaining chapters, so it is essential to cover some rules governing the operation of summation. Three rules are presented here along with several examples. Make sure you understand the basic concepts involved in these rules, for they will be helpful in the coming chapters.

RULE 1. The sum of a constant times a variable is equal to the constant times the sum of the variable. $\Sigma CX = C\Sigma X$.

Arithmetic Example:

Given the following four numbers added together:

$$7 + 3 + 4 + 8 = 22$$

let the constant C be equal to 5 and multiply each number above by the constant:

$$\begin{aligned} (5)7 + (5)3 + (5)4 + (5)8 &= \\ 35 + 15 + 20 + 40 &= 110 \end{aligned}$$

But you can get the same result by multiplying the original sum by the constant 5:

$$22 \times 5 = 110$$

Algebraic Proof:

The sum of a constant times a variable X would be

$$\Sigma CX = CX_1 + CX_2 + CX_3 + \cdots + CX_N$$

But collecting terms and factoring would give

$$\Sigma CX = C(X_1 + X_2 + X_3 + \cdots + X_N) = C\Sigma X$$

So the sum of a constant times a variable (ΣCX) is equal to the constant times the sum of the variable ($C\Sigma X$), or $\Sigma CX = C\Sigma X$.

RULE 2. The sum of a constant is equal to N times the constant, where N is the number of things added. $\Sigma C = NC$.

Arithmetic Example:

Let the constant C be equal again to 5 and be added six times:

$$5 + 5 + 5 + 5 + 5 + 5 = 30$$

But collecting and noting that the constant was added six times ($N = 6$):

$$6 \times 5 = 30$$

Algebraic Proof:

All values of C are the same, and they are added N times, so

$$\Sigma C = C + C + C + \cdots + C = NC$$

RULE 3. The summation sign operates like a multiplier on quantities within parentheses. $\Sigma(X - Y) = \Sigma X - \Sigma Y$.

Arithmetic Example:

Given a number of pairs of X and Y values, subtract each Y from its paired X value and add the columns:

X	Y	$X - Y$
6	3	3
7	2	5
9	2	7
4	2	2
26	9	17

Then note that the sum of $X - Y$ is equal to 17, which is the same result as the sum of $X(26)$ minus the sum of $Y(9)$.

Algebraic Proof:

Given a number of $X - Y$ pairs, add them together:

$$\Sigma(X - Y) = (X_1 - Y_1) + (X_2 - Y_2) + (X_3 - Y_3) + \cdots + (X_N - Y_N)$$

Collect the common terms:

$$\Sigma(X - Y) = (X_1 + X_2 + X_3 + \cdots + X_N) - (Y_1 + Y_2 + Y_3 + \cdots + Y_N)$$

and add the X's separately and Y's separately:

$$\Sigma(X - Y) = \Sigma X - \Sigma Y$$

To the Students—How to Study This Stuff

It is a strange phenomenon, but a surprising number of college students who enthusiastically tackle such diverse college subjects as archaeology, music theory, and metaphysics are afflicted with terminal insecurity when confronted with a required course in statistics. A "weakness" in mathematics is often given as an excuse, and this would be a legitimate reaction to a course that stressed the mathematical theory underlying each concept and the derivation of all formulas for a given statistic. However, this textbook for your current statistics course assumes no mathematical training beyond your first high school algebra course. Also, when it is necessary to manipulate some mathematical symbols, each step is shown and carefully explained.

But there is no denying the fact that a first course in statistics can be difficult, simply because the subject matter is new to most of you. As a result, this book has been organized deliberately to facilitate the understanding of each new concept. The beginning chapters contain the simpler material, and succeeding chapters become progressively more complex. However, there is a logical progression from one chapter to the next, so even the more complex concepts will not prove too difficult if you make certain that you completely understand the material in each chapter before going on.

Teaching by example has been shown to be an effective educational tool, so plan to spend time looking over each example and coordinating it with the text narrative. The *Study Questions* will help you to focus your reading on the important concepts in each chapter. The *Sample Problem* at the end of each chapter should help tie together the concepts presented, and there is an ample supply of *Exercises* that will sharpen your skills on the various concepts in each chapter. The answers to selected exercises are given in the *Appendix*.

Because the material in the earlier chapters is somewhat simpler, you may have a natural tendency to devote less time to these chapters and to neglect the exercises. This could be a serious mistake, since it is the most frequent reason given by students for failing or dropping the course. Because the later chapters will require more time and effort, it is essential not to fall behind in your work at any time in the course. If you keep up with the daily assignments on both the reading and the problems, you will find that the material will be less difficult to learn and much easier to remember.

Concluding Remarks

It would be unrealistic to expect a first course in statistical methods to transform a beginner into a polished researcher or research analyst. If you find that one of your primary interests is conducting research, you will undoubtedly choose one or more additional courses in this area. Since different textbooks and different instructors may use other systems of notation and methods of approach in the advanced courses, it is difficult to write an introductory textbook that will prepare everyone equally well for future courses that they may take. The approach used in this book is to present the material in an intuitive fashion, stressing the understanding of basic concepts, so that there will be maximum transfer to the next course you take. Symbols may differ and formulas may be slightly altered, but if you understand the elementary principles presented you should be able to make the transition to an advanced course with a minimum of discomfort.

STUDY QUESTIONS

1. Why is a pictogram often misleading in representing the number in each of several categories?
2. In general, what is the "task" of statistics?
3. In what ways are descriptive and inferential statistics the same? In what ways are they different?
4. Give an example of each of the four types of scales.
5. What does the symbol X_i indicate?

EXERCISES

1. Indicate whether the tasks described would involve nominal, ordinal, interval, or ratio scaling.
   a. Measuring the time (to the nearest second) for a 10-kilometer run.
   b. Assigning room numbers at a new motel.
   c. Measuring square footage in classrooms at an elementary school.
   d. Assigning grades on a multiple choice test taken by a class in European history.
   e. Ranking a French poodle as the best of breed at a dog show.
   f. Explaining the Celsius temperature scale.
   g. Choosing the best cherry pie from 5 entries at the county fair.
   h. Distributing automobile license plate numbers at a motor vehicle registration office.
2. Classify the following types of data as nominal, ordinal, interval, or ratio.
   a. Fourth graders at an elementary school who have not had chicken pox.
   b. Number of items correct for 27 students on a mathematics aptitude test.
   c. Time required to depress the brake pedal in a driving simulator after consuming 3 ounces of alcohol in 30 minutes.
   d. Tags with pet identification numbers distributed after rabies vaccination.
   e. Daily high temperatures for the month of August in degrees Fahrenheit.
   f. Position at the finish line for the first four runners in a 1500-meter run.
   g. List of the heights of 52 Colorado mountain peaks that are over 14,000 feet.

3. Use summation notation to simplify the following.
a. $X_1 + X_2 + X_3 + X_4$.
b. $X_1^2 + X_2^2 + X_3^2$.
c. $X_3 + X_4 + X_5 + X_6$.
d. $X_1 + X_2 + X_3 + \cdots + X_N$.

4. What summation symbol would you use to express the following operations on a column of test scores (X)?
a. You want to add all the scores from the 7th one through the 15th.
b. You want to add the first 8 scores in the column.
c. You want to add all of the scores in the column.
d. You want to square each of the scores and add all of them.

5. Describe what operations are indicated by the following symbols.
a. $\sum_{i=2}^{20} X_i$.
b. $\sum_{i=5}^{7} X_i$.
c. $\sum_{i=1}^{N} Y_i^2$.
d. ΣX.

6. Use the summation rules to arrive at the final form of the following expressions.
a. $\Sigma(X - C)$.
b. $\Sigma(X + Y + C)$.
c. $\Sigma(CX)^2$.
d. $\Sigma(X + Y)^2$.

CHAPTER 2

Frequency Distributions and Graphical Methods

In the last chapter we noted that one of the tasks of statistics was to describe a mass of numbers in a meaningful way. The mass of numbers may represent such diverse measurements as test scores, survey results, medical records, or reaction times—but they are similar in that the final computer printouts, scoring sheets, or file folders result in a batch of numbers that defy any reasonable attempt to make sense out of them. Consider Table 2.1, which shows the results of a unit exam in a statistics course.

TABLE 2.1. **Statistics Unit Exam Scores**

69	70	72	62	78
71	85	72	73	91
71	61	85	82	82
82	81	74	79	90
66	88	82	86	83
89	94	86	76	75
81	79	93	76	80
68	81	64	87	80
95	75	84	90	92
88	97	86	68	67

Notice how difficult it is to get any meaning out of this collection of scores. With some effort we can find the highest score, 97, or the lowest score, 61. But we would be hard pressed indeed to find out where the concentrations of scores were, or how many students scored above 85, or how many people had a score of 82. In short, this collection of data gives us very little information about the performance of the group on the statistics exam. So, to bring a semblance of order to a group of observations, we resort to a technique that is often used to display this kind of information in a meaningful way—the *frequency distribution*.

Basically, the frequency distribution is simply a table constructed to show *how many times* a given score or group of scores occurred.[†] For example, we could set up a table where the highest score is at the top and the lowest at the bottom, with all possible scores in between, and indicate how often each score occurred. Such a table is called a *simple frequency distribution*; this technique is shown for the statistics scores in Table 2.2.

TABLE 2.2. **Simple Frequency Distribution of Statistics Exam Scores**

Score	*f*	Score	*f*
97	1	78	1
96	0	77	0
95	1	76	2
94	1	75	2
93	1	74	1
92	1	73	1
91	1	72	2
90	2	71	2
89	1	70	1
88	2	69	1
87	1	68	2
86	3	67	1
85	2	66	1
84	1	65	0
83	1	64	1
82	4	63	0
81	3	62	1
80	2	61	1
79	2		

Although the simple frequency distribution may be convenient for a teacher assigning grades, the *pattern* of the distribution does not emerge as it should. It is still difficult to readily see the concentrations of scores in such a distribution.

The most common form of the frequency distribution is the *grouped frequency distribution*, which is shown for the statistics exam scores in

[†]We will often use the term *scores* even though the numbers may represent heights, weights, reaction times, heart rates, or numbers of eyelashes, for instance.

Table 2.3. Note that instead of displaying how often *each* score occurred, the scores have been grouped into intervals (thus the name *grouped* frequency distribution) and the number in the frequency column *f* tells how many scores are in the given interval. For example, in Table 2.3, 8 students had scores from 90 through 99.

TABLE 2.3. **Poor Illustration of a Grouped Frequency Distribution of Statistics Exam Scores**

Scores	*f*
90–99	8
80–89	20
70–79	14
60–69	8
	N = 50

Unfortunately, Table 2.3 has only four intervals, resulting in a frequency distribution that does not tell us very much about the pattern of distribution of our exam scores. It would be better to have a frequency distribution with more and smaller intervals. Such an example is shown in Table 2.4, where there are now eight intervals. Notice that this frequency distribution gives us a better idea of the spread and concentration of our scores. We can easily see the clustering of scores about the center of the distribution, with fewer and fewer scores at the extreme ends of the distribution.

TABLE 2.4. **Better Illustration of a Grouped Frequency Distribution of Scores**

Scores	*f*
95–99	2
90–94	6
85–89	9
80–84	11
75–79	7
70–74	7
65–69	5
60–64	3
	N = 50

Defining Some Terms

Before we can get at the business of constructing frequency distributions, we must define some concepts that are essential to understanding the nature of the data to be displayed in a frequency distribution. The concepts described will all refer to the frequency distribution of the statistics exam scores of Table 2.4.

Real Limits and Apparent Limits

The top interval in Table 2.4 is 95–99, with a frequency of 2. In other words, two students scored somewhere from 95 through 99, the *apparent limits* of this interval. The next interval shows six scores from 90 through 94. However, to preserve the continuity of our measuring system for calculations to be discussed later on, we cannot leave a gap between 94 (the top score in the 90–94 interval) and 95 (the lowest score in the 95–99 interval). For this reason, it is understood that the *real limits* of any interval extend from $\frac{1}{2}$ *unit below the apparent lower limit to* $\frac{1}{2}$ *unit above the apparent upper limit*. Thus the real limits of the 95–99 interval are 94.5 and 99.5, while the real limits of the 90–94 interval are 89.5 and 94.5. The real lower limit is designated L and the real upper limit is U. So for the 60–64 interval, $L = 59.5$ and $U = 64.5$. There will be more said about real limits in a later section on continuous and discrete data.

Midpoints

The exact center of any interval is called its *midpoint*, abbreviated *MP*. The *MP* of any interval is found by adding the apparent upper limit to the apparent lower limit and dividing by 2. In Table 2.4, the *MP* of the 75–79 interval is 77 (75 + 79 divided by 2), and the *MP* for the 80–84 interval is 82.

Interval Size

The *size* of the interval, denoted by the symbol i, is the distance between the real lower limit and the real upper limit. In other words you can determine the size of the interval simply by subtracting L from U. For example, in Table 2.4, using the 70–74 interval, you would calculate $i = U - L = 74.5 - 69.5 = 5$. Note that i is the same for all the intervals in Table 2.4. And, in Table 2.3, you will note that the interval size is 10 ($99.5 - 89.5 = 10$).

Frequency

It should be obvious by now that frequency (the symbol is f) simply indicates how many scores are located in each interval. In Table 2.4, $f = 9$ for the 85–89 interval, indicating that nine students scored between 85 and 89.

Number

As was mentioned in the last chapter, in connection with summation notation, the number of scores in a distribution is denoted by the symbol N. In Table 2.4, $N = 50$. Note that N is the total of all the frequencies for the different intervals; that is, $\Sigma f = N$.

Constructing the Frequency Distribution

The first step in constructing a frequency distribution from any group of data is to locate the highest and lowest score. For the data of Table 2.1, the two extremes are 97 and 61. This gives us $97 - 61 = 36$. Then we add 1 to the difference between the highest and lowest scores ($36 + 1 = 37$), because simple subtraction ignores the real limits of a score. Technically, we should subtract 60.5 from 97.5 to obtain the range, 37. A convenient rule of thumb is to have from 8 to 15 intervals, depending on the size of the group. With only 50 scores 8 intervals will serve very nicely, while for a group of 200 you might get more information from your frequency distribution if you used from 12 to 15 intervals.

After you have calculated the range between the highest and lowest scores, the next step is to determine i, the size of the interval. This you do by dividing the range by the number of intervals you wish to employ. For the statistics exam scores the range of 37 divided by 8 would give 4.6 or 5, which is the interval size used in Table 2.4. *The choice of the number of intervals and the size of the interval is quite arbitrary.* Had you decided to use 10 intervals for the statistics exam scores, the interval size would be $i = \frac{37}{10} = 3.7$ or 4. However, 4 is seldom used for i, and most commonly you will see i values of 3, 5, 10, 25, 50, and other multiples of 10.

Since the choice of the number of intervals and i is arbitrary, your main concern is to have your frequency distribution display as much information as possible concerning concentrations and patterns of the scores. Obviously, the top interval should contain the highest score, and the bottom interval the lowest score. The bottom interval begins with a multiple of the interval size. The lowest score among those of

Table 2.4 might be 60, 61, 62, 63, or 64, but the interval would still begin at 60.

After you have determined the size of *i* and the number of intervals you will use, you simply place the intervals in a column labeled "Scores" at the left side of a worksheet and then begin going through the data, placing a tally mark by the interval in which each score lies. Each entry in the frequency column is simply the addition of these tally marks for each interval. Your original worksheet for the statistics exam scores would look like Table 2.5, and the finished product would, of course, look like Table 2.4.

TABLE 2.5. **Worksheet for Frequency Distribution of Statistics Exam Scores**

Scores	Tally	f
95–99	||	2
90–94	|||| |	6
85–89	|||| ||||	9
80–84	|||| |||| |	11
75–79	|||| ||	7
70–74	|||| ||	7
65–69	||||	5
60–64	|||	3
		N = 50

Continuous Versus Discrete Data

The data that constitute the raw material for studies in education and the behavioral sciences are numbers representing quantities such as reaction time, IQ, age, test scores, and the like. It is convenient at this time to point out that some of the measurements that describe an individual's characteristics are *continuous* and some are *discrete*.

If the precision of a measurement or observation depends upon the accuracy of the measuring instrument, we say we have continuous data. For example, in measuring an individual's reaction time in pushing the brake pedal on a driving simulator in response to a red light, we may get a reaction time of 0.32 seconds. However, if we had a more accurate clock we might measure the same reaction as 0.324 seconds, or even as 0.3241 seconds. Or, in a similar vein, a person's height might be 5 feet

10 inches, but with a better measuring instrument we might find it to be 5 feet 10.3 inches or even 5 feet 10.317 inches.

In any case we usually report just how accurate our measures are. If our reported reaction time is 0.32 seconds, we note that the measure is accurate to the nearest hundredth of a second. If our height is 5 feet 10 inches, we note that it is accurate to the nearest inch. Whenever we work with continuous data, we must be conscious of the *real limits* of the reported values, which extend from one-half unit below the reported value to one-half unit above. In the example above, any reaction time between 0.315 and 0.325 is reported as 0.32 seconds. Or any height from 5 feet 9.5 inches to 5 feet 10.5 inches would be reported as 5 feet 10 inches. If the measurement answers the question "How much?", we are talking about continuous data.

However, if data are of the *frequency* or *counting* type (and answer the question "How many?"), we are speaking of *discrete* data. If there are 47 divorces for 100 marriages, or 17 auto fatalities in December, or 183 cases of Asian flu, then we have discrete data. There are no "in between" values, since you cannot have 0.35 of a divorce, or $\frac{1}{2}$ a case of the flu. This does not stop the statistician from reporting fractional values, and you may have read of an average family size of 4.08 persons, or that the average worker may change jobs 3.8 times before retiring. This, of course, can be meaningful, provided you remember that the basic data consisted of discrete numbers. Later on in this book you will find that for purposes of calculation it will be necessary to assume that, for example, 3 children means between 2.5 and 3.5 children—but deep in our hearts we know that these are discrete data, and the fractions are for computational purposes only.

One very common type of measure deserves special consideration—that of test scores, especially those reported in number of items correct. If a student gets 43 items correct out of a possible 60 items, it sounds as if we are dealing with discrete data, since you cannot get part of an item correct. However, test scores in terms of items correct are considered continuous data, *since the underlying variable that is being measured is assumed to be continuous.* A 60-item test may be measuring mathematics achievement, and it is assumed that the amount of achievement is continuous, with a score of 43 extending from 42.5 to 43.5. We are simply assuming that some students scoring 43 may have slightly less achievement than what is represented by the 43 but still greater than what is represented by a score of 42. Those with slightly more achievement than a score of 43, but not quite enough to obtain a score of 44, would still get a score of 43. So, when the underlying variable is assumed to be continuous, it is meaningful to consider a score of 43 extending from 42.5 to 43.5.

Our noble ancestor who first uttered the now well-worn cliché about a picture being worth a thousand words was undoubtedly looking at the first graph. One can hardly read an assignment in a textbook or scan a newspaper or magazine without running into a graphical presentation of one form or another. Because graphical methods are such an important part of the statistical tools of the educator or behavioral scientist, it is essential that we spend considerable time and energy on this topic. There are many different kinds of graphs, but we will focus our attention on two main types: graphs that display the frequency distribution and graphs that represent a functional relationship between two variables.

By inspecting the frequency distribution of the statistics exam scores in Table 2.4, we can see that a few people made low scores and a few made high scores, but the majority of the scores are concentrated toward the middle of the distribtion. It is difficult, however, to picture the entire distribution as a whole. For example, certain irregularities in the distribution may easily escape a casual glance. Since a pictorial representation of the data enables us to see the pattern of the distribution almost instantaneously, it is desirable to graph the frequency distribution. The most common graphical methods for the frequency distribution are the *histogram* and the *frequency polygon*.

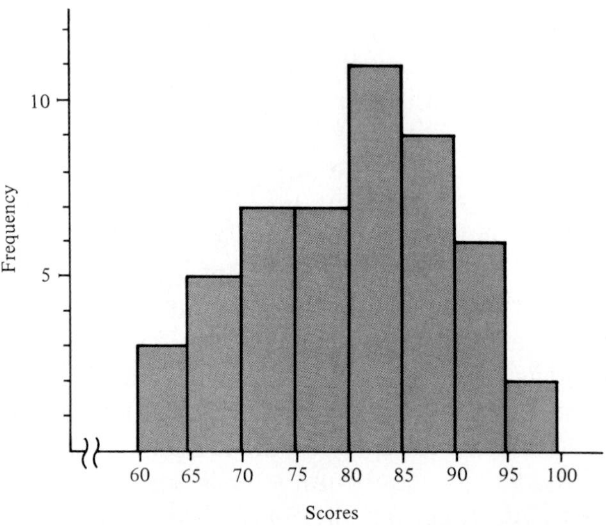

FIGURE 2.1. **Histogram of statistics exam scores.**

Histogram

The histogram is similar to the familiar bar graph that you have seen quite frequently, and it can be interpreted in much the same way. Figure 2.1 shows a histogram based on the frequency distribution of the statistics exam scores of Table 2.4. Note that the height of each column represents the *frequency* in each interval.

When we examine the histogram of the statistics exam scores in Figure 2.1, we see that it is not symmetrical and that the shape is irregular. However, there is a tendency for many scores to fall toward the center of the distribution, with progressively fewer scores as you move in either direction from the "hump." We also see that there is a wide variation in the scores.

Certainly, if we worked at it long enough we could get the above information from the frequency distribution itself, but the graphical method gives us this information at a glance. Also, in a later section, when we get into the topic of the different kinds of shapes that distributions may take, it will be much easier to compare several graphs instead of frequency distributions.

The steps in the construction of a histogram are

1. Lay out an area on a sheet of graph paper that corresponds roughly to the proportions of Figure 2.1. It is a good practice to have the height of the graph about three-fourths of the width. The horizontal line, called the x-axis or the abscissa, is drawn long enough to include all of the scores plus a little unused space at each end. Label this axis ("Scores" in this example) and put a number of scores at appropriate intervals (60, 65, 70, etc., are logical choices here).
2. At the left end of the x-axis draw a vertical line (the ordinate, or *y*-axis). Divide the *y*-axis into units so that the largest frequency will not quite reach the top of the graph. Number these units and label this axis "Frequency."
3. Now you can complete the histogram simply by drawing lines parallel to the x-axis at the height of the frequency for each interval and connecting the lines to the x-axis by vertical lines to the *real* limits of the intervals. For example, as shown in Figure 2.1, three people scored in the 60–64 interval, so the horizontal line above this interval is drawn 3 units up from the x-axis, and vertical lines extend down to the real limits of 59.5 and 64.5.
4. Give the histogram a title, either above or below the figure. The title should be a clear statement of what the histogram represents.

Frequency Polygon

Probably, most pictorial representations of the frequency distribution take the form of the *frequency polygon*, which is a line graph instead of the bar-type graph of the histogram. The frequency polygon for the distribution of statistics exam scores is shown in Figure 2.2.

Note that we get the same information from the frequency polygon as we did from the histogram, that is, the concentrations and spread of the scores. Since most graphs of the frequency distribution are frequency polygons, it might be a good idea to become thoroughly acquainted with the construction of the frequency polygon described below. Especially note that the points plotted in Figure 2.2 are directly above the *midpoint* of each interval and that the polygon drops down to the baseline (zero frequency) at both ends of the graph.

The steps in the construction of the frequency polygon are

1. Lay out the area for the graph with the proper proportions and label the x- and y-axes just as you did for the histogram.
2. Instead of drawing a bar corresponding to the frequency for each interval as you did for the histogram, place a *dot* above the *midpoint* of each interval. For example, three persons scored in the 60–64 interval. $MP = 62$ for this interval, so you would place a dot opposite a frequency of 3 and above the score of 62. After you place the points for all intervals, simply connect the points with straight lines.
3. It is mathematically incorrect and artistically unsatisfying to leave the polygon suspended in midair, so the curve connecting the points

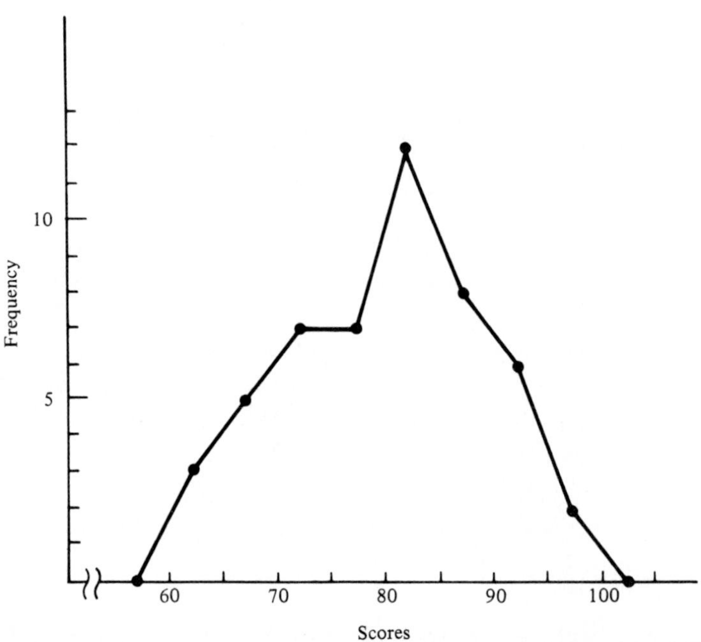

FIGURE 2.2. **Frequency polygon of statistics exam scores.**

drops down to the baseline at the extreme ends of the distribution. The curve touches the baseline at the *midpoint* of the adjacent interval, whose frequency, of course, is 0. For example, on the right side of Figure 2.2 the curve connects the point at 97 (the midpoint of the 95–99 interval) with the x-axis at 102 (the midpoint of the nonexistent 100–104 interval). The same method is used to connect the point at 62 with the x-axis at 57.

NOTE 2.1

WHERE DO THE SCORES LIE?

We have to remember that in gaining a bit of efficiency we have lost a little accuracy. Note that when scores are placed in a frequency distribution (and graphed as a polygon or histogram) the *identity* of an individual score is lost. In the statistics exam data we see that five individuals scored between 65 and 69. Once the data are in the form of a frequency distribution we cannot determine the exact values of the individual scores. We only know that the five scores lie somewhere in the 65–69 interval. Several statistical techniques to be presented later on will make different assumptions regarding the precise location of the scores in an interval. But we will leave that for a later section and for now will note that the frequency polygon shown in Figure 2.2 assumes that the *average* of all the scores in any interval falls at the *midpoint* of that interval.

CAUTIONS ON GRAPHING. A certain amount of care has to be exercised in transforming the ordinary frequency distribution into a frequency polygon. The greatest single error committed by the novice is in using incorrect proportions in drawing the graph. In the last section we noted that the height of the graph should be about three-fourths of the width. It is essential to follow this convention, since most graphs that you will be examining in books and journal articles will have the same general proportion. It is a lot easier to analyze someone else's graph if you have been using a similar layout and format all along.

Figure 2.3 shows the statistics exam scores represented in two incorrectly drawn frequency polygons. Polygon A has the scale of the ordinate too great with respect to that of the abscissa, while polygon B is just the reverse. Although these figures are obvious exaggerations, it must be noted that it does not take much deviation from the three-fourths rule of thumb to produce a distortion that interferes with interpreting the graph.

COMPARING TWO DISTRIBUTIONS. One very common use of the frequency polygon is in presenting data from two frequency distributions with

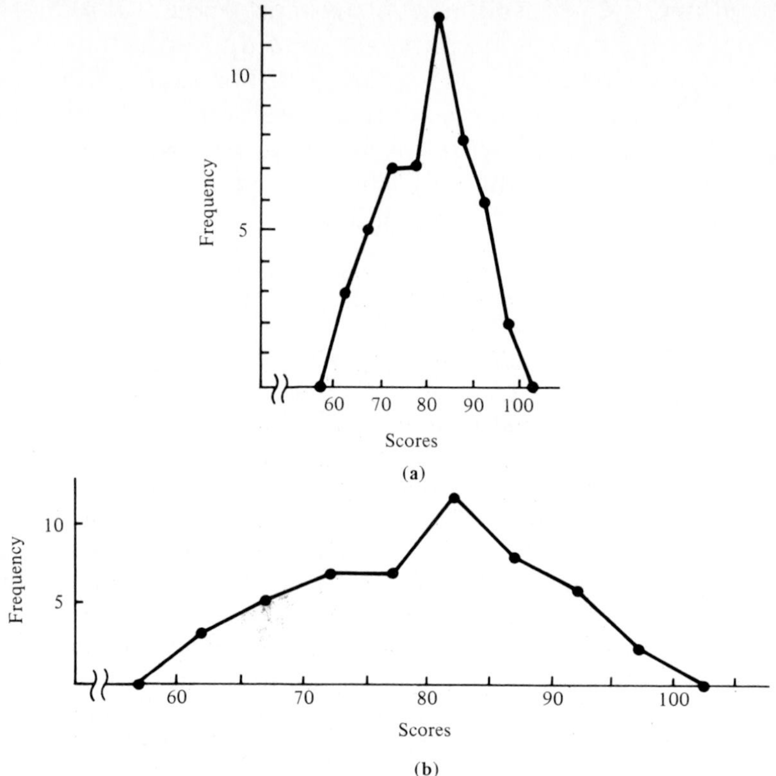

FIGURE 2.3. **Graphical distortion produced by improper proportion in statistics exam scores.**

approximately equal values of N. This procedure permits a comparison of two sets of scores on the same variable, in addition to providing information on each distribution separately. Just such a comparison is shown in Figure 2.4 for coordination test scores for a group of 90 seventh-grade girls and 85 seventh-grade boys. Note the ease with which you can make direct comparisons regarding the relative abilities of the two groups and the spread and concentrations of scores. It is obvious from the graph that the ability of the girls as shown by this particular coordination test is superior to that of the boys.

Typical and Not-So-Typical Frequency Distributions

The Normal Curve

You have all heard of, and probably used, the term *normal curve*. Many physiological measurements (e.g., height, weight, length of nose, and

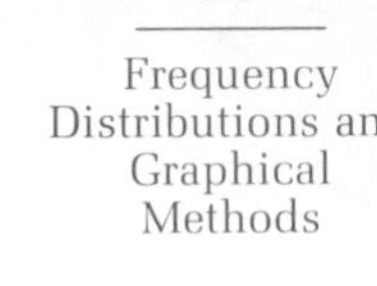

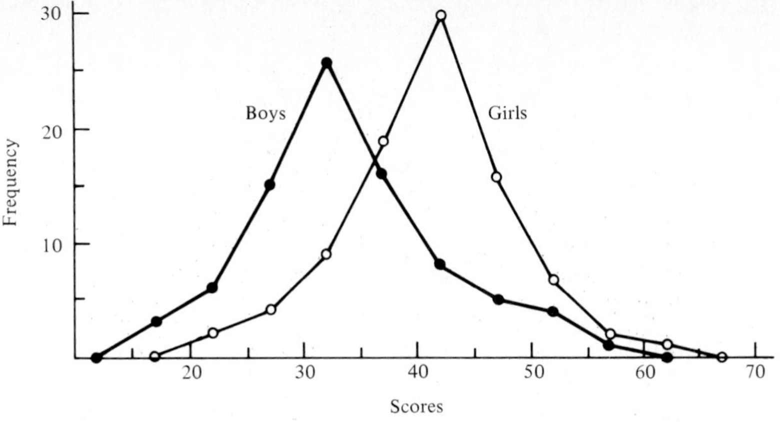

FIGURE 2.4. **Coordination scores for seventh-grade boys and girls.**

number of eyelashes) and behavioral measurements (IQ scores, reaction time, and aptitude test scores) are normally distributed in the population. By "normally distributed" we mean that the frequency polygon for the distribution of measurements closely approximates the mathematical model shown in Figure 2.5. This bell-shaped, symmetrical curve is called a *normal curve*. If this were a frequency polygon of test scores, it would show the typical concentration of scores in the middle of the distribution, with fewer and fewer scores as you approach the extremes. For example, if this curve represented the distribution of the heights of American adult males, the curve would be the highest (greatest frequency) around 5 feet 9 inches and would get progressively lower (smaller and smaller frequency) as you got out toward 6 feet 5 inches or 5 feet 1 inch. The curve would almost touch the baseline (very small frequency of occurrence) at a height of 7 feet 1 inch or 4 feet 8 inches.

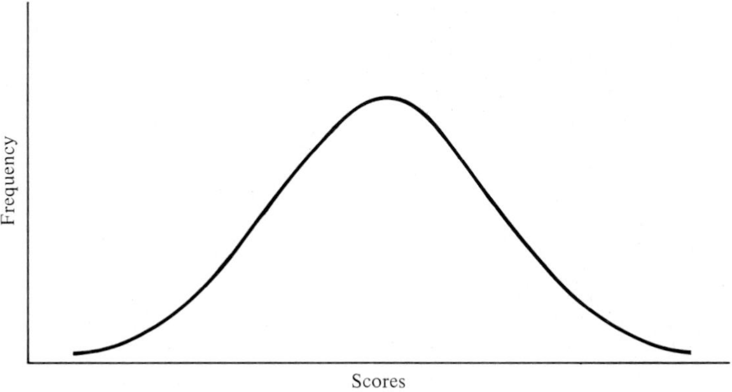

FIGURE 2.5. **A normal curve model.**

The frequency polygon of the statistics exam scores of Figure 2.2 bears only a slight resemblance to the normal curve. When we have only 50 scores we must expect a lack of symmetry in the distribution. If the statistics exam had been given to 500 students, we would expect that some of the irregularities would disappear, resulting in a smoother, more symmetrical curve. If the test scores were obtained from several thousand individuals, the resulting frequency polygon would come even closer in appearance to the normal curve model of Figure 2.5.

Even though the data with which we will most often be working will never be exactly normally distributed, we will find that the normal curve model has some very useful properties in relation to our data. In fact, these properties are a foundation for a very important part of statistical techniques, and we will have occasion to devote an entire chapter to these properties later on in the book.

Skewness

There will be times when a graph of the frequency distribution will not have the typical, symmetric bell shape with the majority of scores concentrated at the center of the distribution. Instead you might find that the majority of the scores are clustered at either the high end or the low end of the distribution. This concentration of scores at one end or the other of the distribution is called *skewness*.

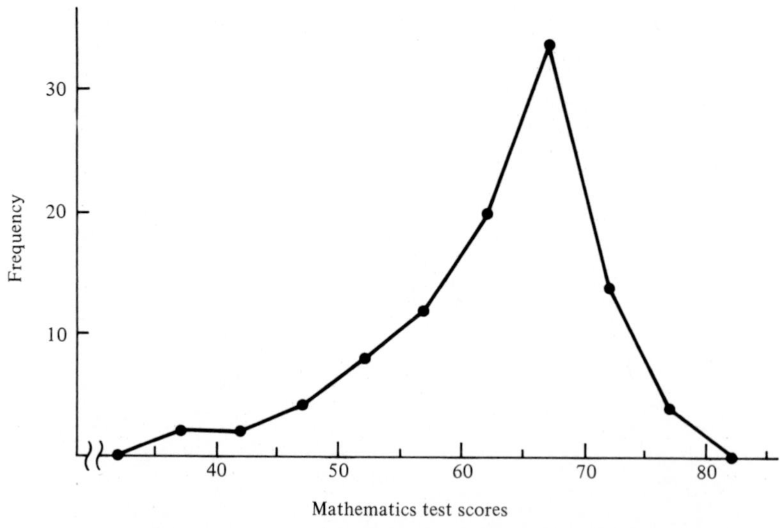

FIGURE 2.6. **Negatively skewed mathematics achievement scores.**

If the scores are concentrated at the upper end of the distribution so the tail of the curve skews to the *left*, we say that the curve is *negatively skewed*. Figure 2.6 shows the results of a national achievement exam in mathematics given to all ninth-graders in a school system enrolled in accelerated classes in mathematics. As you can see, there is a large concentration of high scores with progressively fewer low scores as you go the left. This graph is telling us that the exam was too easy for the majority of the students, which was to be expected, since they are in accelerated classes.

If the scores are clustered at the lower end of the distribution so the tail of the curve skews to the right, we say that the curve is *positively skewed*. Positive skewness is characterized by a preponderance of low scores, such as would occur if a difficult test were administered to an average group or an ordinary test administered to a below-average group. Another set of data that is usually positively skewed is the distribution of earned incomes. Figure 2.7 shows the incomes of 14,000 workers in a survey conducted in a national opinion poll. You can see that there is an obviously positive skew, since the majority of incomes fall between $12,500 and $22,500, which are at the lower end of the distribution, while other incomes extend a considerable distance to the right.

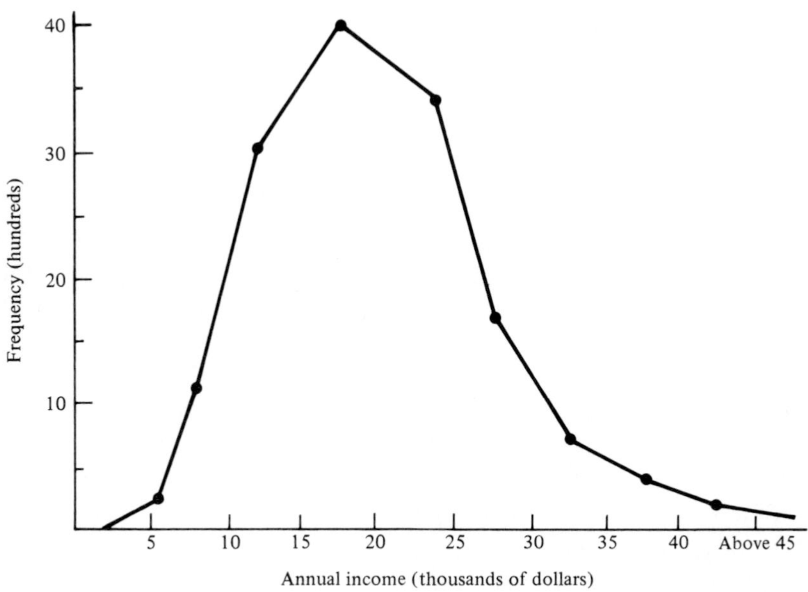

FIGURE 2.7. **Positively skewed distribution of annual incomes of 14,000 workers.**

If we examine the frequency polygon of the statistics exam scores in Figure 2.2, we see that there is a very slight negative skewness, since there are a few more scores near the upper end of the distribution. This type of skewness is quite common in academic situations, especially in college-level classes. Since most college classes represent quite a high level of ability, there usually will be more higher scores than lower scores. To put it another way, there usually are more A and B letter grades than D and F grades.

We have been using the "eyeball" method for examining the skewness of a distribution, and this may be perfectly adequate in most situations. However, if you want a more precise, mathematically defined method for measuring the amount of skewness, see one of the books listed in the References.

NOTE 2.2

REMEMBERING WHICH SKEWNESS IS WHICH

A handy device for determining whether a curve is negatively or positively skewed is to look at the *tail* of the curve (the tail is on the opposite side from the concentration of scores). If you remember your algebra, you know the direction *left* is negative, so if the tail goes to the left, the skewness is negative. If the tail points to the right, since the direction *right* is positive, you have positive skewness. In checking this with Figure 2.6 and 2.7, note that in Figure 2.6 the tail points to the left (negative skewness) while in Figure 2.7 it points to the right (positive skewness).

Kurtosis

Another property of a frequency distribution besides the amount of symmetry (symmetry is the opposite of skewness) is its *kurtosis*. When we speak of kurtosis we are referring to the "peakedness" or "flatness" of a frequency polygon. If the curve has a very sharp peak, it indicates an extreme concentration of scores about the center, and we say that it is *leptokurtic*. If the curve is quite flat, it would tell us that while there is some degree of concentration at the center there are quite a few scores that are dispersed away from the middle of the distribution; we say the curve is *platykurtic*. The curve that represents a happy medium is a *mesokurtic curve*. The normal curve model of Figure 2.5 is mesokurtic. All three degrees of kurtosis are shown in Figure 2.8.

We ordinarily do not worry too much about the type of kurtosis in a set of data from test scores, surveys, and experiments, where we are interested in describing the performance of a group of individuals. How-

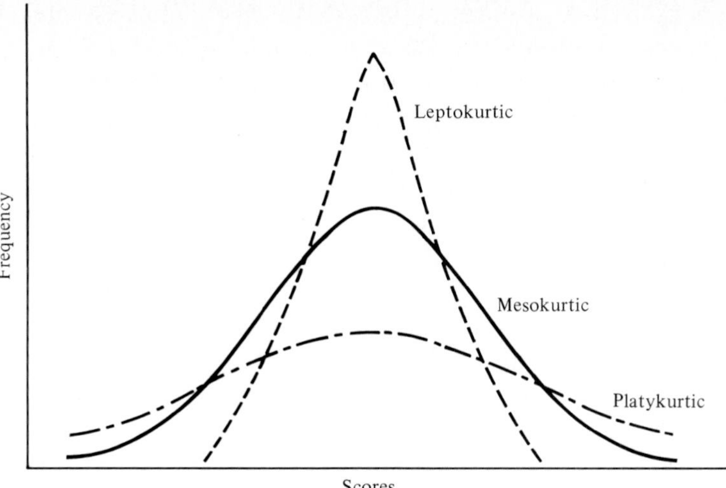

FIGURE 2.8. **Types of kurtosis.**

ever, when we consider theoretical issues in sampling procedures and statistical tests in later chapters, we will come back to the concept of kurtosis.

NOTE 2.3

REMEMBERING WHICH KURTOSIS IS WHICH

A colleague of mine uses some vivid imagery to illustrate the different types of kurtosis. He notes that the platykurtic curve is flat and rounded like the back of a duckbilled platypus. The other type of kurtosis reminds him of a kangaroo, since the leptokurtic curve looks as if it is "lepping" around. He claims that not a single student has forgotten the difference between platykurtic and leptokurtic curves!

J-Curves

Before leaving our discussion of the shapes of frequency distributions, we should make note of a special type of extreme skewness—the J-curve. The J-curve hypothesis of social conformity was first proposed by Floyd Allport and is related to conforming behavior among groups of people. The large majority of scores fall at the end of the scale representing socially acceptable behavior, while the small minority of scores represent a deviation from this social norm.

Nelson and Nilsen (1984) assessed drivers' behavior at a four-way stop sign under two conditions: when a marked police car was parked near the intersection and when no police car was present. The speed with which each driver entered the intersection was tabulated and the results are shown in Figure 2.9.

The J-shaped curve for the drivers' behavior with the police car present is the result of extreme skewness, since the great majority of the drivers either made a "complete stop" or a "rolling stop." However, a few only slowed down slightly and two went through the intersection without slowing down at all, resulting in the J-shaped curve. As you can see, the curve has an entirely different shape when the police car was absent.

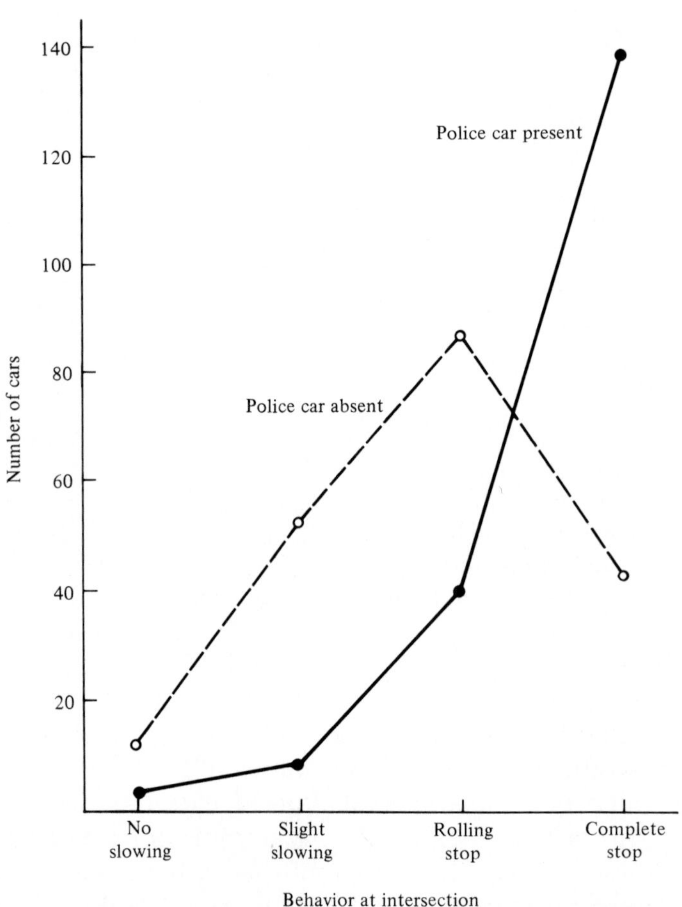

FIGURE 2.9. **Conforming behavior at a four-way stop sign with a marked police car present or absent.**

This type of curve can be found in a number of situations where a group standard can be quantified and deviations from the social norm observed. Some examples might be the length of time drivers will park in a No Parking zone, the time of arrival at a concert or religious service, and the number of drinks consumed during the social hour preceding a banquet.

Graphing the Functional Relationship

After spending considerable time and energy on frequency distributions and the graphical methods for displaying such data, we will have to shift our frame of reference slightly in order to take up the topic of graphing a *functional relationship*. We will be directing our attention not to a pictorial representation of the frequency distribution but rather to a pictorial image of the *relationship* between two variables.

In the broadest sense we might say that the objective of educators and behavioral scientists is to discover functional relationships between variables. An educator might be interested in knowing if the number of siblings (brothers and sisters) that a child has will have an effect on his or her social relationships in nursery school. If an investigator finds that the only child has more difficulty in getting along with others than the child with two siblings, we say that the researcher has demonstrated a *functional relationship* between the two variables of number of siblings and amount of social adjustment.

An economist might be interested in learning what effect the amount of leisure time has on the individual's expenditures for recreational items such as snowmobiles, speedboats, camper trailers, and the like. If she finds that people working a 48-hour week spend 3% of their take-home pay on recreation while those working a 36-hour week spend 7%, we say that she has found a functional relationship between hours worked and leisure expenditures.

In still another example, an engineering psychologist might be concerned with the brightness of a car's brake lights and the reaction time of the driver in the following car. If he finds that the reaction time of 100 drivers decreases as the brightness of the brake light is increased, we would say that he has established a functional relationship between reaction time and brake light brightness.

Independent and Dependent Variables

In all of the examples given we have noted that there is a relationship between two variables—the *independent* variable and the *dependent* variable.

The independent variable is the one that is manipulated in some way by the researcher, who may choose to compare an only child with a child who has two siblings, or workers working a 48-hour workweek with those working a 36-hour workweek, or three different brightness levels of brake lights. The *independent variables* are, respectively, number of siblings, length of workweek, and brake light brightness.

The dependent variable is dependent (hence the name) upon the value of the independent variable. It is always a measurement of some sort—in the examples above it would be the amount of social adjustment, amount of money spent on recreation, and the speed of reaction. It is sometimes called the *response variable*, since it is always the measure of some type of response. If there is a relationship between the independent and dependent variables, we note that the dependent variable changes in response to changes in the independent variable. As you vary the independent variable, you note that the dependent variable, which you are measuring, also changes.

These changes are best shown in the form of a graph such as Figure 2.10. As we mentioned earlier, a researcher might be investigating how the reaction time of a driver to a brake light depends on the brightness of the brake light. Note that the dependent variable (reaction time) is plotted on the *y*-axis while the independent variable (brake light brightness) is plotted on the *x*-axis. As you can see from the graph, when a dim brake light is used as a stimulus, the response time is about 0.35 seconds. The driver's response is somewhat faster when a moderately bright brake light is used, and the response is the fastest (about 0.30

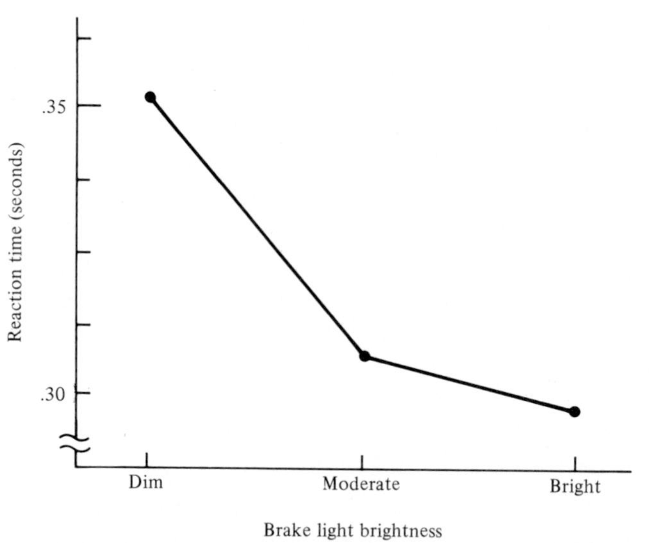

FIGURE 2.10. **Reaction time to three brake light systems.**

seconds) when a bright brake light is used. There is no doubt about it—the researcher has demonstrated a functional relationship between the two variables. Reaction time does indeed depend on the brightness of the stimulus, and the graph displays this functional relationship very efficiently.

The rules for graphing functional relationships are much the same as for graphing the frequency distribution as far as proportion, labeling of axes, and so on. And, as mentioned above, the dependent variable—what the researcher is measuring—always goes on the y-axis. One important difference that should be noted is that the curve does not drop down to 0 at the baseline as was the case with the frequency polygon. If it did this in Figure 2.10, for example, it would imply that we had some measurements at other brightness levels than dim, moderate, and bright. Since reaction times were not measured at "very dim" or "very bright," we cannot extend the curve to those points.

The above discussion was not intended to be a thorough and exhaustive presentation of the topic of functional relationships. A complete treatment of functional relationships and different types of variables more properly belongs in a textbook of experimental design or advanced statistics. Our purpose for considering the topic at this point was twofold: A discussion of graphical methods would not be complete without it, and there will be several occasions in future chapters when we will want to consider ways to measure the *amount* of relationship that exists between two variables.

SAMPLE PROBLEM

The College Life Questionnaire was administered to a random sample of 219 full-time freshmen students at a small midwestern state university. Forty-seven percent or 104 indicated that they had a part-time job either on or off campus. They were asked to list the number of hours they worked each week. The results follow.

2	11	9	6	10	5	10	6	8	5
5	3	11	9	11	5	10	10	10	7
9	5	3	10	12	8	15	6	5	8
10	5	6	4	13	10	22	9	12	9
7	8	13	6	4	16	10	16	12	6
7	10	9	6	8	4	5	6	10	5
18	10	8	6	5	12	4	10	15	10
8	6	20	15	14	18	6	4	25	9
8	15	6	15	8	12	7	9	4	11
6	7	20	5	14	4	12	9	5	4
9	8	10	10						

To construct a frequency distribution from these data, we first note that he highest "score" is 25 hours worked per week and the lowest is 2.

With a range of 25 − 2 + 1 = 24, we could use 9 intervals with $i = 3$. We could use 8 intervals if we began the bottom interval at 2, but the frequency distribution might be more descriptive if we begin with zero.

Once we have decided on the bottom interval of 0–2, we list the remaining intervals in a column and go through the table of raw data, placing a tally mark beside the interval where each score falls. In Table 2.6 the worksheet is shown on the left and the finished product is on the right.

The frequency polygon for this data is shown in Figure 2.11. Note that for easy readability the score values on the horizontal axis are in multiples of 5, even though $i = 3$ for this distribution. Also note that the points are plotted above the midpoints for each interval.

In analyzing the frequency polygon, we note that there is a rather marked degree of positive skewness; that is, the tail skews to the right. This would indicate that most freshmen with part-time jobs work a moderate number of hours (3 to 11 hours per week), but there are a few who are working as much as 15 or more while carrying a full load of course work.

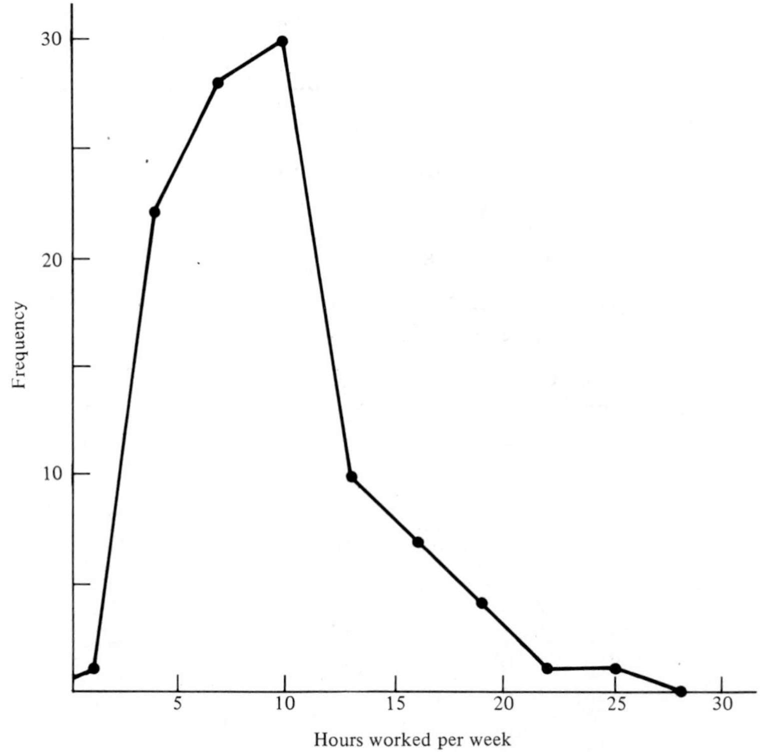

FIGURE 2.11. **Hours worked at a part-time job by college freshmen.**

TABLE 2.6. **Number of Hours Worked per Week at a Part-Time Job**

Scores	Tally	Scores	*f*
24–26	|	24–26	1
21–23	|	21–23	1
18–20	||||	18–20	4
15–17	卌 ||	15–17	7
12–14	卌 卌	12–14	10
9–11	卌 卌 卌 卌 卌 卌	9–11	30
6–8	卌 卌 卌 卌 卌 |||	6–8	28
3–5	卌 卌 卌 卌 ||	3–5	22
0–2	|	0–2	1
		N =	104

STUDY QUESTIONS

1. Why might a frequency distribution with 10 intervals be preferred to one with only 4 or 5 intervals?
2. When might you prefer a frequency distribution where $i = 1$?
3. List the steps you would use in constructing a frequency distribution.
4. What is the difference between discrete and continuous data?
5. What are the steps you would use in constructing a frequency polygon?
6. How do a histogram and a frequency polygon differ?
7. Describe the shape of a distribution that is negatively skewed and give an example. Do the same for a distribution that is positively skewed.
8. What is the difference between the independent variable and the dependent variable? Which goes on the *x*-axis? Which goes on the *y*-axis?

EXERCISES

1. For the following intervals give the real lower and upper limits, the midpoint, and the size of the interval.

a. 50–99	**f.** 80–89
b. 1–2	**g.** 100–124
c. 2.0–2.4	**h.** 50–79
d. 61–63	**i.** 25–29
e. 19.5–19.9	**j.** 0.02–0.04

2. List the real lower and upper limits, the midpoint, and the size of the interval for the following intervals.

a. 120–129	**f.** 100–149
b. 49–51	**g.** 65–69
c. 2.5–2.9	**h.** 70–99
d. 25–49	**i.** 17.0–17.4
e. 0.06–0.08	**j.** 3–4

3. Indicate whether the following statistics represent discrete or continuous data.
   - **a.** Accident rate per 100,000 miles driven on the Interstate Highway System.
   - **b.** Average height of the coaches of teams in the National Basketball Association.
   - **c.** Number of items correct on an American history test.
   - **d.** Incidence of mononucleosis diagnosed at a university health center.
   - **e.** Batting averages of the Chicago Cubs.
   - **f.** Average monthly weight loss by 42 members of the local chapter of a national weight loss organization.
   - **g.** Average finish time for participants in a 10-kilometer run.
   - **h.** Number of errors on a 1-minute typing test.
   - **i.** Average number of cavitites for a group of children using a fluoride mouth wash.
   - **j.** Average distance walked by students in a 24-hour "Walk for Hunger" fund-raising activity.

4. Indicate whether you expect the following to be continuous or discrete data.
   - **a.** Average number of children of mothers aged 30–34 in 3 different family income levels.
   - **b.** Average speed of the first 3 finishers in the Indianapolis 500.
   - **c.** Free throw percentage of the best free throw shooter of the Boston Celtics.
   - **d.** Daily hours of sunshine from September 21 to December 21.
   - **e.** Amount of gasoline consumed by 3 prototype engines in cars driven 1000 miles on a test track.
   - **f.** Proportion of university freshmen working 10 hours or more at a part-time job.
   - **g.** Number of items correct on a biology quiz.
   - **h.** Average monthly rainfall during the growing season in Des Moines, Iowa.
   - **i.** Number of cases of coronary heart disease among males aged 50–59 smoking 3 or more packs of cigarettes a day.
   - **j.** Percent of students at a state university receiving financial aid.

5. The heights of 48 male college students follow. Construct a simple frequency distribution ($i = 1$) from these data.

73	71	74	72	71	73	70	71	70	73
70	72	71	70	74	71	73	70	72	73

72	71	69	74	69	69	71	68	69	72
69	71	68	69	71	68	71	74	69	70
75	69	67	70	73	65	70	74		

6. On a questionnaire conducted by a class in exercise physiology, the weights of 35 women were tabulated with the following results. Construct a grouped frequency distribution with $i = 5$.

124	111	150	145	116	126	121	137	129	120
121	118	124	134	124	117	121	131	130	119
109	132	125	96	109	116	109	150	115	102
108	139	130	115	149					

7. Thirty biology students were given a laboratory quiz with the following results. Construct a frequency distribution with $i = 3$.

25	16	12	22	20	10	9	18	18	2
16	12	15	15	11	19	19	8	14	17
13	25	23	11	15	15	15	14	7	14

8. A final exam in a course in physiological psychology was taken by 39 students with the following results. Construct a frequency distribution with $i = 5$.

60	64	62	58	72	63	68	67	63	61
64	59	67	62	60	57	57	63	55	61
48	66	47	51	55	53	69	43	52	61
65	62	32	54	56	57	60	36	64	

9. A 50-item unit exam was given to 98 members of an introductory psychology class and the number of items correct for each student is shown below.

a. Construct a simple frequency distribution ($i = 1$) for this group of scores.

b. Construct a grouped frequency distribution with $i = 3$. Note that the pattern of scores is more obvious than with $i = 1$.

32	29	35	26	32	22	33	30	27	30
36	29	33	38	32	43	29	39	36	30
18	35	37	38	33	30	39	31	38	32
36	30	19	29	37	35	33	34	30	38
30	39	30	39	30	42	36	22	36	33
39	31	26	30	39	31	42	25	35	30
38	38	33	39	32	42	36	34	35	31
41	44	47	31	39	36	34	37	37	30
46	31	31	41	43	37	44	42	40	37
24	28	25	41	33	30	36	28		

10. Resting heart rates before vigorous physical exercise were measured for a sample of 92 undergraduates at Pennsylvania State University.
    a. Construct a simple frequency distribution ($i = 1$) for these data.
    b. Construct a grouped frequency distribution with $i = 5$. Note how this frequency distribution gives a better picture of the pattern of scores than with $i = 1$.

64	58	62	66	64	74	84	68	68	76
80	92	68	60	62	66	70	68	72	70
66	70	96	62	78	82	100	62	96	78
62	80	62	60	72	62	76	62	54	74
68	72	68	82	64	58	54	70	68	48
88	70	90	78	70	90	92	60	72	68
74	68	84	61	64	94	60	72	58	88
84	62	66	80	78	68	72	82	76	87
78	68	86	76	90	74	88	74	76	84
66	90								

11. A 24-point quiz was given to 95 university students in 2 sections of a course in educational psychology with the results shown below.
    a. Construct a simple frequency distribution ($i = 1$) for these data.
    b. Draw a frequency polygon of this frequency distribution.
    c. What type of skewness is shown here?

20	16	13	18	19	18	18	20	20	18
20	12	19	19	20	17	20	17	20	18
11	20	20	18	19	17	20	19	18	19
21	13	22	21	19	22	18	19	16	21
17	21	14	22	20	23	21	20	21	22
18	22	17	14	24	18	24	17	23	19
20	18	20	21	14	19	21	21	21	22
22	19	18	24	20	15	17	23	23	16
24	24	19	21	21	20	15	23	23	20
19	21	18	15	21					

12. A final exam in African history was administered to 89 college students in 2 sections of the course. The results are shown below.
    a. Construct a grouped frequency distribution with $i = 3$.
    b. Draw a frequency polygon of this distribution.
    c. What type of skewness is shown, if any?

88	82	87	74	72	67	83	78	83	81
84	80	89	82	78	82	85	78	75	82
85	73	66	74	74	81	73	81	62	81
90	84	83	76	76	81	85	68	82	86
82	79	83	86	80	79	84	85	83	80
89	83	77	85	89	88	83	71	84	86
83	83	84	77	75	79	70	72	87	88

85	80	87	86	87	86	74	84	68	78
86	79	85	76	78	87	81	80	83	

13. The College Life Questionnaire mentioned earlier in the Sample Problem also was used to ask 207 college freshmen to estimate the number of hours that they studied each week. A grouped frequency distribution of the scores is shown below.

a. Construct a histogram of these data.
b. Construct a frequency polygon for these data.
c. What type of skewness is shown here?
d. Estimate the number of hours that you study in an average week and locate that number on the frequency polygon. How do you compare with this sample of college freshmen?

Hours Spent Studying	f
41–43	2
38–40	0
35–37	4
32–34	9
29–31	8
26–28	14
23–25	20
20–22	22
17–19	27
14–16	22
11–13	35
8–10	23
5–7	21
	N = 207

14. The College Life Questionnaire was also given to a random sample of students at a different university in an attempt to discover how many students held part-time jobs, and, if so, how many hours they worked. Out of the original sample of 408, 197 (48%) held part-time jobs. The number of hours each worked per week is shown in the grouped frequency distribution below.

a. Construct a histogram from the data.
b. Construct a frequency polygon from these data.
c. What type of skewness is present here?

Hours Worked	f
27–29	1
24–26	3

21–23	1
18–20	12
15–17	9
12–14	9
9–11	49
6–8	55
3–5	51
0–2	7
	N = 197

15. Indicate the shape of the curve (normal, positively skewed, or negatively skewed) that you would be most likely to find in the following data.

a. College students receiving help in reading comprehension skills at the Reading Center are given the verbal subtest of a popular college entrance examination.

b. All the seniors at a local high school are given a manual dexterity test.

c. The teacher of a sixth-grade class accidentally administered an arithmetic test that was intended for the fifth-grade class.

d. All seniors at a large high school who are planning to attend college are given a standardized vocabulary test.

e. A savings and loan institution tabulates a distribution of savings account balances for all their depositors.

f. Reaction times to the onset of a light are measured for a group of driver-training students.

16. Give an example of data that would likely be negatively skewed.

a. Draw a hypothetical frequency polygon that would describe these data and label the axes.

b. Do the same for an example of positive skewness.

17. An investigator is studying the effects of fatigue on body sway. A group of truck drivers is tested for amount of body sway before entering a driving simulator and after 4 and 8 hours of simulated driving. A group of high school students who have just passed their driving exams are also measured for amount of body sway after the same time intervals in the simulator.

a. Graph the results given below.

b. Which is the dependent variable? Which is the independent variable?

c. What is the relationship between body sway and driving time? How is this relationship affected by driving experience?

	Average body sway (mm)		
Group	**0 hr**	**4 hr**	**8 hr**
Students	59	87	102
Truck drivers	57	80	88

18. A manufacturer of computer and word processor keyboards wanted to know the optimum key displacement (i.e., how far down the key must be depressed before making contact) that would result in the fewest errors. The designer had 9 keyboards identical in appearance, but the key displacement varied from 1.2 millimeters (about $\frac{1}{16}$ inch) to 4.4 millimeters (almost $\frac{3}{16}$ inch). A number of clerical personnel used the different machines to type a one-page selection, and the total number of errors for each machine was recorded.

a. What was the dependent variable? The independent variable?

b. The total number of errors for the different key displacements is shown below. Graph these data.

c. What would you conclude about the relationship between the distance that a key has to be depressed and the number of errors made?

Key Displacement	**Errors**
1.2	45
1.6	33
2.0	26
2.4	23
2.8	14
3.2	19
3.6	22
4.0	34
4.4	39

19. The number of homicides each year in the United States from 1968 to 1982 is shown in the following table. (*Source*: Department of Justice, *Uniform Crime Reports for the United States, 1983*.) Draw a graph of these data.

Year	**Homicides**	**Year**	**Homicides**
1968	12,503	1976	16,605
1969	13,575	1977	18,033
1970	13,649	1978	18,714
1971	16,183	1979	20,591
1972	15,832	1980	21,860
1973	17,123	1981	20,053
1974	18,632	1982	19,485
1975	18,642		

20. It is generally known that the total amount of weight that can be lifted by a weightlifter is related to the weight of the weightlifter. Shown below are the weights (in pounds) of the 1984 finalists in a weightlifting competition and the total amount (in kilograms) they lifted. (*Source*: U.S. Weightlifting

Federation, Men's 1984 National Championships.) Graph these results to see if there is a relationship between the weightlifter's weight and the amount lifted.

Weight of Finalist (pounds)	Amount Lifted (kilograms)
114	205
123	233
132	235
148	275
165	300
181	310
198	348
220	360
242	363

CHAPTER 3

Central Tendency

Hardly a day goes by that we do not hear some reference to the concept of "average" performance. Energy-conscious drivers inform us that their cars will average 30 or 40 miles per gallon, while sports fans are forever discussing some ballplayer's batting average or earned run average or their own bowling average. Or you may insist that you are "just an average student" or that your parents back home live in a typical average house. Figure 3.1 suggests the degree to which measures of average performance affect the lives of all of us.

The "average" used so frequently by the layperson is part of the concept of *central tendency.* As you saw in the last chapter, frequency distributions, along with the histogram and the frequency polygon, are valuable devices that enable us to extract meaning from a mass of data. However, we would like more efficient ways of expressing our results than a mere picture of the distribution as a whole. Specifically, we would like to have a statistical method that would yield a *single value* that would tell us something about the entire distribution.

One such single value is called a measure of *central tendency*. It is the single value that best describes the performance of the group as a whole. For example, the average income in a community might be $27,500, or the average weight of the defensive linemen of the Minne-

FIGURE 3.1. The concept of central tendency is even noted in the comic strips.

sota Vikings might be 280 pounds, or the average test score in an educational psychology course might be 83.2. All of these single values cited have one thing in common: They are values that best characterize the group as a whole. No person actually had a score of 83.2 on the educational psychology exam, but the average of 83.2 is the best single value that represents the performance of that group of students.

There are a number of measures of central tendency, all designed to give representative values of some distribution. In this chapter we will concentrate on three of the most commonly used: the arithmetic mean, the median, and the mode. But keep in mind as we discuss the characteristics of these measures of central tendency that our eventual aim will be to find one that best describes the performance of the group as a whole.

Mean

The *average*, referred to a bit earlier, is more correctly called the *arithmetic mean*, or often just the *mean*. I am sure you remember from an arithmetic class of many years ago that to calculate the mean you simply add up all the numbers and divide by how many numbers there are. So, if you paid 69 cents, 49 cents, 98 cents, 59 cents, and 95 cents for 5 items in the supermarket, the average or mean price paid per item would be $3.70 divided by 5, or 74 cents.

TABLE 3.1. **Calculating the Mean: Field Goal Attempts by a College Men's Basketball Team in an NCAA Division 2 Playoff**

X	
6	
9	
10	
7	
9	$\overline{X} = \frac{\Sigma X}{N} = \frac{60}{10} = 6$
2	
4	
5	
6	
2	
$\Sigma X = 60$	

However, statisticians tend to get a little nervous when terms are not precisely defined, so let us be more formal with the terminology. Table 3.1 shows the number of field goal attempts by each of the 10 members of a college men's basketball team in an NCAA Division 2 playoff game, and the terms ΣX and N, which we first met in Chapter 1. We remember that ΣX is the symbol for the sum of the X values, and N is the number of values. In Table 3.1, the addition of the X values yields $\Sigma X = 60$, and N is, of course, 10. The formula for the mean, whose symbol is $\overline{X}$ (say "X bar"), would be

$$\overline{X} = \frac{\Sigma X}{N} \qquad (3.1)^{\dagger}$$

which simply states that we obtain the mean by summing all the X values and dividing by N, the number of observations. So, for the data of Table 3.1, the mean would be

$$\overline{X} = \frac{\Sigma X}{N} = \frac{60}{10} = 6$$

So what does the mean mean? What does $\overline{X} = 6$ in Table 3.1 tell us about the data? In short, we would say that the value of 6 is the best single value which describes this distribution. Since these values were the number of field goal attempts by 10 college basketball players, we would conclude that the mean of 6 is the best single value that represents the number of shots attempted by the 10 team members as a whole.

Another useful feature of the mean is that it makes it possible to compare an individual's performance with the rest of the group. Was his number of shots attempted above the mean or below it? Is he above average or below average? How far above or below is he? We can answer these questions in terms of deviation or distance from the mean. Using the men's basketball data again, we list each player and number of field goals he attempted in Table 3.2. In addition to the number of shots attempted (X) there is another column ($X - \overline{X}$), which shows how far each value is from the mean. The column headed ($X - \overline{X}$) contains deviations, and these are obtained, obviously, by subtracting the mean from each value. For example, player B's deviation would be 3 (9 minus 6), indicating that his performance is located 3 units above the mean. In a similar fashion we calculate that player F, who attempted 2 shots, would have a deviation of -4 (2 minus 6), locating him 4 units below the mean.

†Computational formulas in this book will be identified by a number that indicates the chapter number and the location of the formula in the chapter. Thus formula 3.1 is the first formula listed in Chapter 3.

TABLE 3.2. **Field Goal Attempts Shown as Deviations from the Mean**

Player	X	$(X - \overline{X})$
A	6	0
B	9	3
C	10	4
D	7	1
E	9	3
F	2	−4
G	4	−2
H	5	−1
I	6	0
J	2	−4
	$\Sigma X = 60$	$\Sigma(X - \overline{X}) = 0$

The reason for belaboring the obvious in the previous paragraph is to demonstrate that the algebraic sum of the deviations is always 0. Note that the sum of the $(X - \overline{X})$ column in Table 3.2 is 0. This is the formal definition of the arithmetic mean. It is calculated in such a way that it is directly in the center of these deviations, making the algebraic sum of these deviations 0. In our usual notation, $\Sigma(X - \overline{X}) = 0$.

This is best shown by the illustration in Figure 3.2. Note that the mean is the fulcrum and that the entire group of observations is in balance.

We need to make one last point about the calculation of the mean. In Chapter 1, if you remember, quite a case was made for the necessity of determining what level of precision we have in a group of data, that is, whether the observations are from a nominal, an ordinal, an interval, or a ratio scale. As was mentioned, we need to know this information before we can calculate any statistic based on the data. As you progress through this book, you will note that for each statistic discussed there will always be an accompanying comment on what type of data is appropriate for a given statistic. So with this in mind, we conclude that since the mean takes into account the distances between observations,

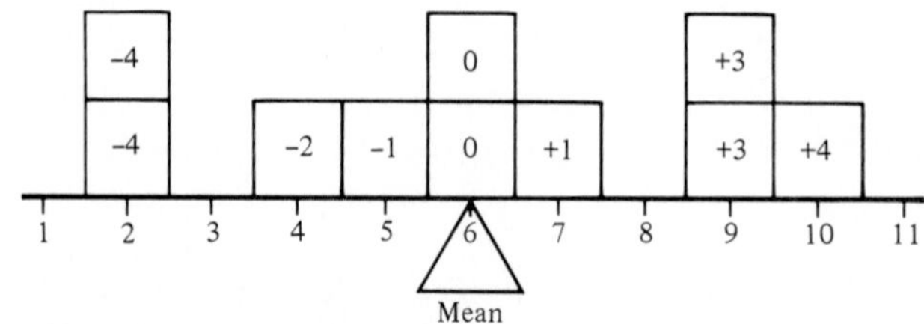

FIGURE 3.2. **Arithmetic mean expressed as a balancing point.**

the measurements from which the mean is calculated must be at least of the interval type. A mean calculated from ordinal data (e.g., mean rank) may be misleading.

Median

The median is probably most familiar to you as the 50th percentile. *The median is the value that exactly separates the upper half of the distribution from the lower half.* It is obviously a measure of central tendency in that the median is the point located in such a way that 50% of the scores are lower than the median and the other 50% are greater than the median. It is important to note that while the mean was the exact center of the *deviations* or distances of the scores from the mean, the median is the exact center of the scores themselves. There are two different methods for calculating the median, depending on whether we have just a "few" scores and all the scores are different, or a larger distribution where some of the scores are identical.

Median: Untied Scores

The calculation of the median for *untied* (a more precise, but somewhat confusing way of saying that all the scores are different) scores is shown in Table 3.3. Personal incomes (in thousands of dollars) are shown for 8 members of a chemical-dependency therapy group and 11 members of a child-abuse therapy group.

TABLE 3.3. **Calculating the Median for Untied Scores: Personal Income (thousands of dollars) for Members of Two Therapy Groups**

Child-Abuse Therapy Group $N = 11$	Chemical-Dependency Therapy Group $N = 8$
$25	$17
24	14
20	12
19	11
18	8
17	6
15	4
12	3
10	
9	
7	
$Med = 17$	$Med = \frac{(8 + 11)}{2} = 9.5$

If any set of data contains an *odd* number of untied scores, such as the incomes for the 11 members of the child abuse group, the median is simply the central score when the scores are *ranked* in order of magnitude. In the example shown, *Med* = $17,000, since 17 is the point that exactly separates the upper and lower halves of the distribution.

If there is an *even* number of untied scores, such as the incomes for the 8 members of the chemical-dependency group, the median is exactly halfway between the two centermost values. Since 8 and 11 are on either side of the center, the value of the median would be 9.5. In other words *Med* = $9500 is the point that exactly separates the upper and lower halves of the distribution.

Median: Distributions with Tied Scores

In practice, we are most likely to be calculating a median for data that contain one or more tied scores. For example, the left portion of Table 3.4 shows the reading comprehension scores of the Peabody Individual Achievement Test for 17 fifth-graders referred to the Reading Clinic in a midwestern school district. In order to calculate a median for these 17 scores, we first group the scores into a *simple* frequency distribution. If you remember, in Chapter 2 we noted that in a simple frequency distribution, $i = 1$, and all possible score values from the lowest to the highest are listed, along with the number of times each occurred in the f column.

TABLE 3.4. **Calculating the Median When Some Scores Are Tied: Reading Comprehension Scores on the Peabody Individual Achievement Test**

Ungrouped Scores	Scores	f	*cum f*
98, 96, 101,	101	1	17
100, 94, 92,	100	1	16
89, 92, 93,	99	0	15
94, 91, 95,	98	1	15
95, 93, 94,	97	0	14
96, 95	96	2	14
	95	3	12
	94	3	9
	93	2	6
	92	2	4
	91	1	2
	90	0	1
	89	1	1
		$N = 17$	

Calculations

(A) 50% of 17 = 8.5

(B) 1 + 0 + 1 + 2 + 2 = 6

(C) $0.5N - cum\ f = 8.5 - 6 = 2.5$

(D) $f_{MED} = 3$

(E) $L = 93.5$

$$\begin{aligned} Med &= L + \left(\frac{0.5N - cum\ f}{f_{MED}}\right) \\ &= 93.5 + \left(\frac{8.5 - 6}{3}\right) \\ &= 93.5 + .83 \\ &= 94.33 \end{aligned}$$

As you can see in Table 3.4, we have added the *cum f* (cumulative frequency) column, which is simply the addition of the frequencies in each interval as you count up from the bottom. For example, there is 1 score of 89, no scores of 90, and 1 score of 91, so the cumulative frequency for 91 is 2, since there are 2 scores of 91 and below. There are 2 scores of 92, so the cumulative frequency of 92 is 4, indicating that there are 4 scores of 92 and below. But remembering the discussion in Chapter 2, we know that technically the 4 scores are below 92.5, the exact upper limit of a score of 92.

The first step in the calculation of the median is to take 50% of *N*. In step A, $0.50 \times 17 = 8.5$. The next step is to find where 8.5 is located in the *cum f* column, since the point where 8.5 scores are below would be the location of the median. In step B, we see that in counting up from the bottom in the *f* column, 1 + 0 + 1 + 2 + 2 = 6 in the *cum f* column. If we went up one more score, we would exceed 8.5, so this tells us that the median is somewhere between 93.5 and 94.5, since there are 6 scores below 93.5 and 9 scores below 94.5.

Now we must determine how far into the 93.5 to 94.5 interval the median is, and we first subtract the frequencies we have just totaled from the value we actually needed. We needed 50% of *N* or 8.5, but we could only count up to 6. So in step C, 8.5 − 6 = 2.5.

We next look at the interval where the median is located and find its frequency. We had previously learned that the median would be between 93.5 and 94.5 (the exact limits for a score of 94) and in Table 3.4 we note that there are 3 scores of 94. We call this f_{MED}, the frequency of the interval containing the median, and in step D, $f_{MED} = 3$.

Finally, we determine the exact lower limit, *L*, for the interval containing the median. Since the score of 94 is really represented by the interval 93.5 to 94.5, the exact lower limit would be $L = 93.5$.

The formula for the median for data in a simple frequency distribution like Table 3.4 is

$$Med = L + \left(\frac{0.5N - cum\ f}{f_{MED}}\right) \tag{3.2}$$

where

L is the exact lower limit of the interval containing the median
$0.5N$ is 50% of the size of the group
$cum\ f$ is the number of scores below the interval containing the median
f_{MED} is the frequency of the interval containing the median

For the reading comprehension scores of Table 3.4,

$$\begin{aligned} Med &= 93.5 + \left(\frac{8.5 - 6}{3}\right) \\ &= 93.5 + \left(\frac{2.5}{3}\right) = 93.5 + .83 \\ &= 94.33 \end{aligned}$$

So the median is 94.33, and we say that 50% of the distribution falls below 94.33, or that 94.33 is the point that divides the distribution exactly in half.

The median is a valuable measure of central tendency for measurements that are only of the ordinal type. Since no assumptions are made concerning distances between observations, it is meaningful to talk about median ranks, for example. So the median is an appropriate measure of central tendency for data that are ordinal or above.

Mode

The French expression *à la mode* literally means "in vogue" or "in style." That is exactly what the mode is—*the score that is made most frequently*, or seems to be "in style." It is classed as a measure of central tendency, since a glance at a graph of the frequency distribution (e.g., Figure 2.2, in the last chapter) shows the grouping about a central point, and the mode is the highest point in the hump, or the most frequent score. The mode is easily obtained by inspection, but it is the crudest measure of central tendency and is not used as often as either the mean or the median. Table 3.5 shows the number of sit-ups completed in 60 seconds by 14 members of a noon-hour fitness class before beginning an 8-week exercise program. We note that the mode is 21, since that score appears most often. However, there are three scores of 19, so it is necessary to distinguish between the *principal mode* and the *secondary mode*. Whenever there are two peaks or concentrations in a frequency distribution, we have what is known as a *bimodal distribution*. For example, if we were to measure the heights of a random sample of high school seniors, we would very likely get a bimodal distribution—one peak corresponding to the concentration about an average for women and another peak for men.

TABLE 3.5. **Determining the Mode: Number of Sit-Ups by a Fitness Class Before Starting an Exercise Program**

X	
24	
23	
22	
21	
21	Principal mode = 21
21	
21	
20	
19	
19	Secondary mode = 19
19	
18	
17	
16	

The chief value of the mode lies in the fact that it is easily obtained by inspection and is useful in locating points of concentration of like scores in a distribution. The mode may be calculated from measurements that are of the nominal type or above.

Comparison of the Mean, Median, and Mode

If a measure of central tendency is a single value that best represents the performance of the group as a whole, which single value should be used? If you compute the mean, median, and mode for the same set of scores, you very rarely will find that all three are identical. Which one will give us the "best single value" that describes the entire distribution?

The answer to this question is by no means a simple one. First of all, we can ignore the mode, since it is a rather crude measure of central tendency. This leaves the mean and median to be considered, and, as you probably have guessed, the mean is most often given as a measure of central tendency. Whenever you see the term *average* used in books, magazines, and newspapers in describing some distribution, most often the author is referring to the mean.

NOTE 3.1

THE EFFECT ON THE MEAN OF ADDING A CONSTANT

If some number, such as 3 (in other words, a constant), is added to all the scores in a distribution, the mean is increased by the amount of the constant. For instance, the mean of the scores 5, 2, 9, 6, and 3 is 25/5 = 5. If the constant 3 were added to each score you now would have 8, 5, 12, 9, and 6, with a mean of 40/5 = 8. Stated symbolically, if a constant C is added to each score in a distribution with a mean $\overline{X}$, the mean of the new distribution will be $\overline{X} + C$. This fact probably does not come as a startling revelation to you, but in later chapters we will have occasion to make use of this peculiar property of the mean. For the mathematically inclined, the proof is very simple. Given the mean of a set of scores:

$$\overline{X} = \frac{\Sigma X}{N} = \frac{X_1 + X_2 + \cdots + X_N}{N}$$

a constant is added to each score:

$$\overline{X} = \frac{(X_1 + C) + (X_2 + C) + \cdots + (X_N + C)}{N}$$

Combining terms, and noting that the sum of a constant is equal to N times the constant:

$$\overline{X} = \frac{\Sigma X + \Sigma C}{N} = \frac{\Sigma X + NC}{N} = \frac{\Sigma X}{N} + \frac{NC}{N}$$

So the mean of the new distribution is

$$\overline{X} = \frac{\Sigma X}{N} + C = \overline{X} + C$$

However, there are many instances where the median is a valuable statistic. It is not affected by extreme or atypical values as much as is the mean, so it is very useful in situations where the distribution is either positively or negatively skewed. For example, suppose we would like to calculate the average income for people living in a small midwestern town. Let us assume that in this community there is one millionaire and the rest of the residents are earning what we would judge to be ordinary incomes. This is a slight exaggeration, but it will serve to illustrate the usefulness of the median when distributions are markedly skewed. The mean, since it takes into account the exact value of each score, would be unduly influenced by the millionaire's income, and this measure of central tendency would be much too high. The median, on the other hand, is the center of the distribution and would not be affected very much by the addition of a single score at the extreme end. Clearly, the median would give the most accurate picture of the income for this situation.

The ability of the median to resist the effect of skewness is demonstrated in Table 3.6 with the personal incomes of the chemical-dependency therapy group mentioned earlier. The original group (left column) has a mean income of $9375 and a median income of $9500. Now suppose a new member joins the group, and let us say he or she has an income of $60,000. Look what happens to the mean and median income of this new group. The mean jumps to $15,000, which is higher than 7 of the 9 incomes. The median is not affected nearly as much, increasing only to $11,000. Again, the median gives a much more representative picture of the entire group.

TABLE 3.6. **The Effect of Skewness on Mean and Median Income**

Original Chemical-Dependency Group	Same Group Plus New Member
$17,000	$60,000
14,000	17,000
12,000	14,000
11,000	12,000
8,000	11,000
6,000	8,000
4,000	6,000
3,000	4,000
$\Sigma X = 75{,}000$	3,000
	$\Sigma X = 135{,}000$
$\overline{X} = \frac{75{,}000}{8} = \$9{,}375$	$\overline{X} = \frac{135{,}000}{9} = \$15{,}000$
Med = $9,500	*Med* = $11,000

The relationship of the mean and median in skewed distributions is demonstrated in Figures 3.3 and 3.4. The negatively skewed distribution of mathematics achievement scores discussed in the last chapter

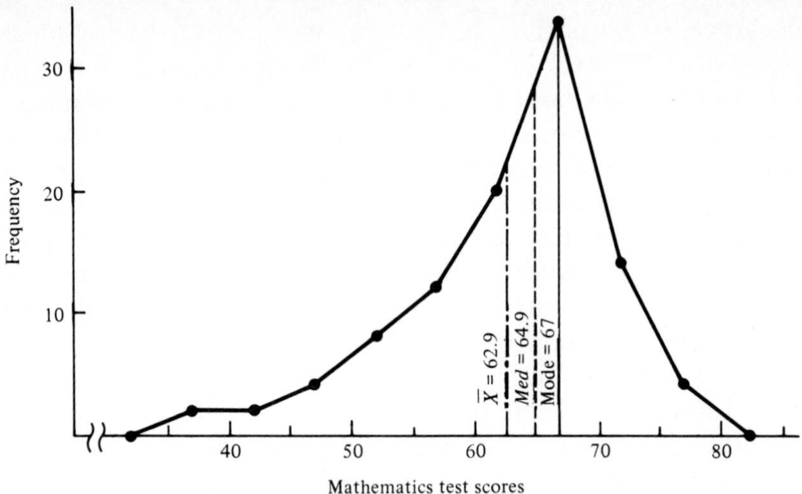

FIGURE 3.3. **Three measures of central tendency in negative skewness.**

is reproduced in Figure 3.3. Note that the mode has the highest value, 67, while the mean is at 62.9, and the median is in between the two at 64.9. This illustrates a general rule: *In negative skewness, the mean is the lowest value of the three measures of central tendency.*

In Figure 3.4 the distribution of incomes of 14,000 workers discussed in the last chapter as an example of positive skewness is reproduced.

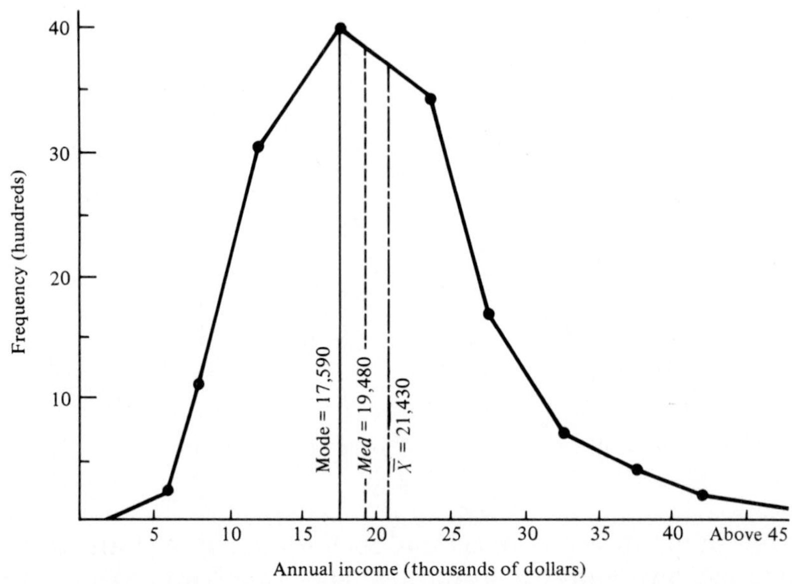

FIGURE 3.4. **Three measures of central tendency in positive skewness.**

Note that the mode is the lowest value, at $17,590, the median is $19,480, and the mean is the highest, at $21,430. *In positive skewness, the mean is the highest value of the three measures of central tendency.*

As you can see from the last two examples, the mean is affected greatly by the extreme or atypical scores in a skewed distribution, since it is "pulled" in the direction of the atypical values. It is for just this reason that the median is the preferred measure of central tendency when there is a marked degree of skewness in the distribution. It is much more likely to be the best single value that describes the distribution as a whole.

Stability

If the median is the preferred measure of central tendency in skewed distributions, why isn't it used all the time for all distributions? The answer to this question is that the *mean* is the most *stable* of the three measures of central tendency. By stability, we mean that in *repeated sampling* the means of the samples will tend to vary the least among themselves.

A complete treatment of *sampling* and *populations* will be coming up in a later chapter, but for the time being we can illustrate this concept by an example. If we took repeated samples of some variable in the population (e.g., IQ scores of 100 10-year-olds, chosen at random) and calculated the three traditional measures of central tendency for various batches of samples, each containing 100 IQ scores, we would find that the means would be most like each other. Since we often want to estimate what the population value is from just one sample, we naturally would want to use the measure of central tendency that is the most reliable or consistent. So the mean is used most often as the measure of central tendency, since it fluctuates the least from sample to sample.

Mean of a Population Versus the Mean of a Sample

The formula $\overline{X} = \Sigma X/N$ that we have been using is technically the formula for calculating the mean of a *sample*. The distinction between sample and population, as mentioned in the preceding paragraph, will be discussed at length in a later chapter, but at this point we need to distinguish between the mean of a sample and the mean of a population.

By *population*, we mean a group of elements that are alike in one or more characteristics as defined by the researcher. A population could be all the college students in the United States, or all the voters in a state, or all the seventh-graders in a school system, or all the dairy cattle in a county. The important thing to remember is that the population is defined by a researcher *for a particular purpose* and *all* the elements satisfying the criteria are members of that population. Size alone does not determine whether or not a group of elements is a population, although in practice most populations turn out to be rather large.

The formula for the mean of a population, μ, is[†]

$$\mu = \frac{\Sigma X}{N} \tag{3.3}$$

which contains the identical components, ΣX and N, as did the *sample* mean. The only difference, as should be obvious by now, is that the calculation of μ is based on *all* the elements of a population, whereas $\overline{X}$ is based on a smaller group of observations called a sample.

NOTE 3.2

ROUNDING OFF MEASURES OF CENTRAL TENDENCY

How many significant digits should you report when you calculate a measure of central tendency? The mean, since it is obtained by dividing ΣX by N, could have a whole string of digits to the right of the decimal point. For example, if $\Sigma X = 93$ and $N = 7$, you could report the mean as 13.3 or 13.29 or 13.286 or 13.2857, depending upon your enthusiasm for division. The problem is that the more digits you include, the more precision is implied—and the precision may be totally unwarranted. For that reason it is conventional to report measures of central tendency to *two more decimal places* than you have in your data. Therefore, for test scores that are whole numbers, such as 84, 72, and so on, your mean might be 58.62. Also, if your data consisted of reaction time measured in hundredths of seconds, such as 0.37, 0.29, and so on, your mean might be 0.3261. The median is treated in a similar fashion.

Calculating a Mean for Several Groups

There will be occasions when you will want to calculate a mean from a number of other means. For example, if the mean reading test score for a group of fourth-graders is 31.3, and for a group of fifth-graders it is 42.7, what is the mean for the combined group of fourth- and fifth-

[†]To avoid confusion, we will use Greek letters anytime we are talking about population values. The Greek letter μ, pronounced "mew," is the symbol for the population mean.

graders? With only this information, *you cannot calculate a combined mean.* You cannot simply "average" the two means to get a combined mean for the two groups (except in the unlikely event that the values of N for the two groups are identical). In order to calculate the combined mean for a group of means, you must know the size of each sample, or N. Then the formula is

$$\overline{X} = \frac{N_1\overline{X}_1 + N_2\overline{X}_2 + N_3\overline{X}_3 + \cdots}{N_1 + N_2 + N_3 + \cdots} \qquad (3.4)$$

where
- N_1 is the size of sample 1
- $\overline{X}_1$ is the mean of sample 1
- N_2 is the size of sample 2
- $\overline{X}_2$ is the mean of sample 2,

and so on.

Table 3.7 shows the means for a motor coordination test for three groups of 10–, 11–, and 12-year-olds, and our objective is to find the combined mean for all 80. Note that in the calculation of the combined mean the mean of each separate group is multiplied by its own N. The sum of these products is then divided by the combined N, giving a mean of 21.13 for all three groups together. This is obviously different from what you would have obtained if you had incorrectly "averaged" the three means themselves.

TABLE 3.7. **Calculation of a Combined Mean for Motor Coordination**

10-year-olds	11-year-olds	12-year-olds
$N = 10$	$N = 30$	$N = 40$
$\overline{X}_1 = 17$	$\overline{X}_2 = 20$	$\overline{X}_3 = 23$

$$\begin{aligned}\overline{X} &= \frac{N_1\overline{X}_1 + N_2\overline{X}_2 + N_3\overline{X}_3}{N_1 + N_2 + N_3} \\ &= \frac{10(17) + 30(20) + 40(23)}{10 + 30 + 40} \\ &= \frac{170 + 600 + 920}{80} = \frac{1{,}690}{80} \\ &= 21.13\end{aligned}$$

NOTE 3.3

WHAT THE COMBINED MEAN FORMULA REALLY MEANS

The formula for finding a mean of means is very easy to remember if you stop and note that $N\overline{X} = \Sigma X$. In other words, all that we are doing is converting the data for each group back to the original sum of the scores. After we have done this for all the groups, we add the individual sums together to get one grand sum of the scores, or a combined ΣX. This is then divided by the combined N for all groups, and the result is the ordinary mean

$$\frac{\Sigma X}{N}$$

By the way, the process of multiplying each $\overline{X}$ by its own N is called *weighting* the mean by its N. This makes sense if you refer to Table 3.7 and note that there are only ten 10-year-olds, while there are forty 12-year-olds. Since there are many more 12-year-olds, their mean should have a greater *weight* in the calculation of the combined mean. Or, stated another way, the mean of 23 is weighted by an N of 40, while the mean of 17 is weighted by only 10. We will have occasion to refer to the process of weighting in later chapters.

Some Concluding Remarks

There should be no doubt in your mind what the purpose of a measure of central tendency is. Quite clearly, it is an attempt to find the best single value that represents the performance of the group as a whole. A measure of central tendency of 57.5 would be the best single value that describes the performance of a group of high school seniors in terms of number of items answered correctly on a college entrance exam, or a group of white rats in terms of number of seconds taken to run a maze, or a group of job applicants in terms of number of pegs correctly placed in a manual dexterity test.

The problem arises when a decision must be made concerning *which* measure of central tendency is the best one to use for a given set of data. The mean, because it is the most stable measure of central tendency, is preferred in most cases. Since, as we indicated in Chapter 1, we are often interested in making inferences concerning the population, the mean is preferred because it is the most reliable. However, as we noted in the case of skewed distributions, the mean may not be the most representative value, and for this reason the median may give a better picture of the distribution as a whole.

SAMPLE PROBLEM

The sex-typing of occupations by children was studied by means of test of memory. Thirty fourth-graders were presented with flashcards containing pairs of proper names and occupations. Each received 12 cards with traditional sex-typed pairs (e.g., Carpenter-Bob, Nurse-Sue). Each flashcard was shown for 3 seconds with a 2-second interval in between. One minute after all cards had been viewed, the children were asked to recall as many of the pairs as they could. The number of errors made by each child is shown below (perfect recall = 0 errors).[†] Calculate the mean and median for the error scores.

0	2	2	2	0
1	0	0	1	6
4	8	2	0	0
4	2	2	0	0
2	2	0	2	2
2	2	2	4	0

Mean: $\overline{X} = \frac{\Sigma X}{N} = \frac{54}{30} = 1.80$ errors

Median: Since the group of scores above contains a number of tied scores, it is necessary to group them into a simple frequency distribution.

X	*f*	*cum f*
8	1	30
7	0	29
6	1	29
5	0	28
4	3	28
3	0	25
2	13	25
1	2	12
0	10	10
	N = 30	

$$Med = L + \left(\frac{0.5N - cum\,f}{f_{MED}}\right)$$

$$= 1.5 + \left(\frac{15 - 12}{13}\right)$$

$$= 1.5 + \tfrac{3}{13}$$

$$= 1.73 \text{ errors}$$

Mode: The most frequent error score was 2.

[†]Adapted from Blaske, D. (1984). Occupational sex-typing by kindergarten and fourth-grade children. *Psychological Reports*, 54, 795–801.

1. What is meant by the concept of central tendency?
2. How does the mean differ from the median?
3. What does $\Sigma(X - \overline{X}) = 0$ mean?
4. When might the median be preferred to the mean?
5. The mean is the most stable measure of central tendency. What does this mean?
6. In calculating a mean for several groups, each individual mean is weighted. What does this mean?

EXERCISES

1. Eighteen university freshmen applied for positions as peer tutors in the English Department. The selection committee makes its choice partly on the basis of the students' performance on the English Usage Test of the ACT (American College Testing Program). Their test scores are as follows. Calculate the mean and also demonstrate that $\Sigma(X - \overline{X}) = 0$.

27	23	22	28	26
23	25	25	30	21
29	26	26	28	27
24	25	24		

2. A school psychologist is attempting to measure students' intellectual ability without any racial or cultural biases and administers both forms of the Culture Fair Test to 20 students. Calculate the mean for this distribution and demonstrate that $\Sigma(X - \overline{X}) = 0$.

68	50	54	52	52
68	54	60	64	59
63	64	62	71	55
67	71	58	63	71

3. A recent research project funded by the National Institutes of Health studied weight loss in a sample of women. One group of 17 women used a combination of both diet and exercise and their weight loss (in pounds) over 15 weeks follows. Calculate the mean and median weight loss.

30	18	23	17	10
18	19	20	26	5
18	22	15	13	20
15	17			

4. Another group of 34 women, also in the NIH research project described in Exercise 3, used only a diet plan to lose weight. Their weight loss over the same 15-week period is shown below. Calculate the mean and median for this data and compare with Exercise 3.

14	10	15	8	11
11	17	16	10	12
7	9	3	15	12
16	7	8	12	4
13	9	12	11	13
9	5	10	8	16
10	7	10	7	

5. A study comparing the salaries of men and women was initiated to see if there were discrepancies in salaries for men and women holding identical jobs. The mean salaries for a sample of 64 women in 3 of the job categories are shown below. Calculate a combined mean to obtain the average salary for all 64 women.

Attorney	**Chemist**	**Architect**
$N = 45$	$N = 12$	$N = 7$
$\overline{X} = \$34{,}700$	$\overline{X} = \$38{,}500$	$\overline{X} = \$28{,}750$

6. The study described in Exercise 5 showed the following results for a sample of 337 men in the same 3 job categories. Calculate a combined mean salary for all 337 men.

Attorney	**Chemist**	**Architect**
$N = 255$	$N = 38$	$N = 44$
$\overline{X} = \$39{,}500$	$\overline{X} = \$43{,}100$	$\overline{X} = \$34{,}300$

7. A seventh-grade mathematics teacher administered a unit exam to one class and found out that the majority of the items were much too easy. Calculate the mean and median for the following test scores. What kind of skewness is illustrated here?

49	48	46	46	47
45	44	47	49	48
48	45	40	48	45
38	49	48	44	48
49	48	41	39	49
47	42	49	47	

8. The Sample Problem at the end of this chapter describes a research project where fourth-grade children were tested on their memory for flashcards containing sex-typed proper names and occupations such as "Secretary-Nancy." In another part of the experiment an additional 30 children were tested with flashcards containing *non* sex-typed pairs, such as "Plumber-Alice" or "Nurse-Greg." These cards were presented in the same manner as the others, and the number of errors made by each child is shown below (perfect recall = 0 errors). Calculate the mean and median for these error scores and compare them with the mean and median for the Sample Problem. What would you conclude?

4	8	4	6	8
5	4	8	5	6
2	0	1	8	10
6	8	8	6	6
4	9	0	4	6
4	7	6	6	4

9. An opinion poll of college students at a midwestern university was designed to find out their "major concerns in the nation today." Their concern about the issues of crime, energy, nuclear war, pollution, inflation, jobs, and taxes was assessed by a seven-point rating scale that asked the amount of their agreement or disagreement with the statement "I believe Congress should go all out to solve the problem of ______." They were asked to circle the number that best matched their opinion on each issue on the following scale:

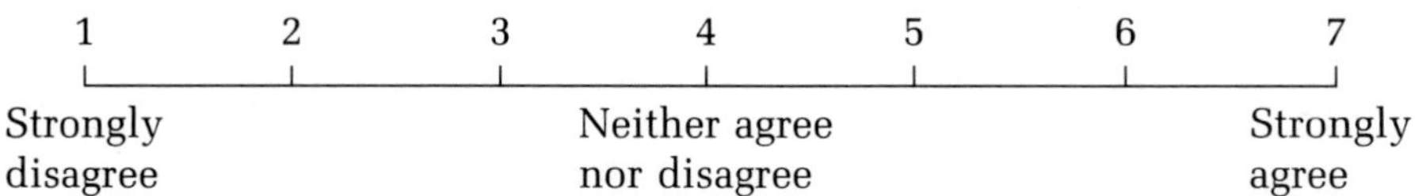

The results below show the ratings of a sample of 33 senior psychology majors on the issue of nuclear war. Calculate the mean and median for these ratings. Is there skewness in this distribution? If so, what kind?

7	7	7	7	7
7	7	6	7	7
7	7	6	4	1
6	5	7	7	7
1	5	6	7	4
2	7	7	7	6
7	5	7		

10. The same sample of psychology majors in Exercise 9 was also asked to rate the importance of taxes. Calculate the mean and median ratings for the following data and compare them with their rating of the issue of nuclear war.

6	2	4	5	1
6	2	5	6	5
4	6	5	5	5
5	3	4	5	6
4	5	6	6	4
2	6	5	6	6
4	6	5		

11. Type A behaviors (e.g., time urgency, competitiveness, and hostility) were studied in a group of college students. (More information about this research is presented in the Sample Problem at the end of Chapter 5.) In one sample, 15 Type A students were identified and their college grade-point averages (GPA) follow. Calculate the mean GPA for these students.

3.00	2.80	3.05	3.20	2.90
3.50	3.50	3.50	3.00	3.10
3.90	3.38	3.30	3.30	3.40

12. In the same study reported in Exercise 11, 20 Type B students (i.e., those lacking many of the traits that are characteristic of Type A students) were identified. Their GPAs are shown below. Calculate the mean for these Type B students and compare with the mean that you calculated for the Type A students in Exercise 11.

3.00	2.70	3.30	3.80	2.00
3.90	3.40	1.50	3.10	1.90
2.00	2.00	3.40	3.10	3.20
3.45	3.30	2.70	3.20	3.00

13. Many medical authorities recommend a maximum daily salt intake not to exceed 5 grams (approximately 1 teaspoon). Twenty adults enrolled in a wellness program filled out a nutrition questionnaire on the first day of class. Shown below is the estimated daily salt intake in grams for each class member. Calculate the mean and median for these data.

16	13	14	12	14
12	15	7	13	11
14	14	15	5	8
10	13	11	14	13

14. One of the wellness class members in Exercise 13 volunteered to survey the sodium content of a sample of popular dry breakfast cereals. She selected 10 cereals from a supermarket display and from information given on the carton recorded the sodium content in milligrams (mg) of an average serving with a half cup of whole milk. The results for the 10 cereals are listed in milligrams. Calculate the mean sodium content.

60	460	190	255	70
345	380	250	280	370

CHAPTER 4

Percentiles and Norms

In the last two chapters we have been talking about scores and groups of scores, raw scores, and frequency distributions. We have taken for granted that a score describes some property of an individual—reading readiness, IQ, clerical aptitude, or reaction time—but we have neglected to emphasize that a score has meaning only in relation to the rest of the group. The entire distribution provides a frame of reference for interpreting the individual score.

If little Brian excitedly tells his mother that he got a score of 35 on a nationwide achievement test in arithmetic, how should his mother react? Obviously, she would want more information before she decides to give him money for an ice cream sundae or send him to his room without TV for the evening. Clearly, what Mom wants to know is just where her Brian stands in relation to the rest of his class or in comparison with boys of similar age all over the nation.

It is for just this purpose that *derived scores* have been developed. A derived score is one that allows us to infer where it is located in some score distribution. As you saw in the last paragraph a raw score does not give us this information. In the present chapter we will be concerned with several kinds of derived scores; others will be discussed in later chapters. Just remember that the use of a derived score is an attempt to make a single score more meaningful by providing a frame of reference for purposes of comparison.

How Norms Are Constructed

Before test developers promote a new test and market it as a nationwide standardized test, they must first construct a set of *norms* based on the population for which it is intended. They may administer the test to a representative sample of several thousand third-graders, high school sophomores, government data entry operators, or army personnel—depending on the type of test. Although we cannot go into the intricacies of how these samples are chosen, let us say that the group of individuals on which the norms are based is an accurate, representative sample of the population that will eventually be taking the test. *It is the set of scores from this representative sample that constitutes the norms for a particular test.*

For example, the publisher of a well-known college ability test contacted the psychology department of a large number of colleges and universities and asked them to choose 20 freshmen and 20 sophomores at random from the student body and administer the examination. These results were forwarded to the publisher and the score distribution became the norms for this standardized test of college ability. A student taking the test in the future could now be compared with her fellow college students from all over the nation.

NOTE 4.1

WHAT IS STANDARD IN A STANDARDIZED TEST?

There are many misconceptions about the term *standard* in regard to a standardized test. The most common, no doubt, is that the norms accompanying a standardized test somehow form a standard or ideal against which performance is compared. It is as if the classroom teacher or a test examiner is the judge and jury for each person's test result and will determine if each examinee is performing up to par. Although this may be the case when certain test results are used for guidance purposes (e.g., in counseling the underachiever or the overachiever), the term *standardized* simply refers to the physical conditions under which the test is administered. In order to make certain that the examinee's performance can be accurately compared with the set of norms, it is essential that the test be administered under very precise, controlled, standard conditions. The examiner's manual is very detailed in this respect, demanding a comfortable testing room, adequate lighting, instructions that must be followed to the letter, precise, accurate timing of portions of the test that have time limits, and a host of other controlled conditions. All are factors that must be equivalent from administration to administration to ensure the accuracy of the set of norms to be used in evaluating an examinee's performance.

In another case the norms for one of the early intelligence tests were constructed from the test results of 3000 children between the ages of 6 and 12. Since it was believed that the occupational level of the fathers was a variable that was related to the children's test performance, the occupational level of the children's fathers in this group of 3000 was checked against 6 occupational classifications of males in the latest U.S. Census. These groups were professionals, semiprofessionals, businessmen, farmers, skilled laborers, and slightly skilled or unskilled laborers. Since there were differing proportions of men in these groups, the percentage of children with fathers of a certain occupation in the sample had to match that of the general population. For example, if 5% of all employed males were in the professional group, then 5% of the children in the sample must have fathers in the professional group. As a result of this rather elaborate sampling procedure the IQ norms were expected to be accurate and representative of children from ages 6 to 12 on this standardized intelligence test.

When norms are used to evaluate an individual score, the comparison often is made in either age or grade equivalent scores, or in percentile rank. These are described in detail in the sections to follow.

Age–Grade Norms

When a test developer uses the biographical data included on the test's answer sheet to determine the average grade in school or average chronological age for a particular distribution of test scores, the norms are called, respectively, grade norms or age norms. That is, for any given grade in school or age in years, the set of norms states the *average* raw score achieved by the sample. So, in Table 4.1 showing age–grade equivalents on a standardized arithmetic test, a raw score of 25 on the arithmetic achievement test was the average for students in the sample who were in the fifth month of the third grade or who were 8 years, 10 months of age.

When this test is used to evaluate a student's arithmetic achievement, the norms are interpreted in just this way. We enter the norms with the raw test score and determine what grade equivalent score or age equivalent score corresponds to that raw score. Again, using the example of Brian with his score of 35, we find that his arithmetic achievement score is comparable to that of the "average" 10-year-old, or the "average" fourth-grader during the seventh month of the fourth grade. All that is left to do in order to interpret Brian's raw score is to find out his age and grade. If he has just turned 9 years of age, or if he is just beginning the fourth grade, we would say that he is definitely above average in his grasp of arithmetic.

However, caution must be exercised in interpreting age–grade norms. It would be tempting to say that since Brian is performing as well as the average 10-year-old, he would have the same knowledge of arithmetic fundamentals that the average 10-year-old has. This is *not* necessarily true, and this is the reason quotation marks appeared around

TABLE 4.1. **Age–Grade Equivalents on an Arithmetic Achievement Test**

Raw Score	Grade	Age	Raw Score	Grade	Age
50	8.0	13–4	30	4.0	9–4
49	7.6	13–0	29	4.0	9–3
48	7.3	12–9	28	3.8	9–2
47	7.1	12–6	27	3.8	9–0
46	6.8	12–2	26	3.6	8–11
45	6.5	11–11	25	3.5	8–10
44	6.2	11–7	24	3.5	8–8
43	6.0	11–5	23	3.3	8–7
42	5.8	11–2	22	3.3	8–6
41	5.6	10–11	21	3.1	8–5
40	5.4	10–8	20	3.0	8–4
39	5.2	10–6	19	3.0	8–3
38	5.0	10–4	18	2.8	8–2
37	4.9	10–3	17	2.8	8–1
36	4.7	10–1	16	2.7	8–0
35	4.7	10–0	15	2.5	7–11
34	4.5	9–10	14	2.3	7–9
33	4.3	9–9	13	2.2	7–8
32	4.3	9–7	12	2.1	7–7
31	4.1	9–6	11	2.0	7–6

the word *average* in the preceding paragraph. Brian's superior score, which was equal to that of the average 10-year-old, is very likely due to his superior mastery of the material at his own level. He picks up more points at this level than would the average student who is a year older, but he would not do as well on more advanced material that is completely foreign to him. For this reason age-grade norms should be interpreted with caution.

Percentile Norms

Probably even more familiar to the average layperson are percentile norms. The test publisher administers the test to a large number of examinees to obtain norms, and an individual score is located in regard to the *percentage of the distribution falling below it*. Many tests have several sets of norms, depending on the purpose of the test. For example, a test of mechanical skills might have norms based on the performances of a group of diesel mechanics at a trade school, a group of high school senior home economics students, or a class of college sophomores in a predentistry curriculum.

The percentile is a way of expressing the location of a particular raw score in a distribution. In the last chapter we found that the median is that point in a distribution below which lie 50% of the scores. In exactly the same way we could calculate points below which lie 20%, 43%, 68%, or any percentage of the scores. These points are called *percentiles* and are usually denoted by the symbol P_p, where P is a percentile and the subscript p is the percentage of the cases below that point. P_{20} (read "a percentile of 20" or the "20th percentile"), for example, is the point below which lie 20% of the scores. Similarly, P_{68} would be the point below which lie 68% of the scores. Obviously, the median would be P_{50}.

To use percentile norms, all we have to do is enter the table of norms with the test score and find what percentage of the distribution (the standardization sample) falls below. Table 4.2 shows a set of norms for a particular college entrance examination developed by the American College Testing Program (ACT). A student who obtains a test score of 18 would be positioned in the distribution in such a way that 44% of the college-bound high school students have scores lower than hers. Stated another way, P_{44} is a score of 18.

You will often run across the term *percentile rank (PR)* in a discussion of percentiles. To avoid confusion, just remember that if you want to know the *score* below which a *given percentage* of the distribution falls, you are talking about a percentile. Thus the 44th *percentile* in Table 4.2 is a score of 18. If you are interested in knowing what *percentage* falls below a *given score*, you are dealing with the score's per-

TABLE 4.2. **ACT Percentile Ranks (Composite Score) for College-bound High School Students**

Standard Score	Percentile Rank	Standard Score	Percentile Rank
33	99.9	19	50
32	99.8	18	44
31	99.4	17	38
30	98.7	16	33
29	97	15	27
28	95	14	22
27	92	13	17
26	88	12	13
25	84	11	9
24	79	10	6
23	74	9	4
22	68	8	2
21	62	7	1
20	56		

From *Using ACT on the Campus* (Iowa City, Iowa, 1974), p. 12. Copyright 1974 by the American College Testing Program. Reprinted by permission of the publisher.

centile rank. Therefore, in Table 4.2, the percentile rank of a score of 18 is 44.

It frequently happens that the norms established for a nationwide standardized test may not be appropriate for a particular sample of examinees. For example, if a private college has rather strict admissions policies, the personnel deans may be interested in knowing how a high school student's entrance exam score compares to the scores obtained by students who were admitted to the college the year before. For this reason the school may construct its own norms, called *local* norms, based on its own student body. In this way the admissions board would know how well the individual compares with students already at the institution. Just such a set of local norms is shown in Table 4.3, for the ACT college entrance examination. Compare these local norms with the national norms of Table 4.2, to see how a selective admissions policy would alter the norms used to describe an individual's performance. For example, a score of 20 has a percentile rank of 56 on the national norms but a rank of only 36 on the local norms. Thus a student with this score would find that, nationally, 56% of the distribution have scores lower than hers while, at a more selective school, only 36% are lower than hers.

Sometimes percentiles are used to compare an individual's own performance on two or more tests. The percentile rank is valuable here because it is impossible to compare raw scores directly. Knowing that John Jones scored 27 on a reading test, 54 on an arithmetic achievement test, and 128 on a mechanical aptitude test does not help us know his relative strengths and weaknesses. However, if we know that his three scores were at the 76th, 53rd, and 83rd percentiles, respectively, we

TABLE 4.3. **ACT Percentile Ranks for a Local Institution (Based on a freshman class, N = 1078)**

Standard Score	Percentile Rank	Standard Score	Percentile Rank
31–36	99	20	36
30	98	19	29
29	97	18	23
28	94	17	17
27	90	16	12
26	86	15	7
25	79	14	5
24	72	13	4
23	64	12	2
22	55	5–11	1
21	45		

know quite a bit about his performance in relation to that of the rest of the examinees.

Percentile norms can also be used to compare an individual's performance on two or more parts of the *same* test. Table 4.4 shows the percentile ranks for the four subtests of the ACT: English, Mathematics, Social Studies, and Natural Sciences. A student who scored 21, 14, 25, and 18, respectively, on these 4 subtests would have percentile ranks of 70, 26, 80, and 39. This information could be valuable for the student and his or her adviser in planning a college schedule or future career.

TABLE 4.4. **ACT Subtest Percentile Ranks for College-Bound High School Students**

Standard Score	English	Mathematics	Social Studies	Natural Sciences
32	99.9	98	99.4	98
31	99.7	97	98.5	96
30	99.5	95	97	93
29	99.0	92	95	89
28	98	89	93	84
27	97	84	89	80
26	96	80	85	75
25	93	76	80	72
24	89	73	73	69
23	84	69	68	65
22	78	66	62	61
21	70	63	57	57
20	62	59	52	51
19	54	55	48	45
18	46	50	44	39
17	39	44	41	33
16	33	38	38	27
15	28	31	35	21
14	24	26	32	16
13	21	21	29	12
12	18	18	25	8
11	15	15	21	6
10	12	12	17	5
9	9	9	14	4
8	7	7	11	3
7	4	6	8	2
6	3	4	5	1

From *Using ACT on the Campus* (Iowa City, Iowa, 1974), p. 12. Copyright 1974 by the American College Testing Program. Reprinted by permission of the publisher.

Although not always highly publicized, it is a fact of life that a test score is subject to error. We will spend considerable time on errors of measurement in Chapter 14, on the subject of test construction, but, for now, let us assume that a test score, like any measurement of height or volume or weight, is subject to some degree of error. Occasionally, test publishers admit the possibility of error, and instead of listing a single percentile rank, they identify each raw score with a *percentile band*. This is to be interpreted according to the following definition: "Two out of three times, the individual's *true* score (i.e., with no measurement error) lies within the limits of the percentile band." Additional explanations of this concept of the probable value of a true score will be forthcoming in Chapter 14.

Table 4.5 shows the national norms for the Cooperative English Test, and, as you notice, the test scores are accompanied by a percentile band. For example, a test score of 157 has a band extending from a percentile rank of 29 to a rank of 52. This means that two out of three times the individual's *true* score would lie in such a way that from 29% to 52% of the distribution is below that point. Such a range of values may seem outrageously imprecise when we are trying to assess an individual's performance, but a percentile band is only admitting a very real fact of life—there are errors in measuring performance—and stating the norms in this manner is one way of recognizing these errors.

Although percentile bands may at first be confusing to work with, it is helpful to realize that the *most likely* location of an individual's true score is near the middle of the percentile band. For example, in Table

TABLE 4.5. **Total Score Percentiles on the Cooperative English Test**

Score	Percentile Band	Score	Percentile Band
180–182	99–100	158–159	36–59
178–179	97–99.8	156–157	29–52
176–177	95–99.4	154–155	23–44
174–175	91–99	152–153	19–36
172–173	86–97	150–151	14–29
170–171	80–95	148–149	10–23
168–169	74–91	146–147	7–19
166–167	67–86	144–145	5–14
164–165	59–80	142–143	3–10
162–163	52–74	140–141	2–7
160–161	44–67	138–139	1–5

From *Technical Report, Cooperative English Test* (Princeton, N.J., 1960), p. 27. Copyright 1960 by the Educational Testing Service. Reprinted by permission of the publisher.

4.5, an observed score of 158 could have a true score that falls between the percentile ranks of 36 and 59, but it is more likely to be near a percentile rank of 45 or 50 than near the extreme limits of 36 or 59.

The concept of a percentile band should remind us that we need to use caution in comparing two individuals with slightly differing scores. To say, for example, that Joe's score of 169 indicates poorer English achievement than Carol's score of 171 violates what is known about errors of measurement. We simply cannot make this fine a discrimination without additional evidence.

The theory behind the construction of percentile bands presented above has been necessarily brief; additional concepts remain to be covered in future chapters. It is hoped that everything will fall into place when the topic of test construction is covered in Chapter 14.

A Final Word About Norms

Earlier in the chapter you noted the use of sampling procedures to obtain norms against which an individual's score could be compared. Samples of a population constitute most norms (e.g., 485 ninth-grade industrial arts students or 1730 government clerk-typists), but there is an increasing use of entire test populations as normative groups. The use of computers and associated data-handling devices has made it possible to tabulate percentile ranks for an entire year's crop of test results. Although much smaller samples could give almost the same degree of accuracy, it is not unusual to have the norms accompanying a test based on huge numbers of examinees. For example, the college entrance exam norms of Table 4.2 are calculated from test results of over 2 million examinees. This seems to be a trend among test publishers whose tests are widely used and very popular. You must admit that a "sample" of over 2 million is impressive indeed!

Graphical Methods for Percentiles and Percentile Ranks

Although a table of norms is often used to determine percentiles or percentile ranks, graphical methods are sometimes employed to permit a quick estimate of the numerical values. Graphical methods are also used when we would like to display the results of two or more distributions at the same time. Before proceeding, let us review the distinction between a *percentile* and a *percentile rank*. If you wish to know what score is the point below which a given percentage of the distribution falls, you are interested in a *percentile*. If you wish to know what percentage of the distribution falls below a given score, you are concerned with *percentile rank*. If you have trouble discriminating, remember:

1. *Percentile:* Given the percentage, find the score.
2. *Percentile Rank:* Given the score, find the percentage.

It is hoped that the sections to follow will clear up any confusion between the two concepts. In keeping with our attention to the type of measurement scale with which a given statistic can be used, we note that percentiles and percentile ranks can be determined from data that are ordinal or above in nature, and that the percentiles and percentile ranks themselves are ordinal values.

Constructing the Ogive

The graph that allows us to read percentiles and percentile ranks directly is called a *cumulative percentage curve* or *ogive* (say "o-jive"). The graph is constructed from a slightly modified grouped frequency distribution such as is shown in Table 4.6. This table contains scores on a reading readiness test administered to 50 first-graders.

TABLE 4.6. **Cumulative Percentage Distribution of Reading Readiness Scores**

Scores	*f*	*cum f*	*cum percentage*
85–89	1	50	100
80–84	3	49	98
75–79	6	46	92
70–74	15	40	80
65–69	12	25	50
60–64	8	13	26
55–59	3	5	10
50–54	2	2	4
	N = 50		

To the grouped frequency distribution that we discussed in Chapter 2 we have included two more columns: a *cumulative frequency* (*cum f*) column and a *cumulative percentage* (*cum percentage*) column. The *cum f* column is simply the addition of the frequencies in each interval as you count up from the bottom. For example, there are 2 scores in the 50–54 interval and 3 in the 55–59 interval, so the cumulative frequency for the 55–59 interval is 5. There are 8 scores in the 60–64 interval, so the cumulative frequency for the 60–64 interval is 13, since in *that interval and below* there are 13 scores. Technically speaking, the entry in the *cum f* column gives the number of scores *below the exact upper limit of the interval*. For example, the 13 in the *cum f* column opposite the 60–64 interval indicates that there are 13 scores below a score of 64.5. Similarly, there are 46 scores below a score of 79.5. This is an important point because the accuracy of your graph depends on this fact.

The *cum percentage* column contains the same information as the *cum f* column, except that the entries have been converted to percentages. Instead of saying that 13 scores are below 64.5, we can say that 26% of the group are below 64.5. To obtain the entries in the *cum percentage* column, you divide each entry in the *cum f* column by the total number of scores, N, and multiply by 100. So, for the entry opposite the 80–84 interval, you would calculate $(49/50) \times 100 = 98\%$. You know that 98% of the scores fall below 84.5.

After the cumulative percentage distribution has been prepared as shown in Table 4.6, the steps in the construction of the ogive are as follows:

1. Lay out the area on a sheet of graph paper with the usual proportions and label the x- and y-axes as shown in Figure 4.1.
2. Place a dot above the *exact upper limit* opposite the cumulative percentage on the y-axis for all the intervals. For example, in Figure

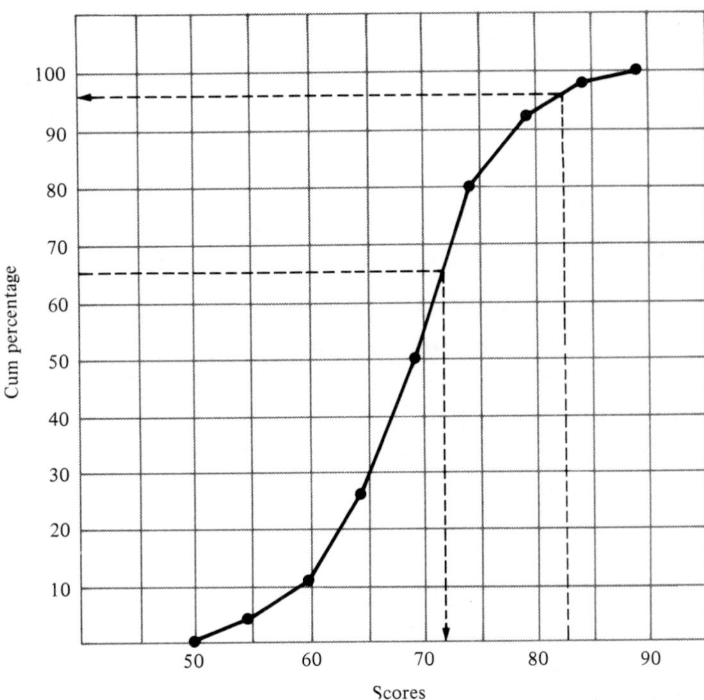

FIGURE 4.1. **Ogive of the reading readiness scores.**

4.1 there is a point above 64.5 and opposite a percentage value of 26.

3. Include a point at the exact lower limit of the bottom interval opposite a percentage value of 0. In Figure 4.1 the curve drops to 0 at a score of 49.5.
4. Connect the dots by straight lines.

Reading Percentiles and Percentile Ranks from the Ogive

Once you have drawn the ogive (or you see the finished product in a book or professional journal), it is a simple matter to determine percentiles or percentile ranks directly from the graph.

If you want to find a particular *percentile*, you simply draw a horizontal line from the cumulative percentage scale on the *y*-axis until it intersects the curve. At the point of intersection you drop a vertical line perpendicular to the x-axis and read the score where the perpendicular line touches the x-axis. For example, in Figure 4.1 a percentile of 65, P_{65}, is found by the above method to be a score of 72.

The reverse procedure is used to determine the *percentile rank* of a particular score. You draw a vertical line from the score until it intersects the curve, and a horizontal line from that point of intersection over to the cumulative percentage scale. That point is the percentile rank of the score. For example, in Figure 4.1 the percentile rank of a score of 83 is 96. Any percentile or percentile rank can be found in this manner, but the accuracy of the procedure depends on how precisely the graph is drawn. Remember that any point on the curve tells us *what percent of the distribution scored that value or below*.

Comparing Ogives for Two Different Distributions

A very useful technique that is often employed with the ogive is a comparison of two frequency distributions. In Figure 2.4 in Chapter 2 there was a comparison of the frequency polygons of motor coordination test scores for a group of seventh-grade boys and girls. The distributions have been converted to ogives and are shown in Figure 4.2.

It is obvious from Figure 4.2 that the coordination test performance of seventh-grade girls was superior to that of their male counterparts. But by using the techniques described above in reading percentiles and percentile ranks directly, we can gain a great deal of specific information as well and can do such things as:

COMPARING MEDIANS. A horizontal line from the 50th percentage value intersects the boys' curve at a score of 33 and the girls' curve at a score

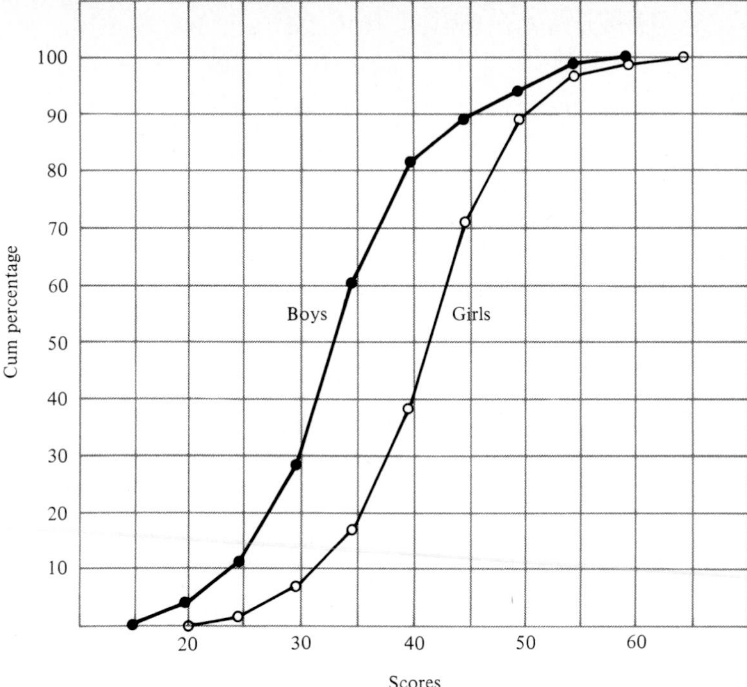

FIGURE 4.2. **Ogives for coordination scores for seventh-grade boys and girls.**

of about 41.2. The "average" girl's performance is a good 8 points higher than the "average" boy's.

COMPARING SCORES. If the test manual states that anyone with a score of 27 or below should be given remedial exercises, how do boys and girls compare? A vertical line from the score of 27 intersects the girls' curve at about 4% and the boys' at about 20%. In other words only 4% of the girls but 20% of the boys would score below 27, or in the "remedial" range.

COMPARING PERCENTILES. Just as when comparing medians, you might decide to compare other percentiles. For example, what score has to be made to place a boy or girl in the top 25% of his or her group? The value that you want is, of course, the 75th percentile, P_{75}, and a horizontal line from the 75th percentage value intersects the boys' curve at a score of 38 and the girls' curve at a score of about 45.5. Comparatively speaking, the examinee would need a coordination score of at least 46 to be in the top 25% of the girls' group but only 38 to be in the top 25% of the boys' group.

It will come as no great surprise to you to find that percentiles are based on a system of 100 units and that theoretically a percentile can take any value between 0 and 100. You should note that although the distribution is usually broken into pieces of 100 there are other systems of dividing a distribution as well. You could conceivably divide the distribution into equal halves, quarters, tenths, or whatever fraction suited your purpose. There are certain divisions that are more common than others:

QUARTILES. The distribution is divided into four equal parts, with 25% of the distribution in each part, so obviously Q_1 is the same as P_{25}, Q_2 equals P_{50}, or the median, and Q_3 equals P_{75}.

DECILES. The distribution is divided into 10 equal parts, with 10% of the distribution in each part, and D_1 equals P_{10}, D_5 equals P_{50}, and so on.

There are other divisions besides those just mentioned, but they are not commonly used. Regardless of the type of division used, it must be remembered that P_{72}, or Q_3, or D_7 are points in the distribution, not intervals. You may score *at* the 72nd percentile, or *at* the 3rd quartile, or *at* the 7th decile, but not *in* the 3rd quartile or *in* the 7th decile. The reason for the confusion is that we tend to think of quartiles as quarters and deciles as tenths, and obviously you can score *in* the upper quarter or *in* the bottom tenth of a distribution. But, again, percentiles, quartiles, deciles, and other "iles" are *points*, and you score either at a given point (e.g., exactly Q_3) or in between two points (e.g., between D_7 and D_8).

STUDY QUESTIONS

1. What does an age–grade norm tell us about a child's performance?
2. What is meant by the term *percentile rank*?
3. What is the difference between a percentile and a percentile band?
4. A student taking the ACT scores 25 on the English subtest. What would he have to score on the Natural Sciences subtest to have the same percentile rank? (Use Table 4.4.)
5. What is the purpose of the cumulative percentage curve?
6. Why is it incorrect to say that someone scored "in the first quartile?"

SAMPLE PROBLEM

In Exercise 13 at the end of Chapter 2, a grouped frequency distribution of the number of hours spent studying by a sample of 207 college students is shown. To assist us in analyzing this data further, we need to calculate a cumulative frequency column and a cumulative percentage column so we can construct an ogive. Then we can draw some conclusions from the information shown in the ogive such as

a. What is the average (median or P_{50}) number of hours spent studying?
b. How many hours would we have to study in order to tell our academic adviser that 75% of the students spend less time studying than we do?
c. What percent of the students study 10 hours or less?

Table 4.7 shows the frequency distribution of hours spent studying per week, and we have calculated the *cum f* and *cum percentage* columns using the instructions in the chapter.

TABLE 4.7. **Cumulative Percentage Distribution of Hours Studied per Week**

Scores	*f*	*cum f*	*cum percentage*
41–43	2	207	100
38–40	0	205	99
35–37	4	205	99
32–34	9	201	97
29–31	8	192	93
26–28	14	184	89
23–25	20	170	82
20–22	22	150	72
17–19	27	128	62
14–16	22	101	49
11–13	35	79	38
8–10	23	44	21
5–7	21	21	10
	N = 207		

Plotting the ogive results in the graph shown in Figure 4.3. The median (P_{50}) would be found by dropping a vertical line across from the 50th percentage value and seeing it touch the x-axis at about 17 hours. Doing the same with the 75th percentage value, the vertical would touch the x-axis at about 23 hours. Finally, to find what percent of the students reported studying 10 hours or less, we would find the percentile rank of a "score" of 10 by noting that a horizontal line from the curve above 10 touches the *y*-axis at about 19. So we would conclude that about 19% of the students reported studying 10 hours or less per week.

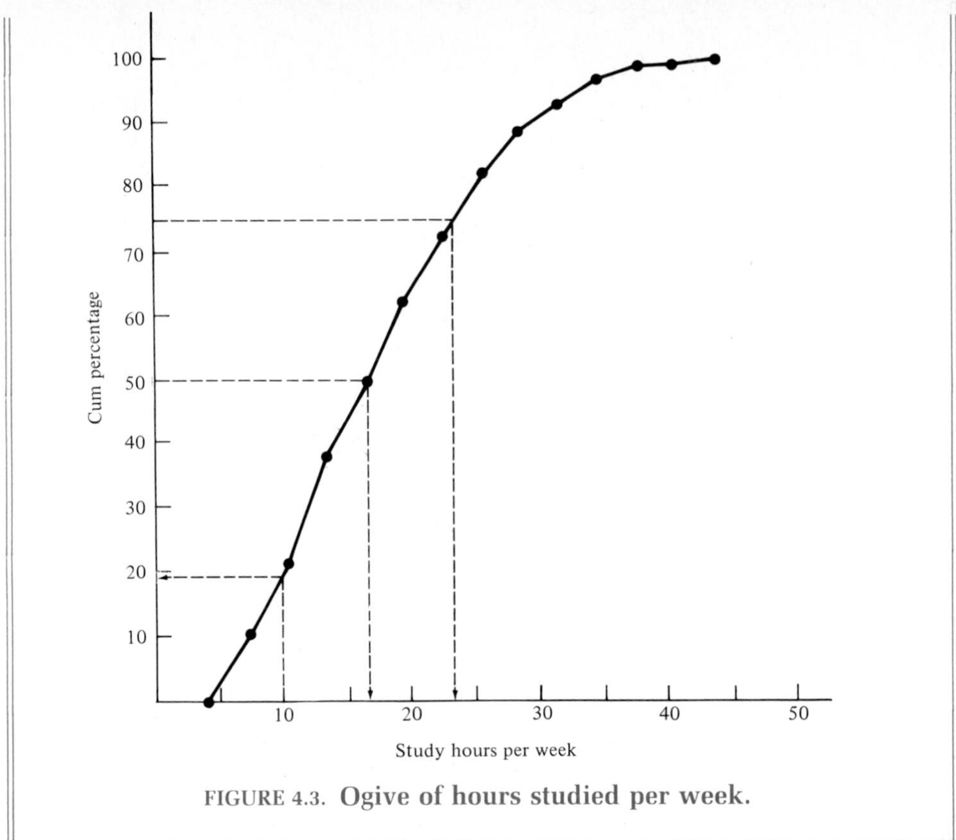

FIGURE 4.3. **Ogive of hours studied per week.**

EXERCISES

1. A student whose age is 9 years and 2 months scores 42 on the Arithmetic Achievement Test. Use the norms of Table 4.1 to determine her age–grade placement. Why would it be incorrect to say that her performance is comparable to that of the average 11-year-old?

2. Dave has just had his 9th birthday. What grade would we expect him to be in? If his mathematics ability is about average for his age, what score would we expect him to achieve on the Arithmetic Achievement Test? (Use Table 4.1.)

3. Lori's composite score on the ACT was 18. Use the national norms in Table 4.2 to determine where she is located in the norm group. Now do the same using the local norms of Table 4.3. What might be the reason for the difference in where she is located in the two sets of norms?

4. Compare the national ACT norms of Table 4.2 with the local norms of Table 4.3. In most cases the percentile ranks of the local norms for a given standard score are lower than the national percentile ranks. What does this mean?

5. In Table 4.4 note that a standard score of 18 is at the median (P_{50}) for the norm group taking the Mathematics subtest of the ACT. What would you have to score (approximately) to be at the median of the Natural Sciences subtest?

6. Use Table 4.4 to evaluate the Social Studies percentile rank of an individual with a standard score of 11. What would he or she have to score to obtain the same percentile rank on the Natural Sciences subtest?

7. Stephanie scores 159 on the Cooperative English Test. Use the norms of Table 4.5 to evaluate her performance in comparison with the norm group.

8. Jeff scores 156 on the Cooperative English Test. Why can't we say definitely that Jeff has poorer English skills than Stephanie with her score of 159?

9. Use the ogive of the reading readiness scores of Figure 4.1 to answer the following.
   a. What was the median (P_{50}) for this group of children?
   b. Twenty percent of the children had scores below what value?
   c. What percent of the children had scores of 67 or above?

10. Use the ogive of the boys' and girls' coordination scores in Figure 4.2 to determine the following.
    a. The top 20% of the girls are considered "highly coordinated." What coordination test score would be necessary to be in this group?
    b. What percent of the boys are at this score or above?
    c. What percent of the girls scored higher than the boys' median of 33?

11. Ninety-four students in an introductory psychology course are given a 24-item quiz with the results shown below. Calculate the values for a *cum f* column and *cum percentage* column and use a sheet of graph paper to draw an ogive for these quiz scores. (*Hint:* Remember to plot each point at the exact "upper limit" of each interval: 10.5, 11.5, etc.)

Score	f	**Score**	f
24	1	16	12
23	4	15	7
22	2	14	5
21	8	13	5
20	9	12	2
19	10	11	1
18	14	10	2
17	12		N = 94

    a. What was the median (P_{50}) performance on this quiz?
    b. The instructor announces that the bottom 20% of this group will receive a failing grade. This would include scores of ____ or below.
    c. Scores of 22 or above were necessary to receive a grade of A. According to the ogive what percent of the group received A grades?

12. Exercise 14 in Chapter 2 described a study of the number of hours worked by college students who held part-time jobs. The grouped frequency distribution of hours worked per week is reprinted below. Calculate the *cum f* and *cum percentage* columns and use graph paper to draw an ogive of the data.

Hours Worked	f
27–29	1
24–26	3
21–23	1
18–20	12
15–17	9
12–14	9
9–11	49
6–8	55
3–5	51
0–2	7
	N = 197

a. What was the median (P_{50}) number of hours worked per week by those students with part-time jobs?
b. A dormitory counselor believes that students working 15 or more hours per week will have serious academic difficulties. According to this study what percent of the students will be affected?
c. The bottom 60% of the students work ____ hours or less per week.

13. One of the purposes of the State-Trait Anxiety Inventory (STAI) is to measure the amount of anxiety that is the relatively stable and consistent anxiety-proneness that is characteristic of an individual's personality. This trait anxiety was investigated in a group of general medical–surgical patients and a group of neuropsychiatric patients by the author of the inventory (Spielberger, 1983). The ogives from a similar distribution of scores are shown in Figure 4.4.†
a. Compare the STAI median for the general medical–surgical patients with the median for the group of neuropsychiatric patients.
b. A score of 40 on the STAI was equaled or exceeded by what percent of the general medical–surgical patients?
c. What percent of the neuropsychiatric patients equaled or exceeded this same score?

14. From the ogives of the State-Trait Anxiety Inventory in Figure 4.4, determine the following.

†Adapted from Spielberger, C. D. (1983). *Manual for the State Anxiety Inventory*. Palo Alto, Calif.: Consulting Psychologists Press, Inc.

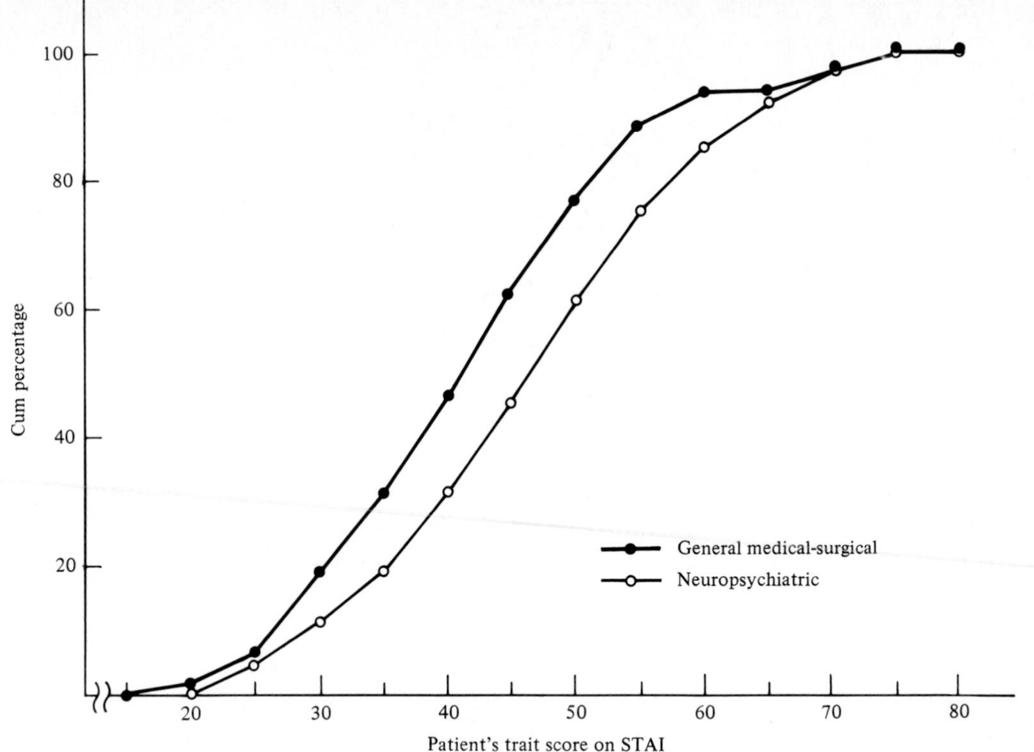

FIGURE 4.4. **Ogive of trait scores on the State-Trait Anxiety Inventory by two types of patients.**

a. Determine P_{30} for both the general medical–surgical patients and the neuropsychiatric patients.

b. There is a difference of about 5 points between the 2 patient groups in part (a). What does this mean?

c. What is the percentile rank of an STAI score of 50 for both patient groups? What does this mean?

CHAPTER 5

Variability

If one thing is obvious from a casual observation of human behavior, it must be the notion of variability. Some people are short, some are tall, others are in between, One 5-year-old might be forward, aggressive, and noisy, while her playmate is withdrawn, passive, and quiet. A teacher giving a test may find that a few of his sixth-graders get top scores, a few get very low scores, and most scores cluster in the center of the group. You notice that your car averages 21 miles per gallon, but you know that on a long trip you can squeeze 26 out of it while in downtown traffic the mileage drops to a mere 13.

Not only is there variation among different people and among different things, but the same person or thing may vary from time to time in certain characteristics. Although Tom Kite and Greg Norman are capable of shooting a subpar 66, both can also score a disastrous 77 in a PGA golf tournament. Similarly, a B student will get A's on occasion and also a few C's.

We said earlier that the task of statistics was to reduce large masses of data to some meaningful values. In Chapter 3 you saw how a measure of central tendency yielded the best *single* value that described the performance of the group as a whole. It was a value that best represented the entire group of observations. But as you noted in the first couple of paragraphs of this chapter, there is more to describing a group of observations than noting the *average* performance, since a measure of central tendency tells you nothing concerning the variation about the average. In preparing frequency distributions and calculating measures of central tendency in earlier chapters, you noted that some observations fell below the mean while some were above the mean. This fluctuation of scores about a measure of central tendency is called variability.

In short, to describe a set of observations accurately we need to know not only the central tendency but also the variability of the observations.

For example, let us suppose that we have two seventh-grade math teachers. Jones and Smith, who give an identical national achievement test in mathematics to their two classes. They happen to meet in the hall one day after the tests are scored, and Jones states that her class averaged 72. Smith stops short and remarks that it certainly is a coincidence because his class average was also a 72.

A naive observer would tend to say that the abilities of the two classes must be quite similar, since they were equal on the achievement test. But note that nothing has been said about the variability of the two groups. It is possible that Jones has an ordinary class with several very bright students, a few slow ones, and the rest run-of-the-mill, average seventh-graders. Smith, on the other hand, could have a peculiar class that has only "average" students if the bright students have been assigned to an accelerated ability group and the slow students have a special class of their own.

The frequency polygons shown in Figure 5.1 illustrate this somewhat contrived, but not uncommon, example. The scores of Jones' students are spread out over a wide range (approximately 42 to 97), since her group contains a few excellent students and several poor students. The scores from Smith's class, on the other hand, show much less dispersion (scores from 51 to 85), which is what you would expect when those students likely to make the high scores and the low scores are in special

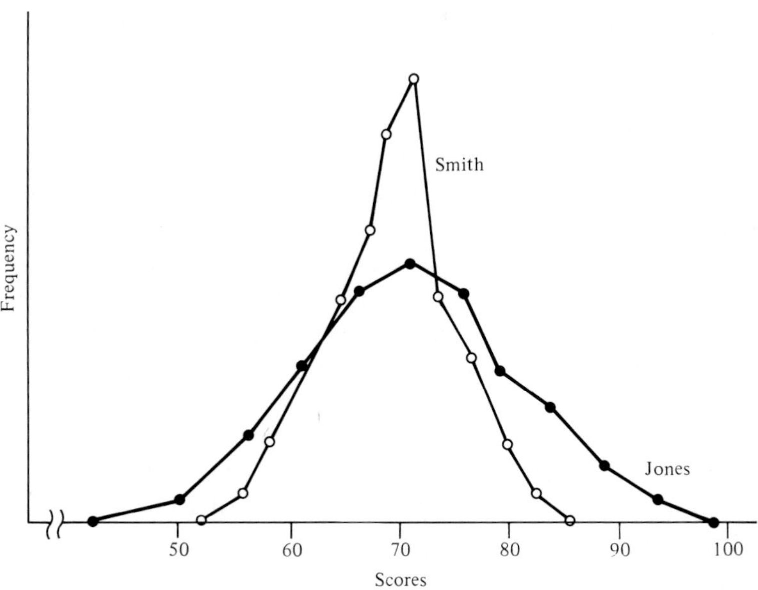

FIGURE 5.1. **Frequency polygons for two distributions of arithmetic achievement scores.**

classes somewhere else in the building. In Figure 5.1 the greater variability of Jones' class is obvious, and there is no doubt that despite the fact that the two classes have equal means, the scores of Jones' students are different from those of Smith.

Clearly, a measure of central tendency is not enough to describe a distribution accurately. Some unkind stories concerning statisticians have arisen from just this last point—you may have heard such things as "A statistician drowned in a river whose average depth was 4 feet" or "A statistician is one who with feet in the oven and head in the refrigerator is, on the average, quite comfortable.

Since a measure of central tendency does not give information about the variability of a group of observations, it is necessary to have some way to measure precisely the amount of variability (scatter, dispersion, spread, or variation) that is present in a distribution of scores. We would like to have a convenient and concise measure that tells us instantly something about how much the scores are spread out. The frequency polygons in Figure 5.1 tell us this, of course, but a single measure would be much more efficient. This measure would do for the characteristic of variability what the mean or median did for central tendency. Let us examine several ways of measuring variability.

Range

One of the simplest and most straightforward measures of variability is the *range*. This statistic can be calculated for measurements that are on an interval scale or above. The range is simply the difference between the *highest* and the *lowest* scores in a distribution plus 1. For example, if a distribution of weights of 100 college men showed the heaviest at 225 pounds and the lightest at 132 pounds, the range would be 225 − 132 + 1 = 94 pounds. (Remember that we add 1 to the difference between the highest and lowest scores, since technically we should subtract 225.5 − 131.5 = 94.) Or, in the example described previously in Figure 5.1, Jones' class had a range of 97 − 42 + 1 = 56, while Smith's class was 85 − 51 + 1 = 35. Obviously, this single value tells us something about the scattering or spread of scores, and the range of 56 for Jones' class indicates considerably more variability in the mathematics achievement scores than the range of 35 for Smith's.

Although the range is a handy preliminary method for determining variability, it has two serious weaknesses. First of all, one extreme value can greatly alter the range. Again looking at Figure 5.1, note that the lowest score in Jones' class is a 42. If the student making that score had been sick that day, the next lowest score might have been a 50. In this case the range would be 97 − 50 + 1 = 48, which is considerably less than the original range of 56, and all because of a flu bug.

The second, and most serious, drawback is that since the range is based on only two measures, the highest and the lowest, it tells us nothing about the *pattern* of the distribution. A good illustration of this drawback is shown in Table 5.1.

TABLE 5.1. **Two Distributions with Equal Ranges but Dissimilar Patterns of Dispersion**

Group A		Group B	
40		40	
33		39	
33	$\overline{X} = \frac{325}{10} = 32.5$	39	$\overline{X} = \frac{325}{10} = 32.5$
33		38	
33		38	
32	Range = 40 − 25 + 1 = 16	27	Range = 40 − 25 + 1 = 16
32		27	
32		26	
32		26	
25		25	
ΣX = 325		ΣX = 325	

Both group A and group B have identical means, 32.50, and identical ranges, 16, but a glance at the *pattern* of the distributions shows how dissimilar they are. The scores of group A range from 25 to 40, with most of the other values clustered tightly around the mean of 32.50. Group B's scores also range from 25 to 40, but note that most of the scores are grouped at the extreme ends of the distribution. In this case the range is misleading as a measure of variability, since it says that groups A and B have equal variability although it is obvious that there are rather severe differences between the two.

It is this last point that is important for out attempt to measure variability in an accurate way. We need a method that depends on *all* the measures in a distribution, not just the two extremes. If we define variability as the fluctuation of the scores about some central point, we need a method that will tell us *how much* these scores are fluctuating about that point, or how far they are from the mean. In Table 5.1 the majority of the scores of group A do not deviate very far from the mean, while those of group B are scattered quite a distance from the mean. Clearly, we need a method that will be sensitive to how far each individual observation deviates from the mean, so let us take a look at one method that might be of help in measuring the amount of deviation.

Average Deviation

You will recall from Chapter 3 that the mean was defined as the center of the deviations, and to prove it we subtracted the mean from each score. For example, if a distribution consisted of five scores, 14, 12, 9, 17, and 8, the mean would be $\frac{60}{5} = 12$. Subtracting the mean from each score $(X - \overline{X})$, we would get values of 2, 0, −3, 5, and −4.

It should be obvious that if we want a measure of variability that takes into account how far each score deviates from the mean, the deviation score $(X - \overline{X})$ would give us this information. If the values of $(X - \overline{X})$ are relatively small, it would indicate that there is not much variability and that the scores are near the mean, while if the values of $(X - \overline{X})$ are large the scores must be scattered farther from the mean.

To construct a measure of variability, it would be necessary only to add the $(X - \overline{X})$ values to get an overall picture of how much variation there was. However, we noted in Chapter 3 that the sum of the deviations about the mean, $\Sigma(X - \overline{X})$, is always 0. We can easily overcome this problem by ignoring the sign of the deviation score (in other words, using the *absolute* value) and then dividing this sum by N to get the average deviation. The formula for the average deviation is

$$AD = \frac{\Sigma\,|(X - \overline{X})|}{N} \tag{5.1}$$

where

$\Sigma\,|(X - \overline{X})|$ is the sum of the absolute values of the deviations from the mean

N is the number of observations

For the group of 5 scores listed earlier:

X	$(X - \overline{X})$	$\|(X - \overline{X})\|$
14	2	2
12	0	0
9	−3	3
17	5	5
8	−4	4
$\Sigma X = 60$	$\Sigma(X - \overline{X}) = 0$	$\Sigma\,\|(X - \overline{X})\| = 14$

$$\overline{X} = \frac{\Sigma X}{N} = \frac{60}{5} = 12$$

$$AD = \frac{\Sigma\,|(X - \overline{X})|}{N} = \frac{14}{5} = 2.8$$

From an intuitive point of view the average deviation makes good sense. The value of 2.8 in the example would tell us that on the average each observation deviates 2.8 units from the mean. So the average deviation is a sensible, easily understood, accurate measure of variability.

Unfortunately, it is hardly ever used. One thing that you will become aware of should you go on to more advanced courses in statistics is that all of the statistical techniques are linked together by certain theoretical foundations—and that the average deviation simply doesn't fit into this framework. So it is doubtful that you will ever see any reference to the average deviation again. The only reason that it was brought up at this point was to emphasize the use of the deviations from the mean $(X - \overline{X})$ as a way of indicating the amount of variability. These deviations will be an important part of the measure of variability to be discussed in the next section.

Standard Deviation

The most widely used measure of variability is the *standard deviation*. This statistic makes use of the deviation of each score from the mean, but the calculation, instead of taking the absolute value of each deviation, squares each deviation to obtain values that are all positive in sign. You will remember from elementary algebra that when two numbers of the same sign are multiplied together the product is always positive. So when we square each of the deviations, we get positive numbers whether our original deviations were positive or negative. We add these squared deviations to obtain $\Sigma(X - \overline{X})^2$ (read "sum of the squared deviations"), divide the sum by N to obtain a sort of average, and then take the square root of that result in order to get back to our original units of measurement. The standard deviation may be calculated from any set of measurements that are at least of interval nature.†

To summarize the steps just described, we can state the formula for the standard deviation as

$$S = \sqrt{\frac{\Sigma(X - \overline{X})^2}{N}} \qquad (5.2)$$

†The meaning of the standard deviation will become clearer when it is interpreted in terms of the normal curve later in this chapter. It is simply the distance on the x-axis between the mean and the steepest part of the normal curve, (e.g., between μ and -1 or μ and $+1$ in Figure 5.2).

where
S is the standard deviation
$\Sigma(X - \overline{X})^2$ is the sum of all the squared deviations from the mean
N is the number of observations

This formula for the standard deviation is called the *deviation formula* because it is calculated from the deviations themselves. In the next section formulas will be used that simplify the calculation of S by eliminating the time-consuming step of subtracting the mean from each score. But before we get ahead of ourselves, let us use the deviation formula to calculate S for the set of data shown in Table 5.2. Substituting into formula (5.2), we obtain

$$S = \sqrt{\frac{\Sigma(X - \overline{X})^2}{N}} = \sqrt{\frac{30}{8}}$$
$$S = \sqrt{3.75}$$
$$S = 1.94$$

Our standard deviation for the data of Table 5.2 is $S = 1.94$.

TABLE 5.2. **Calculation of the Standard Deviation (Deviation Method)**

X	$(X - \overline{X})$	$(X - \overline{X})^2$
6	1	1
8	3	9
2	−3	9
4	−1	1
4	−1	1
3	−2	4
7	2	4
6	1	1
$\Sigma X = 40$	0	$\Sigma(X - \overline{X})^2 = 30$

$$\overline{X} = \frac{\Sigma X}{N} = \frac{40}{8} = 5$$
$$S = \sqrt{\frac{\Sigma(X - \overline{X})^2}{N}} = \sqrt{\frac{30}{8}}$$
$$S = \sqrt{3.75}$$
$$S = 1.94$$

The example shown in Table 5.2 uses the deviation formula to obtain S. Although this method was used here to show you how S is based on squared deviations from the mean, it is not often used in practice because its calculations are time-consuming and laborious. It is necessary to subtract the mean from each score, square these deviations, average them by dividing by N, and extract the square root. However, some of the concepts involved in this method, especially the sum of squares, $\Sigma(X - \overline{X})^2$, will be useful to us in a later chapter. As was mentioned earlier, the deviation method illustrates the concept of variability very well in that each score is treated as a deviation from the mean.

The method described above for calculating S employed the deviation of each score from the mean, but it was mentioned that there was a more efficient method. It does not take much thought to realize that the calculation in Table 5.2 would have been much more time-consuming and prone to error if ΣX had been 41 instead of 40. Then the mean would have been $\frac{41}{8} = 5.125$. Consider for a moment subtracting 5.125 from each score and then squaring the deviation. Instead of pondering this plodding, time-consuming process, let us consider two alternate formulas for the standard deviation. One requires the mean of the distribution, and the other requires only ΣX and ΣX^2, which makes it especially suited for pocket calculators.

Table 5.3 shows the number of field goal attempts by a college basketball team that we first examined in Table 3.1 in Chapter 3. Both formulas are demonstrated for this distribution.

TABLE 5.3. Calculating ΣX and ΣX^2 for Raw Score Formulas: Basketball Data

X	X^2	
6	36	
9	81	
10	100	
7	49	
9	81	$\overline{X} = \frac{60}{10} = 6.00$
2	4	
4	16	
5	25	
6	36	
2	4	
$\Sigma X = 60$	$\Sigma X^2 = 432$	

Standard Deviation. Mean Formula

The formula for calculating the standard deviation using the mean is

$$S = \sqrt{\frac{\Sigma X^2}{N} - \overline{X}^2} \qquad (5.3)$$

where

S is the standard deviation
ΣX^2 is the sum of the squared scores
$\overline{X}$ is the mean of the distribution
N is the number of observations

As you can see in Table 5.3, the scores are placed in the X column, each score is squared and the square entered in the X^2 column, and the two columns are summed to obtain ΣX and ΣX^2. The mean of X is also calculated. The appropriate values from Table 5.3 are substituted into the formula

$$S = \sqrt{\frac{\Sigma X^2}{N} - \bar{X}^2} = \sqrt{\frac{432}{10} - (6)^2}$$
$$= \sqrt{43.2 - 36} = \sqrt{7.2}$$
$$S = 2.68$$

This formula is algebraically equivalent to the deviation formula (formula 5.2) and is much easier to compute because you are working with the original raw scores instead of the deviations. Notice in the above formula that the ΣX^2 refers to the sum of the squares of the *raw* scores, whereas in the deviation method you added the squared *deviations*, $\Sigma(X - \bar{X})^2$.

Standard Deviation. Machine Formula

Since most of you own or have access to a pocket calculator, there is an alternative to formula 5.3 that might be more convenient, since it does not require the separate calculation of the mean. The *machine formula* (the term dates back to the days when desk calculators were large and cumbersome and were called machines) for S is

$$S = \frac{1}{N}\sqrt{N\Sigma X^2 - (\Sigma X)^2} \qquad (5.4)$$

where
- ΣX^2 is the sum of the squared scores
- $(\Sigma X)^2$ is the square of the sum of the scores
- N is the number of observations

Note the difference between ΣX^2 and $(\Sigma X)^2$. The symbol ΣX^2 indicates that you are to square each individual score and then sum these squared scores. The term $(\Sigma X)^2$ indicates that you are to add all the scores first and then square that sum. To clarify these terms, let us calculate S for the data of Table 5.3 using the machine formula.

$$S = \frac{1}{N}\sqrt{N\Sigma X^2 - (\Sigma X)^2} = \frac{1}{10}\sqrt{10(432) - (60)^2}$$

$$= \frac{1}{10}\sqrt{4320 - 3600} = \frac{1}{10}\sqrt{720} = \frac{26.83}{10}$$

$$S = 2.68$$

Your result, S = 2.68, is the same as that obtained with formula 5.3. Since pocket calculators are readily available, we will use this machine formula throughout the rest of this book.

NOTE 5.1

CALCULATING THE STANDARD DEVIATION WITH A POCKET CALCULATOR

If your pocket calculator has memory and square root functions, the machine formula for the standard deviation is especially easy to calculate. After you have calculated ΣX in the usual way, the quantity ΣX^2 can be obtained by squaring each number and entering the square in your M+ storage. For example, in Table 5.3 you would multiply the first score, 6, by itself to obtain 36 and enter this in the memory by pushing the M+ key. The second score, 9, would be squared and entered into M+ and so on until you reached the last score, 2, and entered its square, 4, into M+. You would then recall memory to obtain $\Sigma X^2 = 432$.

You now can perform the operation called for in the machine formula,

$$S = \frac{1}{N}\sqrt{N\Sigma X^2 - (\Sigma X)^2}$$

You would first obtain $N\Sigma X^2$ by multiplying 10×432 and storing the 4320 in M+. Next you would calculate $(\Sigma X)^2$ by multiplying $60 \times 60 = 3600$ and subtracting this from memory by pushing the M− key. The next step would be to recall the result, 720, from memory and press the square root key to obtain 26.83. The last step, dividing by N, 10, would yield 2.68, the standard deviation for the data of Table 5.3.

Now that we have spent the time and effort on the calculation of S, what does it do? What does it tell us about a distribution of scores? Obviously, it tells us in a relative fashion *how much* the scores in a distribution deviate from the mean. If S is small, there is little variability and the majority of the observations are tightly clustered about the mean. If S is large, the scores are more widely scattered above and below the mean. One of the primary uses of S is to compare two or more distributions with respect to their variability. Let us repeat Table 5.1, where we noted that groups A and B had equal ranges but the scores were distributed differently in the two distributions. The results are shown in Table 5.4.

TABLE 5.4. **Comparing the Standard Deviations of Two Different Distributions**

Group A		Group B	
X	$\boldsymbol{X^2}$	***X***	$\boldsymbol{X^2}$
40	1,600	40	1,600
33	1,089	39	1,521
33	1,089	39	1,521
33	1,089	38	1,444
33	1,089	38	1,444
32	1,024	27	729
32	1,024	27	729
32	1,024	26	676
32	1,024	26	676
25	625	25	625
$\Sigma X = 325$	$\Sigma X^2 = 10{,}677$	$\Sigma X = 325$	$\Sigma X^2 = 10{,}965$

Group A:

$$\overline{X} = \frac{\Sigma X}{N} = \frac{325}{10} = 32.5$$

$$S = \frac{1}{N}\sqrt{N\Sigma X^2 - (\Sigma X)^2}$$

$$= \frac{1}{10}\sqrt{10(10{,}677) - (325)^2}$$

$$= \frac{1}{10}\sqrt{1{,}145}$$

$$= \frac{33.84}{10}$$

$$S = 3.38$$

Group B:

$$\overline{X} = \frac{\Sigma X}{N} = \frac{325}{10} = 32.5$$

$$S = \frac{1}{N}\sqrt{N\Sigma X^2 - (\Sigma X)^2}$$

$$= \frac{1}{10}\sqrt{10(10{,}965) - (325)^2}$$

$$= \frac{1}{10}\sqrt{4{,}025}$$

$$= \frac{63.44}{10}$$

$$S = 6.34$$

There can be no doubt that S is sensitive to the pattern of scores in a distribution. The scores of group A are tightly clustered about the mean and S is 3.38. Compare this with 6.34 for group B, where the scores are more widely dispersed from the mean. We have accomplished what we said we would do, that is, develop a measure of variability that reflects the distance each score deviates from the mean. S is just this measure. The importance of S cannot be overemphasized, and, if there are some concepts that are still not clear, it would be a good idea to review the preceding sections before going on to the next topic.

Variance

Before leaving the section on measures of variability, we should take a quick look at a measure of variability that will be of considerable importance in later chapters. This measure is called the *variance;* it is simply the square of the standard deviation. In Table 5.2, you will note that the deviation formula† for the standard deviation is

$$S = \sqrt{\frac{\Sigma(X - \overline{X})^2}{N}}$$

and so the formula for the variance is simply

$$S^2 = \frac{\Sigma(X - \overline{X})^2}{N}$$

For the data of Table 5.2, S is 1.94, and the variance, S^2, is 3.75.

NOTE 5.2

ROUNDING OFF THE STANDARD DEVIATION

We noted in Chapter 3 that the mean is conventionally rounded off to two more decimal places than we have in our set of raw data. The same convention applies to the standard deviation, so for whole numbers such as the test scores for Group A shown in Table 5.4, we report a mean of 32.50 and a standard deviation of 3.38. Also, if you had a set of reaction time scores measured in hundredths of seconds, your mean might be 0.3261 with a standard deviation of 0.0942.

We will have occasion to discuss the concept of variance in several future chapters, but, for the time being, consider variance (S^2) as just another method for describing the amount of variation in a set of scores. For example, in Table 5.4, S^2 for group A is $(3.38)^2 = 11.45$, while for group B it is $(6.34)^2 = 40.25$. Obviously, there is greater variability in the scores of group B than in those of group A.

†In Chapter 7 we will point out the important difference between dividing $\Sigma(X - \overline{X})^2$ by N or by $N - 1$ to obtain the variance.

Calculating the Variability of a Population Using σ and σ^2

Just as we distinguished between the mean of a sample, $\overline{X}$, and the mean of a population, μ, back in Chapter 3, we must do the same for measures of variability. The symbol σ (lowercase Greek letter sigma) and σ^2 (say "sigma squared") represent the population standard deviation and population variance, respectively. The formulas for σ and σ^2 are

$$\sigma = \sqrt{\frac{\Sigma(X - \mu)^2}{N}}$$

$$\sigma^2 = \frac{\Sigma(X - \mu)^2}{N}$$

It is obvious that these two formulas are very much like the formulas for S and S^2. All we have to remember is that σ and σ^2 are calculated using *all* the elements of a population, while S and S^2 are calculated using a smaller group of observations.

The Standard Deviation and the Normal Curve

It was noted earlier that frequency distributions of much of the data in education and the behavioral sciences approximate the normal curve. The normal curve will be discussed in detail in the next chapter, but at this point it will be helpful to see the relationship of the standard deviation and the normal curve. If we can assume that a variable (height, weight, IQ score, coordination score, etc.) approximates the shape of the normal curve, we can even better understand the concept of variability. Figure 5.2 shows a normal curve, with its mean, μ, and standard deviation, σ, units on the baseline.

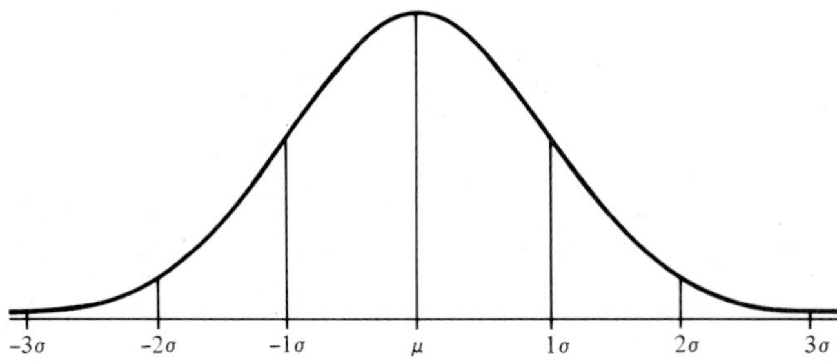

FIGURE 5.2. Normal curve with mean (μ) and standard deviation (σ) units.

In the normal curve of Figure 5.2, the mean is erected from the baseline, and this vertical line divides the distribution into two equal parts. In other words 50% of the scores lie below the mean (to the left) and 50% above the mean (to the right). Vertical lines are also erected from the baseline corresponding to the different σ units so that the area under the curve (e.g., number of scores) between 1σ unit below the mean and 1σ unit above the mean is approximately 68% of the total area. Similarly, approximately 95% of the distribution lies between -2 and $+2\sigma$

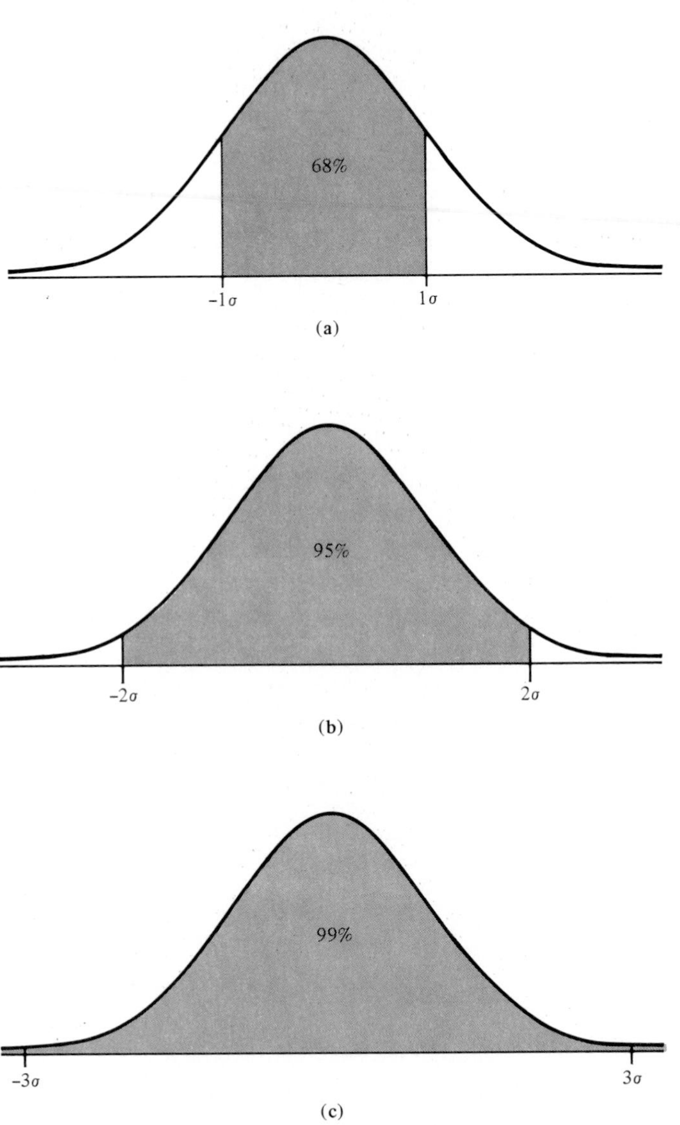

FIGURE 5.3. **Percentage of area under the normal curve.**

units from the mean, and about 99% of the distribution lies between -3 and $+3\sigma$ units from the mean. These statements are represented graphically in Figure 5.3.

Let us apply this approach to a real-life setting in Figure 5.4. The mean of a well-known intelligence test is 100 with a standard deviation of 16. If this variable is normally distributed in the population, we can make some inferences using the normal curve model shown in Figures 5.2 and 5.3. With what we know about the normal curve, we are able to make statements to the effect that approximately 68% of the population would have tested IQ scores between $\mu \pm 1\sigma$, or 100 ± 16, or between 84 and 116. As you can see from Figure 5.4, intervals could be constructed to describe IQ scores between $\mu \pm 2\sigma$ or $\mu \pm 3\sigma$ that would include 95% and 99% of the population, respectively. In the next chapter we will spend considerable time on the normal curve; it was introduced at this point to show the interpretation of the standard deviation when variability is being studied in a variable that approximates the normal curve.

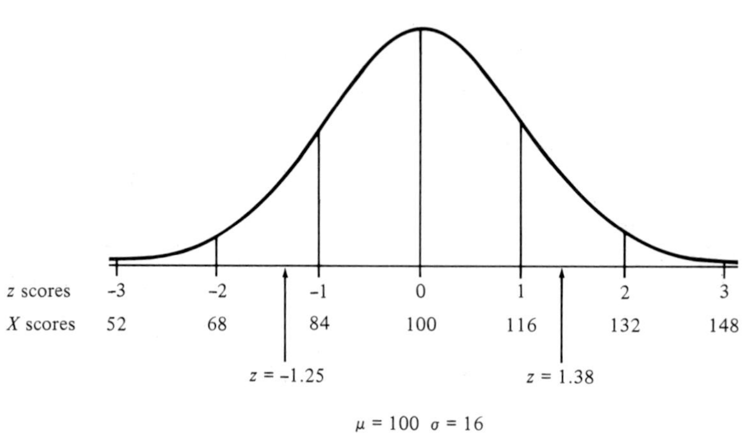

FIGURE 5.4. **Population distribution of IQ scores.**

It is a little awkward in discussing a score or observation to have to say that it was "2 standard deviations above the mean" or "1.5 standard deviations below the mean." To make it a little easier to pinpoint the location of a score in any distribution, the z score was developed. The z score is simply a way of telling how far a score is from the mean in *standard deviation units*. In Table 5.3 we noted the number of field goal attempts and reported the mean to be $\overline{X} = 6.00$ and calculated a standard deviation of $S = 2.68$. Let us use this group of observations to demonstrate the calculation of the z score.

The formula for converting any score (X) into its corresponding z score is

$$z = \frac{X - \overline{X}}{S} \tag{5.5}$$

where

z is the z score
X is the observed score
$\overline{X}$ is the mean of the distribution of scores
S is the standard deviation of the distribution

From Table 5.5 what would be the z score for an observed value of 9? Using formula 5.5, we obtain

$$z = \frac{X - \overline{X}}{S} = \frac{9 - 6}{2.68} = \frac{3}{2.68} = 1.12$$

Again, in Table 5.5, what would be the z score for an observed score of 2 field goal attempts?

$$z = \frac{X - \overline{X}}{S} = \frac{2 - 6}{2.68} = \frac{-4}{2.68} = -1.49$$

Note that when a z score is positive, it is located *above* the mean, and when negative it is *below* the mean. Observe the location of these last two z scores in Table 5.5. If you have difficulty understanding just what a z score is, remember that it is a way of telling how much a score deviates from the mean in standard deviation units. The score of 2 just calculated is 1.49 standard deviations below the mean.

TABLE 5.5. **Calculation of Selected *z* Scores for Field Goal Attempts Data**

X	
6	
9	$X = 9, z = ?$
10	$z = \frac{X - \bar{X}}{S} = \frac{9 - 6}{2.68} = 1.12$
7	
9	
2	
4	$X = 2, z = ?$
5	$z = \frac{X - \bar{X}}{S} = \frac{2 - 6}{2.68} = -1.49$
6	
2	
$\bar{X} = 6.0$	
$S = 2.68$	

A z score may also be used to find the location of a score that is a normally distributed variable. Using the example again of a population of IQ test scores shown in Figure 5.4, we may want to determine the location of an IQ score of, say, 80 or 122. Since the mean is now μ and the standard deviation is σ, the formula for the z score is changed accordingly to

$$z = \frac{X - \mu}{\sigma} \tag{5.6}$$

where
z is the z score
X is the observed score
μ is the population mean
σ is the population standard deviation

Let us calculate z scores for IQs of 122 and 80 and locate them on the normal curve of Figure 5.4.

$$z = \frac{X - \mu}{\sigma} = \frac{122 - 100}{16} = \frac{22}{16} = 1.38$$

$$z = \frac{X - \mu}{\sigma} = \frac{80 - 100}{16} = \frac{-20}{16} = -1.25$$

We can observe, both graphically and algebraically, that a score above the mean results in a positive z score and a score below the mean results in a negative z score.

Another compelling reason for using a z score is to make comparisons between different distributions. Knowing that John got a score of 78 on a mathematics achievement test, 115 on a natural science aptitude test, and 57 on an English usage exam tells us nothing about his performance in relation to the rest of the group. But if these three variables are normally distributed in the population (a reasonably safe assumption), we can make direct comparisons by using the z score approach. For example, Table 5.6 shows the means and standard deviations for the three tests just mentioned.

TABLE 5.6. **Means, Standard Deviations, and John's Scores (X) on Three Tests**

Mathematics	Natural Science	English
$\mu = 75$	$\mu = 103$	$\mu = 52$
$\sigma = 6$	$\sigma = 14$	$\sigma = 4$
$X = 78$	$X = 115$	$X = 57$

In comparing z scores for the three tests you would first calculate the z scores for the three tests as follows:

$$\text{Mathematics:} \quad z = \frac{X - \mu}{\sigma} = \frac{78 - 75}{6} = \frac{3}{6} = 0.5$$

$$\text{Natural Science:} \quad z = \frac{115 - 103}{14} = \frac{12}{14} = 0.86$$

$$\text{English:} \quad z = \frac{57 - 52}{4} = \frac{5}{4} = 1.25$$

So we find that John, in terms of the rest of the group, did best on the English test and poorest (though still above average) on the mathematics exam.

Because a z score can be interpreted in relation to the rest of the distribution without knowledge of the observed scores themselves, z scores will be used frequently throughout the rest of this book. Their relation to the normal curve, which you were introduced to in Figure 5.4, will be explained further in the next chapter.

NOTE 5.3

FINDING THE OBSERVED SCORE WHEN z IS GIVEN

There will be times when you will have to calculate the observed score when the z score is known. If you are not particularly proficient in algebra, the z score formulas (5.5 and 5.6) can be reworked so that you solve for X. By appropriate algebraic manipulation,

$$z = \frac{X - \overline{X}}{S} \qquad \text{becomes} \qquad X = z(S) + \overline{X}$$

and

$$z = \frac{X - \mu}{\sigma} \qquad \text{becomes} \qquad X = z(\sigma) + \mu$$

For example, in Figure 5.4 what IQ score corresponds to a z of -1.75? Solving for X, we get

$$X = z(\sigma) + \mu = -1.75(16) + 100 = -28 + 100 = 72$$

Thus 72 is the IQ score that corresponds to a z of -1.75. Notice that your choice of $X = z(S) + \overline{X}$ or $X = z(\sigma) + \mu$ depends on whether you are working with a *sample* of scores or making a statement about a normally distributed variable in the *population*.

Other Standard Scores

The z score is a *derived* score, since it is derived from the original score units. (Remember that the percentile ranks discussed in the last chapter were also derived scores.) It is also referred to as a *standard score*, since it is based on standard deviation units. However, there are several minor disadvantages in the use of the z score. There are negative values for any scores below the mean, the mean of the z score distribution is 0, and the z scores are decimal fractions—all of which results in a certain amount of computational complexity.

A number of other standard score systems have been devised that do not have these disadvantages. These are listed (along with the familiar z scores for comparison) in Table 5.7.

The standard scores of Table 5.7 are only a small sample of possible standard score systems in current use. However, they are all related to the z score, since they still tell us the basic fact: how far any score deviates from the mean in standard deviation units.

TABLE 5.7. **Typical Standard Score Systems**

z scores:	$\mu = 0$	$\sigma = 1$
T scores:	$\mu = 50$	$\sigma = 10$
General Aptitude Test Battery (GATB):	$\mu = 100$	$\sigma = 20$
College Entrance Examination Board (CEEB):	$\mu = 500$	$\sigma = 100$

The z score formula is used to determine how many standard deviation units a given score in one of these distributions deviates from the mean. Where in the CEEB distribution, for example, does a score of 437 lie?

$$z = \frac{X - \mu}{\sigma} = \frac{437 - 500}{100} = \frac{-63}{100} = -0.63$$

Thus a CEEB score of 437 lies 0.63 of a standard deviation below the mean of 500. Similarly, a T score of 62 yields a z of 1.2, which indicates it is 1.2 σ units above the mean. Figure 5.5 shows the relationship of various derived scores to the normal curve.

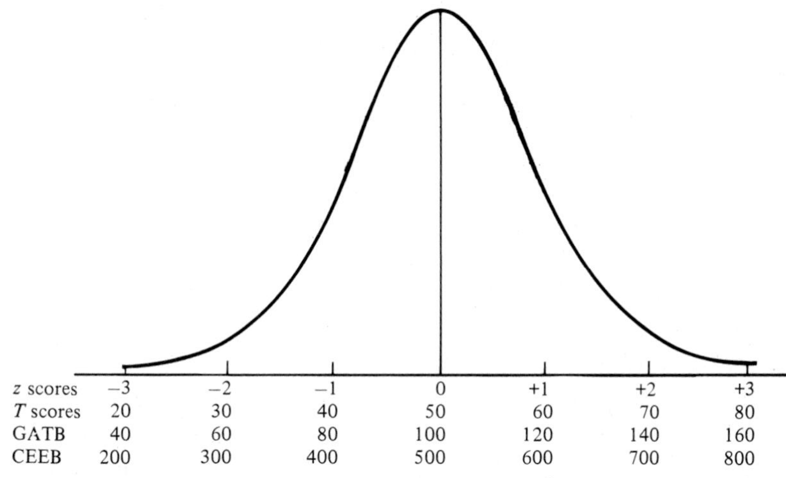

FIGURE 5.5. **The normal curve and standard scores.**

Negative Numbers Under the Radical Sign

In computing S using the raw score formulas, you may get a negative number under the radical sign after you have subtracted $\overline{X}^2$ from $\Sigma X^2/N$ in the formula using the mean, or $(\Sigma X)^2$ from $N\Sigma X^2$ in the machine formula. If this should happen, *you have made a mistake*. You must recheck your work, perhaps the addition of the X^2 column, to find the error. Whether you are using the mean formula or the machine formula, a negative number under the radical sign indicates an error in your calculations.

How to Estimate a Standard Deviation— A Rough Check for Accuracy

In Figures 5.3 and 5.4 you noted that $\mu \pm 3\sigma$ units just about covered the entire range of scores under the normal curve, since 99% of the total area lies between -3 and $+3\sigma$. We can make use of this fact to exercise a rough check on the accuracy of our calculations of S for a group of scores. For an approximately normal distribution, S should be about one-sixth of the range, *when N is quite large*.

But, when you are working with only 50 or 100 scores, you will find that S is about one-fourth or one-fifth of the range. So, when you calculate S for a distribution containing 72 scores, where the lowest score is 25 and the highest is 88, you would know that S should be about 13 to 16 for the range of 64. If your S turned out to be 2.53 or 27.69, you would probably want to recheck your work for possible errors. This will not work every time because some distributions will have peculiar, nonnormal shapes, but for most situations the one-fourth-to-one-fifth rule of thumb will be helpful.

Effect of Adding a Constant to Each Score

It can be shown that when a constant is added to each score in a distribution, S remains unchanged. The addition of a constant to each score does not change each score's relative position with respect to the new mean, and, since S is based on deviations from the mean, the deviations themselves would remain unchanged. Table 5.8 shows distribution A with its mean and S and distribution B, where the constant 3 has been added to each score of distribution A. The deviation method for calculating S is used so you can see that the deviations remain unchanged.

TABLE 5.8. **Adding a Constant Does Not Affect the Standard Deviation**

Distribution A

X	$(X - \overline{X})$	$(X - \overline{X})^2$
9	3	9
7	1	1
6	0	0
5	−1	1
3	−3	9
$\Sigma X = 30$	0	$\Sigma(X - \overline{X})^2 = 20$

$$\overline{X} = \frac{\Sigma X}{N} = \frac{30}{5} = 6$$

$$S = \sqrt{\frac{\Sigma(X - \overline{X})^2}{N}}$$

$$= \sqrt{\frac{20}{5}} = \sqrt{4}$$

$$S = 2$$

Distribution B (A + 3)

X	$(X - \overline{X})$	$(X - \overline{X})^2$
12	3	9
10	1	1
9	0	0
8	−1	1
6	−3	9
$\Sigma X = 45$	0	$\Sigma(X - \overline{X})^2 = 20$

$$\overline{X} = \frac{45}{5} = 9$$

$$S = \sqrt{\frac{20}{5}} = \sqrt{4}$$

$$S = 2$$

As you remember from Chapter 3, the addition of a constant to every score *did* alter the mean in that the mean increased by an amount equal to the constant. However, this addition of a constant, as shown in Table 5.8, does not change the standard deviation.

Semi-Interquartile Range

There are times when it is convenient to be able to examine the variability of a group of observations about the *median* rather than the mean. Such a measure of variability is the *semi-interquartile range* (sometimes called the quartile deviation, or Q), which can be used with any data that is appropriate for the median, that is, with data that is of an ordinal nature or above.

The semi-interquartile range is one-half the distance between Q_3 and Q_1 (or P_{75} and P_{25}). The formula for Q is

$$Q = \frac{P_{75} - P_{25}}{2} \qquad (5.7)$$

In a normal distribution we would expect that about 50% of the observations would be covered in the range $Med \pm Q$. For example, in the Sample Problem at the end of this chapter, the median is 7.3. If P_{75} were 9.5 and P_{25} were 4.5, Q would be calculated as

$$Q = \frac{P_{75} - P_{25}}{2} = \frac{9.5 - 4.5}{2} = \frac{5}{2} = 2.5$$

We would then assume that approximately 50% of the distribution would be between 7.3 ± 2.5, or between 4.8 and 9.8.

The semi-interquartile range has not been a popular statistic in recent years, but you still may run across it in older books and journal articles. Because of its limited use, we will not have any more to do with it.

SAMPLE PROBLEM

A clinical psychologist was studying the prevalence of Type A behavior in college students and used the Jenkins Activity Survey (JAS). The JAS assesses the extent of Type A behaviors such as aggression, competitive achievement striving, sense of time urgency, and potential for hostility that have been shown to be linked to coronary heart disease. The JAS items survey such everyday behaviors as eating rapidly, hurrying to get places even when there is plenty of time, and the ability to control one's temper. Scores on the student version of the JAS can range from 0 to 21 where a score of 21 would indicate extreme Type A behavior. The psychologist administered the JAS to a sample of 52 college women chosen at random from courses in introductory psychology. Calculate the mean, range, and S for this distribution.

X	X^2	X	X^2	X	X^2
12	144	3	9	4	16
7	49	8	64	6	36
7	49	3	9	12	144
12	144	4	16	5	25
8	64	8	64	12	144
7	49	11	121	7	49
3	9	3	9	9	81
8	64	8	64	11	121
6	36	5	25	1	1
7	49	2	4	1	1
8	64	10	100	8	64
6	36	8	64	13	169
8	64	11	121	1	1
14	196	5	25	6	36

10	100	14	196	6	36
3	9	9	81	9	81
3	9	6	36	1	1
11	121				

Summary: $\Sigma X = 370 \quad \Sigma X^2 = 3270 \quad N = 52$

Mean: $\overline{X} = \frac{\Sigma X}{N} = \frac{370}{52} = 7.12$

Range: Range $= 14 - 1 + 1 = 14$

Standard Deviation:

$$S = \frac{1}{N}\sqrt{N\Sigma X^2 - (\Sigma X)^2} = \frac{1}{52}\sqrt{52(3270) - (370)^2}$$

$$S = \frac{1}{52}\sqrt{170{,}040 - 136{,}900} = \frac{1}{52}\sqrt{33{,}140}$$

$$S = \frac{182.0440}{52} = 3.50$$

As a rough check on the accuracy of our calculation of S, let us see if our value of 3.50 is about one-fourth to one-fifth of the range. One-fourth of the range of 14 would be 3.5 while one-fifth would be 2.8, so our value of 3.50 appears reasonable.

If Ann scored 5 on this test, what would be her z score? And if Sally was told that her z score was 1.39, what was her score on the test?

Ann: $X = 5, z = ?$

$$z = \frac{X - \overline{X}}{S} = \frac{5 - 7.12}{3.5} = \frac{-2.12}{3.5} = -0.61$$

Sally: $z = 1.39, X = ?$

$$X = z(S) + \overline{X} = 1.39(3.5) + 7.12 = 12.0$$

STUDY QUESTIONS

1. Give a definition of the term *variability*.

2. What would be an example of the definition of variability?

3. Why is the range inadequate as a measure of variability?

4. What is meant by the operation $\Sigma(X - \overline{X})^2$?

5. How is the quantity $\Sigma(X - \overline{X})^2$ related to the concept of variability?

6 What is the difference between ΣX^2 and $(\Sigma X)^2$? Demonstrate the difference in these two operations by choosing 5 numbers and calculating both values for each.

7. What is the difference between S and σ?

8. What does $z = 1.32$ indicate? What about $z = -2.43$?

EXERCISES

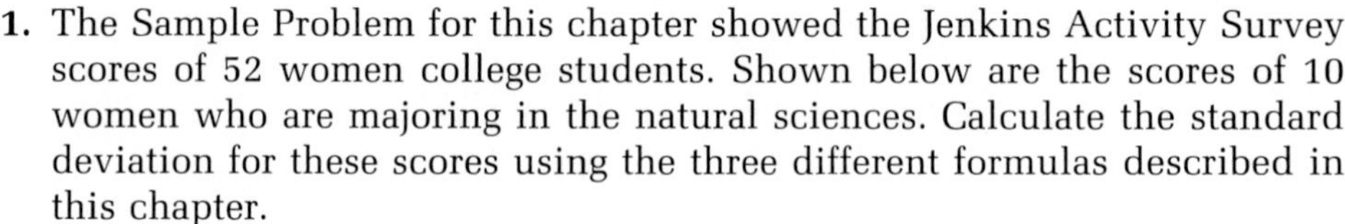

1. The Sample Problem for this chapter showed the Jenkins Activity Survey scores of 52 women college students. Shown below are the scores of 10 women who are majoring in the natural sciences. Calculate the standard deviation for these scores using the three different formulas described in this chapter.

7	8
4	5
12	7
8	11
10	9

2. The tournament program for the NCAA Division 2 men's basketball play-offs listed the rosters of participating teams. Shown below are the heights (in inches) of a midwestern private college team. Calculate the standard deviation for these heights using the three different formulas described in this chapter.

78	77
81	78
76	75
72	76
73	79
77	76

3. The sodium content of 10 breakfast cereals first listed in Chapter 3 is repeated here. Calculate the range and the standard deviation for this data. Why would the range be an inferior measure of variability in this instance?

460	255
380	250
370	190
345	70
280	60

4. The grade-point averages (GPAs) of a sample of 15 Type A college students are shown below. Calculate the range and the standard deviation. Why could the range be a misleading measure of variability in this instance?

3.90	3.50	3.30	3.10	3.00
3.50	3.40	3.30	3.05	2.90
3.50	3.38	3.20	3.00	2.80

5. The National Institutes of Health research project mentioned in Chapter 3 studied weight loss in a sample of 17 women using a combination of both diet and exercise. Shown below is the weight loss in pounds over a 15-week period. Calculate the mean and standard deviation for these data.

30	18	23	17	10
18	19	20	26	5
18	22	15	13	20
15	17			

6. The weight loss study mentioned in Exercise 5 also included 34 women who used diet alone, and their weight loss in pounds for the same 15 week period follows. Calculate the mean and standard deviation for these data.

14	10	15	8	11
11	17	16	10	12
7	9	3	15	12
16	7	8	12	4
13	9	12	11	13
9	5	10	8	16
10	7	10	7	

7. Eighteen university students applying for positions as peer tutors in the English Department were given the English Usage Test and their scores are as follows. Calculate the mean and standard deviation for these scores.

27	23	22	28	26
23	25	25	30	21
29	26	26	28	27
24	25	24		

8. The Culture Fair Test is a test instrument that attempts to measure intellectual ability without racial or cultural biases. Twenty college students in an educational measurement class took the test with the following results. Calculate the mean and standard deviation for these data.

68	50	54	52	52
68	54	60	64	59
63	64	62	71	55
67	71	58	63	71

9. The average female college student is 5 feet 4 inches tall with a standard deviation of 2.48 inches (i.e, $\mu = 64$ inches, $\sigma = 2.48$ inches). Redraw Figure 5.4 marking off the heights corresponding to $\mu \pm 3\sigma$.

10. The average male college student is 5 feet 10 inches tall with a standard deviation of 2.52 inches (so $\mu = 70$ inches, $\sigma = 2.52$ inches). Redraw Figure 5.4 marking off the heights corresponding to $\mu \pm 3\sigma$.

11. Means and standard deviations for the English, Mathematics, Social Studies and Natural Sciences subtests of the ACT (American College Testing Program) are shown below. Judith scored 23, 24, 25 and 26 respectively on the 4 subtests. Assume that these test scores are normally distributed in the population and calculate Judith's corresponding z scores. On which test did she do the best? The worst?

	English	Mathematics	Social Studies	Natural Sciences
μ	17.7	18.7	18.3	20.4
σ	5.6	7.2	7.3	6.4

12. Andrew scored 24 on all 4 subtests of the ACT described in Exercise 11, and claims he has equal ability in the 4 areas. Use the z score method to show why you would dispute his claim.

13. Jayne, Jean, and Joan scored 26, 11, and 21 on the English subtest of the ACT described above in Exercise 11. Sketch a normal curve similar to Figure 5.4 and locate their z scores on the normal curve.

14. Tom, Dick, and Harry scored 17, 11, and 30 on the Natural Sciences subtest of the ACT described in Exercise 11. Sketch a normal curve similar to Figure 5.4 and locate their z scores on the normal curve.

15. Use the $\frac{1}{4}$ to $\frac{1}{5}$ rule of thumb to estimate the standard deviation for the weight loss data of Exercise 5. Do the same for Exercise 6. How do these estimates compare with the calculated values?

16. Use the $\frac{1}{4}$ to $\frac{1}{5}$ rule to estimate the standard deviation of the English Usage Test data of Exercise 7. Do the same for the Culture Fair Test data of Exercise 8. How do the actual calculated values compare with these estimates?

CHAPTER 6

The Normal Curve

We have had several occasions in previous chapters to refer to the normal curve in a cursory fashion, but it is necessary at this point to devote an entire chapter to this much maligned and misunderstood statistical concept. Technically, this chapter is not consistent with the rest of the text as far as presenting techniques for either describing a set of data or making inferences about the nature of the population. Rather, it is a necessary interruption in order to introduce a mathematical concept that is a foundation for many statistical concepts, and it is essential to insert the material at this point. This chapter is somewhat similar to the interruption of an auto mechanic's on-the-job training with a study unit on the theory of automotive electronics. Comparatively speaking, we will temporarily put aside our statistical wrenches and screwdrivers for a short dissertation on the properties of the normal curve.

In previous chapters we noted that many physiological and psychological measurements were "normally" distributed; that is, a graph of the measurements took on the familiar symmetrical, bell-shaped form. The graph of IQ test scores shown in Figure 6.1 is but one of the countless examples of such a distribution.

The Normal Curve as a Model

One of the reasons for the tremendous progress of the sciences over the last few decades has been their utilization of *mathematical models*. Without getting bogged down in technical jargon, we can say that scientists are overjoyed when the data of whatever they happen to be observing "fit" a particular mathematical model that they have chosen. Astronomers noted very early that the orbital path of planets seemed to be elliptical in form (an oval is one type of an ellipse), and by using this model they were able to predict such things as eclipses and the

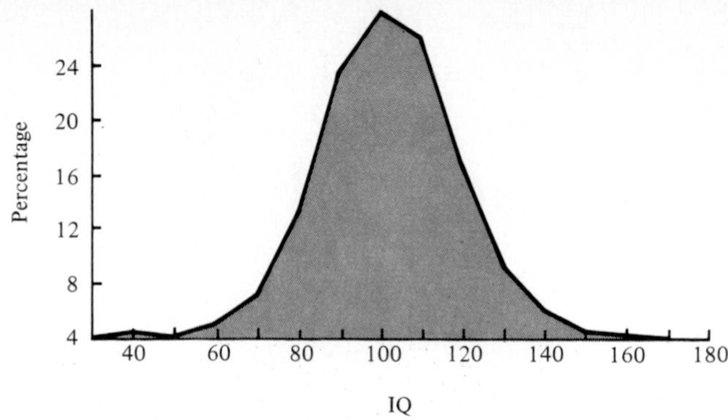

FIGURE 6.1. **Stanford-Binet IQ Scores of 1937 Standardization Group.** (From L. M. Terman and N. Merrill, *Stanford-Binet Intelligence Scale* (Boston, 1960), p. 18. Copyright 1960 by The Riverside Publishing Company. Used by permission of the publisher.)

existence of, at that time, still undiscovered planets. In a similar way, physicists and other scientists have used such mathematical models as exponential or logarithmic functions, wave motion, and ballistic mechanics to study the behavior of electrons, light, body motion, and other natural phenomena. And—here is the point of the whole discussion—mathematical models were used to help scientists *picture* or *visualize* the nature of the data and to enable them to *predict* the outcome of data to be gathered in the future.

Applications of the normal curve were developed in much the same way. Borrowing from the work of several earlier mathematicians (see Note 6.1), Adolphe Quetelet (1796–1874), a Belgian mathematician and astronomer, was the first to note that the distribution of certain body measurements appeared to approximate the normal curve. He found that the heights of French soldiers and chest circumferences of Scottish soldiers yielded frequency distributions that fit the characteristic bell-shaped curve. Sir Francis Galton (1822–1911) continued with Quetelet's approach and studied the distributions of a wide variety of physical and mental characteristics. Quetelet, Galton, and countless researchers in innumerable investigations since those early days have made good use of the normal curve model to describe a set of data and to predict the results of future investigations.

So, why are we so excited (or, at least, mildly enthusiastic) over the possibility of using the normal curve as a model for data in education and the behavioral sciences? Simply because, if we can assume that the *population* of measurements from which we draw our sample is normally distributed, we can make use of the known properties of the normal curve both to visualize our data and to make predictions based on our sample. Since prediction is such an important part of statistics, we shall examine in detail the properties of this mathematical concept, the normal curve, in the remainder of this chapter.

NOTE 6.1

ORIGIN OF THE NORMAL CURVE

The noble art of mathematics was very early applied to problems in probability posed by gamblers. One very knotty problem concerned such questions as "What is the probability that in 10 flips of a coin heads will result 6 times?" or "What is the probability that in 20 throws of a die a '2' will show 7 times?" Although solutions to these simple problems occurred quite early, it remained for a general application to cover the impossible calculations involved in answering a question such as "What is the probability that in 12,000 tosses of a coin heads will occur 1876 times?" In 1733 Abraham de Moivre (1667–1754) developed a mathematical curve that would predict the probabilities of events associated with such things as flipping coins and rolling dice. The curve that de Moivre developed was, of course, what we now call the normal curve.

You may also have seen this curve referred to as the "Gaussian curve," after C. F. Gauss (1777–1855), who developed the normal curve independently of de Moivre in 1809. His interest was primarily in errors of measurement or observation in astronomy, and he noted that the curve described a distribution of errors. But, as was mentioned earlier, it was probably Quetelet who first applied the curve to a variety of physical data.

Some Characteristics of the Normal Curve

Since the normal curve is a mathematical function, it has a formula, which is

$$y = \frac{1}{\sigma\sqrt{2\pi}} e^{-\frac{(X - \mu^2)}{2\sigma^2}}$$

where

y = height of the curve
σ = standard deviation of the population
π = pi, or 3.14
e = natural logarithm, approximately 2.718
X = any score value
μ = the population mean

Admittedly, the formula is an imposing one, and fortunately we will have very little to do with it as such. It is mentioned here so that we are aware of the fact that the normal curve is indeed a mathematical function, and because we must call attention to some of the terms in the formula.

There is a whole family of normal curves, depending on the values for the population mean (μ) and the population standard deviation (σ). That is, we can change the position of the curve on a scale of measurement by altering the mean, or we can change the spread of the curve by altering the standard deviation. We could have a normal curve by substituting into the formula a mean of 10, 50, or 100 and a standard deviation of 2, 17, or 29, or any value that suited our fancy.

In Figure 6.2 a normal curve with a mean of 0 and a standard deviation of 1 is shown. This is called the *unit normal curve,* and it is the one we will be working with throughout this chapter. It is called the unit curve because the *area* under the curve is exactly equal to 1 square unit. As we shall see later, using this area of exactly 1 will simplify our computations greatly.

Note that the units on the x-axis of Figure 6.2 are the familiar z scores that we dealt with in the last chapter. As you discovered then, a z score is a way of specifying a location on the scale that holds for whatever mean or standard deviation you have. The z score simply tells where a particular value lies in terms of standard deviation units from the mean. For example, in Figure 6.2 a z of 2.5 is obviously 2.5 standard deviations to the right of a mean of 0. (If having a mean of 0 bothers you, consider the mean as a point that is 0 standard deviations from the mean!)

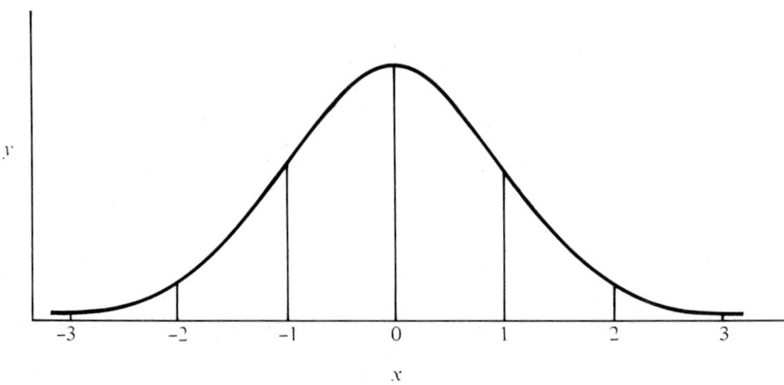

FIGURE 6.2. The unit normal curve.

The last characteristic of the normal curve that we will be concerned with is an important one indeed—that of determining the amount of *area* under the curve. We will find ourselves using this technique throughout the remaining chapters in the book, and it might be an understatement to say that any student of statistics *must* show at least a minimum level of competency in working with normal curve areas.

Some of the areas under the curve are obvious, and a quick glance at Figure 6.2 should reveal that, if the total area under the curve is 1.00, then 0.50 must be the area to the left of the mean and 0.50 the area to the right of the mean. Also, as you may remember from the last chapter, we found that about 68% of the total area (0.68 in decimal terms) is under the curve between a z of -1 and a z of $+1$, 95% between z scores of -2 and $+2$, and about 99% between z scores of -3 and $+3$.

However, we are usually interested in the "in between" values, and we must resort either to some high-level mathematics or a table of normal curve areas to determine the answers to such questions as "How much of the area under the curve lies between the mean and a z of 1.53?" or "How much of the area under the curve lies between a z of -2.38 and a z of 1.92?" Fortunately, we have Table B in Appendix 4, which makes it very simple to answer a variety of questions relating to normal curve areas. Note that the left side of the pairs of columns in Table B lists the z score, and the entry opposite gives the area under the normal curve *between that z score and the mean*. So to answer the question "How much of the area lies between the mean and a z of 1.53?" you simply look up the z value of 1.53 in Table B and note that 0.4370 or 43.7% of the total area is included under the curve between those points. The shaded area in Figure 6.3a illustrates this graphically.

The z scores of Table B are all positive, but since the curve is perfectly symmetrical we use the same tabled values for negative z scores. For example, if we wanted to know the area under the curve between the mean and a z of -1.53, we would get the same answer as above: 43.7%. The only thing that would be different would be the location of the shaded area in Figure 6.3a, and we would see that the area under consideration would be to the *left* of the mean, between -1.53 and the mean.

Two techniques will be demonstrated in the following sections, with variations of each illustrated in Figures 6.3 through 6.5. Basically, we have two approaches: we either know the z score and use Table B to find an area, or we know the area and use Table B to determine the appropriate z score.

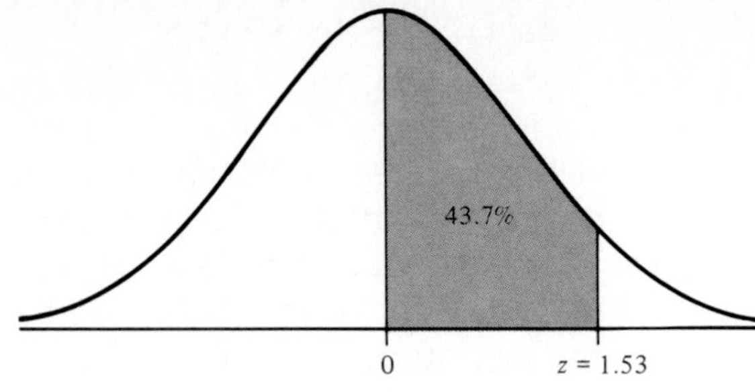

(a) 43.7% between the mean and a z of 1.53.

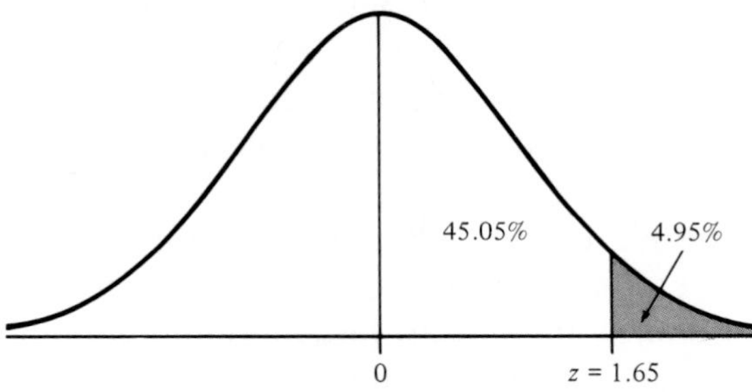

(b) 4.95% above a z of 1.65.

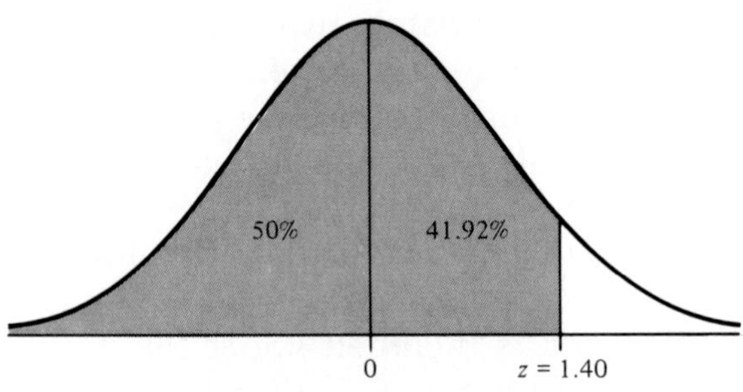

(c) 91.92% below a z of 1.40.

FIGURE 6.3. **Determining areas under the normal curve.**

Given a z Score, Find the Area

In the preceding section Table B was described as giving a direct answer to finding the area under the curve between a z score and the mean, and the answer was illustrated in Figure 6.3a. The following examples show how to determine the area *above* or below a z score or *between* z scores.

FIGURE 6.3b. *How do you find the area under the curve* above *a positive z value* (or below *a negative* z)? You cannot read the value directly from

Table B, since the areas in Table B are *between z and the mean.* In this case you must *subtract* the tabled value from 0.5000. For example, if you want to find what percentage of the area under the curve lies above a z of 1.65, the tabled value from Table B is 0.4505. But this value is the area between z and the mean, so you must subtract 0.4505 from 0.5000 to obtain 4.95%, which is the area above a z of 1.65. Figure 6.3b demonstrates this graphically. A similar approach would be used to find the area below a negative z value.

FIGURE 6.3c. *How do you find the area below a positive z score?* In this case it is necessary to add the tabled value to 0.5000, since the value in Table B gives only the area from the z value down to the mean. Figure 6.3c shows how this is done to answer a question such as "What percentage of the area under the curve lies below a z of 1.40?" Table B shows the area from a z of 1.40 down to the mean to be 0.4192, which, when added to 0.5000, would give 91.92% of the area below a z of 1.40.

To determine the percentage of the area *between* two z scores, it is necessary to use one of two techniques, depending on whether the z scores are on the same side of the mean or on opposite sides.

FIGURE 6.4a. *How do you determine the area under the curve between two z scores on different sides of the mean?* Figure 6.4a shows how you would find the area between a negative z and a positive z, for example, −1.20 and 1.96. Since Table B gives the area from the z score to the mean, it is only necessary to add the two areas to arrive at the answer. As you can see, the area between the mean and a z of −1.20 is 0.3849, and the area between the mean and a z of 1.96 is 0.4750, for a total area between the two of 0.8599 or 85.99%.

FIGURE 6.4b. *How do you determine the area under the curve between two z scores on the same side of the mean?* You simply subtract the smaller area from the larger area. For example, if we wish to determine the area between a z score of 1.25 and 2.20, we look up the tabled value for 2.20 and find that 0.4861 of the total area lies between the mean and a z of 2.20. We then look in Table B for the area between the mean and a z of 1.25 and find the value to be 0.3944. We then subtract 0.3944 from 0.4861 to show that 0.0917 or 9.17% of the area lies between a z of 1.25 and 2.20. This is illustrated graphically in Figure 6.4b.

Given an Area, Find the z Score

The preceding section just about exhausts all the possible ways in which Table B can be used when the z score is known and we wish to find the area. However, there are many occasions when we will have a

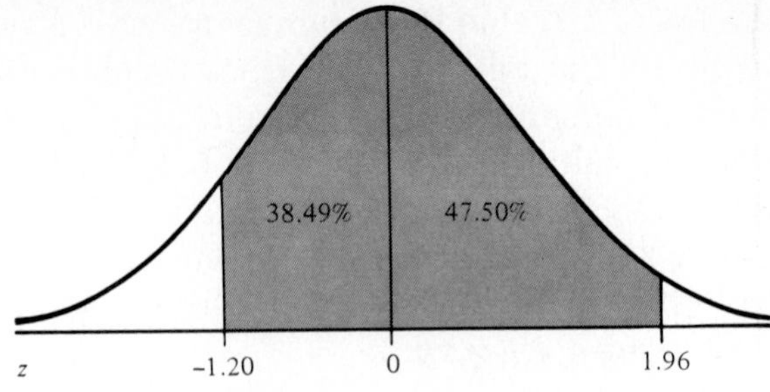

(a) 85.99% between z scores of -1.20 and 1.96.

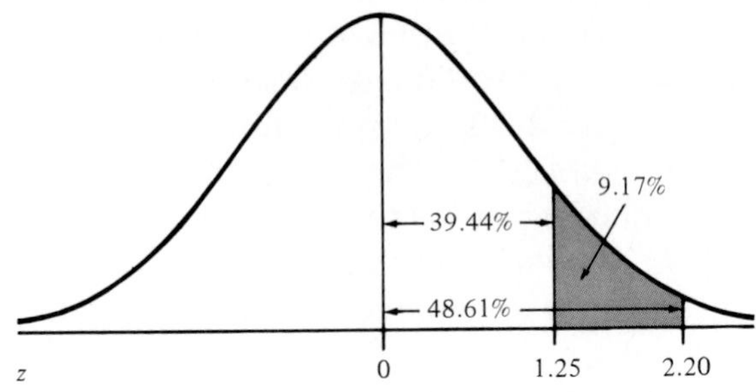

(b) 9.17% between z scores of 1.25 and 2.20.

FIGURE 6.4. **More areas under the normal curve.**

known percentage of area under the curve and will use Table B to find the appropriate z score. Such occasions are obviously the reverse of the problems described in the preceding section, but the same general logic is used. Only the procedure is changed, in that you go to Table B with a percentage of area and look for a z. Figure 6.5 shows several examples of how Table B can be used in this way.

FIGURE 6.5a. *How do you find what z scores form the boundaries of a centrally located area?* All you have to do is divide the area by 2 (since Table B gives values only for one-half of the curve) and look up the z score that corresponds to that value. For example, the middle 95% of the area is between what two z scores? Figure 6.5a shows that one-half of the area, or 47.5%, would be on either side of the mean. We then go to Table B and look in the *area* column to locate 0.4750 and find that the z score that marks off 47.5% of the area from the mean is 1.96. Since the curve is symmetrical, we know that a z score of −1.96 would mark off 47.5% of the area below the mean. So we conclude that z scores of −1.96 and 1.96 are the boundaries enclosing the middle 95% of the distribution.

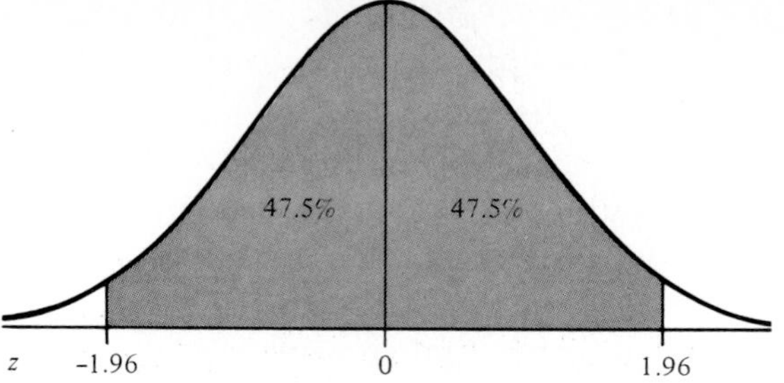

(a) z scores of −1.96 and 1.96 enclose the middle 95%.

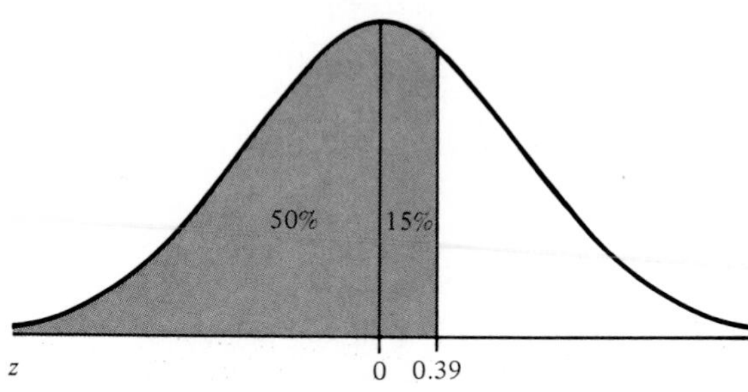

(b) P_{65} falls at a z score of 0.39.

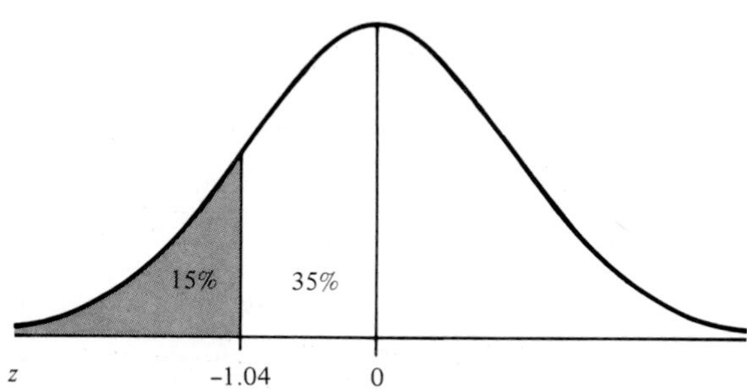

(c) P_{15} falls at a z score of −1.04.

FIGURE 6.5. **Still more areas under the normal curve.**

How do you find the z score corresponding to a given percentile? In an earlier chapter we noted that a percentile is a point below which lies a given percentage of the distribution. So to answer the question "What is the z score that corresponds to the 65th percentile, or P_{65}?" we note that in Figure 6.5b, since 50% of the distribution is below the mean, 15% would be above. We then look up 15% in the *area* column of Table B, because the areas shown in Table B are between the mean and the z value. We find that the exact value of 0.1500 is not listed in Table B, so we take the value *closest* to 0.1500, which is 0.1517, and see that the corresponding z score is 0.39. So the z score at P_{65} is 0.39.

FIGURE 6.5c. *What if the percentile is below the mean?* As you can see, the situation is a little different if the percentile falls below the mean, at P_{15}, for instance. Since the areas of Table B are between the mean and the z score, it is necessary to subtract 0.1500 from 0.5000 to get 0.3500 with which to enter Table B. The nearest value in Table B is 0.3508 and the corresponding z score is 1.04. Since the percentile in question is below the mean, the z value for P_{15} is -1.04.

It should be evident by now that in any problems involving normal curve areas, it is a good idea to draw a rough sketch of the curve, both so you can follow the procedure (i.e., use Table B correctly) and so you get a pictorial representation that helps you visualize the reasoning behind the procedure. In some of the practical applications to follow, we will find such a pictorial approach very helpful in visualizing variables that otherwise would be difficult to understand.

Some Practical Applications

The preceding section was basically a set of instructions on the use of a normal curve table, at best a mechanical and mathematical procedure. We have already seen that the scientist is looking for mathematical models for understanding and predicting, so we now turn to a very important activity of behavioral scientists—applying the normal curve to real life situations.

In Chapter 2 we saw that much of the data of a psychological and physiological nature is normally distributed in the population. Variables, such as height, weight, and test scores, approximate this normal curve model. But in the remaining sections of this chapter we must keep in mind the distinction between the mathematical model and the distribution of observed data. The mathematical characteristics of the normal curve printed in Table B (z scores and corresponding areas under the curve) apply to the model only, and the accuracy of our predictions regarding the real world depends on how close the population "fits" the normal curve model.

With these reservations in mind, let us use the normal curve model in a practical setting to make predictions about the variable of tested IQ scores. We noted in the last chapter that according to one popular IQ test the mean IQ is 100 with a standard deviation of 16. With only this information (and the assumption that IQ is normally distributed in the population) we can make a number of predictions based on the normal curve model.

FIGURE 6.6a. *If an individual with a tested IQ of 70 or below is considered mentally retarded, what percentage of the population would be classified as mentally retarded?* The first step in answering this ques-

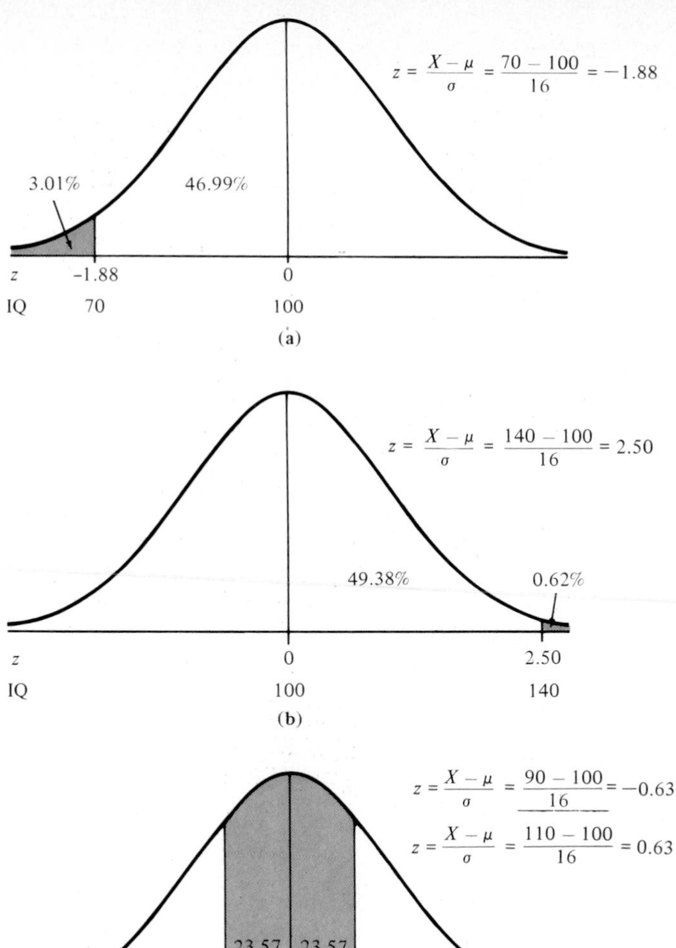

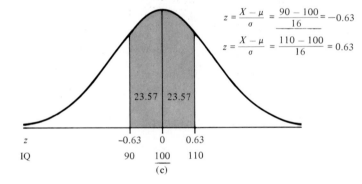

FIGURE 6.6. **Normal curve applications with IQ test scores.**

tion is to convert the score of 70 to a z score. The formula given in Chapter 5 for a z score in a population was $z = (X - \mu)/\sigma$. So a score of 70 in the distribution of IQ scores would have a z of

$$z = \frac{X - \mu}{\sigma} = \frac{70 - 100}{16} = \frac{-30}{16} = -1.88$$

Referring to Figure 6.6a and Table B of Appendix 4 you can see that a z of −1.88 is below the mean, and the area between z and the mean is 0.4699. But the question asked concerned the population *below* 70, so we must subtract 0.4699 from 0.5000 to get 0.0301. Thus we would conclude that 3.01% of the population has a tested IQ of 70 or below.

FIGURE 6.6b. *If an IQ of 140 or above is necessary for a person to be considered mentally gifted, what percentage of the population is mentally gifted?* Since we need a z score to enter Table B we find that the z score corresponding to an IQ of 140 is

$$z = \frac{X - \mu}{\sigma} = \frac{140 - 100}{16} = \frac{40}{16} = 2.50$$

Using Table B and Figure 6.6b, we note that 0.4938 of the area is between the mean and the z of 2.50, leaving 0.0062 or 0.62% of the area above the z score. So we conclude that 0.62% of the population would have tested IQ scores above 140.

FIGURE 6.6c. *If the range of "average" IQ scores is 90 to 110, what percentage of the population is considered average in intelligence?* Converting IQ scores to z scores, we get

$$z = \frac{X - \mu}{\sigma} = \frac{90 - 100}{16} = \frac{-10}{16} = -0.63$$

$$z = \frac{X - \mu}{\sigma} = \frac{110 - 100}{16} = \frac{10}{16} = 0.63$$

We see that the area of the curve in question is that between z scores of −0.63 and 0.63. Referring to Table B, we find that 0.2357 of the area lies between the mean and a z of 0.63. Since the curve is symmetrical, we know that the same area lies between the mean and a z of −0.63, for a total area of 0.2357 + 0.2357, or 0.4714. So we conclude that 47.14% of the population would have IQ scores between 90 and 110.

The careful reader will note that the three examples above were similar in that a decision had to be made about the location of a score in the distribution, the score was converted to a z score, and Table B was used to find the percentage of area above, below, or between the z scores. In terms of the earlier section on the use of Table B, we are again saying, "Given a z, find the percentage." The reverse procedure—"Given a percentage, find the z"—can also be used in practical applications, and Figure 6.7 illustrates this fact.

FIGURE 6.7a. *The middle 95% of the population have IQ scores between what two values?* Dividing the distribution into two equal parts on each side of the mean gives 47.5% on each side. Table B shows us that the z scores corresponding to these areas are −1.96 and 1.96, so we know that the middle 95% have IQ scores between z values of −1.96 and 1.96. However, the question is stated in terms of IQ scores, not z scores, so we must convert our z values to IQ scores, and we find:

$$X = z\sigma + \mu = -1.96(16) + 100 = -31.36 + 100 = 68.64$$
$$X = z\sigma + \mu = 1.96(16) + 100 = 31.36 + 100 = 131.36$$

So, as Figure 6.7a indicates, the middle 95% of the population have scores between 69 and 131.

FIGURE 6.7b. *How high does an individual's IQ have to be to place among the top 10% of IQ scores?* By drawing a quick sketch, you can see at a glance that we want the score that corresponds to the 90th

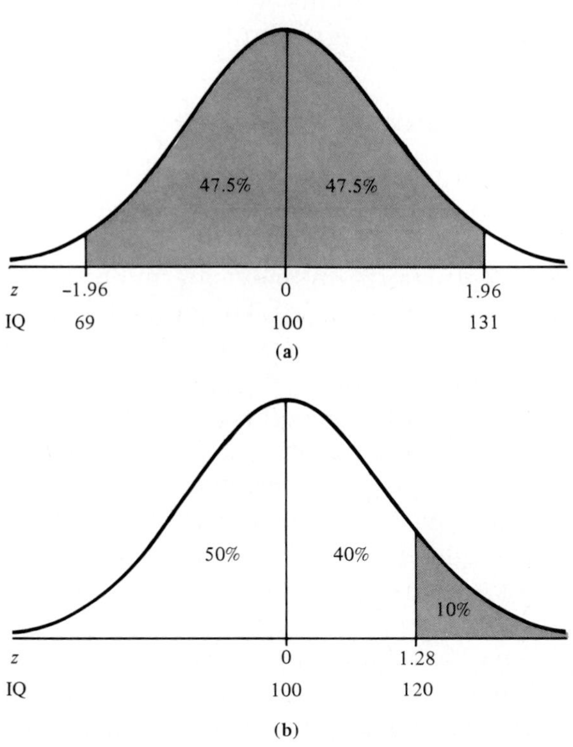

FIGURE 6.7. **More normal curve applications with IQ test scores.**

percentile, since that would divide the bottom 90% from the top 10%. As Figure 6.7b shows, we use Table B to find the z score that marks off 40% of the area, and we note that the closest value to 40% is 0.3997 and its corresponding z score is 1.28. Converting the z score to an IQ score, we find:

$$X = z\sigma + \mu = 1.28(16) + 100 = 20.48 + 100 = 120.48$$

Thus we conclude that an individual must have an IQ score of at least 120 to be considered in the top 10% of the population.

Probability and the Normal Curve

As we noted earlier in the chapter, especially in Note 6.1, the development of the normal curve was directly linked to some practical applications involving probability—mostly questions that gamblers might ask on the probability of the flip of a coin or the roll of a die. Although a complete treatise on probability is beyond the intended scope of this book, it will be helpful to become acquainted with some of the terms used in probability, since we must be aware of just how the normal curve can assist us in making probability statements.

Coins, Cards, Dice, and People

A formal discussion of probability would make the subject seem unfamiliar to most of you, but, whether you are aware of it or not, we are constantly surrounded by probability statements. In a single evening the news, weather, and sports on television might all contribute statements directly related to probability. The news anchor might comment on the increase in car insurance rates, the weather reporter will report a 30% chance of rain for the next 24 hours, and the sports editor will note that the boxer being interviewed is a 5-to-1 favorite for tomorrow's middleweight match. Also if, during a TV commercial, the reigning Dairy Princess spins a wheel to determine the lucky winner of a prize Holstein cow, we will have seen demonstrated all the different types of probability in a single evening! With these examples in mind, let us sort out two different types of probability.

CLASSICAL. This type of probability is undoubtedly the most familiar to us, since we all have flipped coins, rolled dice, drawn a playing card from a deck, and maybe even, like the Dairy Princess, had a chance to spin a roulette wheel. We define probability, *p*, in this case by the following formula:

$$p = \frac{\text{number of ways the event in question can occur}}{\text{number of events possible}}$$

So the probability of getting heads in the flip of a coin is $\frac{1}{2}$, since the 1 event is getting heads on a normal coin, and the total number of events possible (represented by the 2) is a head and a tail. Similarly, the probability of rolling a "4" on one toss of a die is $\frac{1}{6}$, since only one side of a die has four spots while there is a total of six possible sides on the die. The probability of drawing an ace of spades from a well-shuffled deck would, of course, be $\frac{1}{52}$, while the probability of drawing *any* ace from the deck would be $\frac{4}{52}$. The typical roulette wheel has 38 numbered spaces and 1 blank space, so the probability that your number will come up on a fair spin of the wheel is $\frac{1}{39}$.

Although we have been using *fractions* to illustrate these probability situations, it is conventional to state probabilities in terms of *proportions*. So, instead of $p = \frac{1}{2}$, $p = \frac{1}{6}$, or $p = \frac{1}{52}$, it is more common to see $p = .50$, $p = .17$, or $p = .02$. A quick glance at the general probability formula above should demonstrate that the probability of a "certainty" or "sure thing" is 1.0. Thus the probability of flipping *either* a head *or* tail is $\frac{2}{2}$ or 1.0.

One lucky by-product of stating probabilities in terms of proportions is that they are so easily converted to percentages when we talk about the frequency of occurrence in the "long run." For example, if the probability of rolling a "4" in one throw of a die is $p = .17$, we can expect a "4" to appear approximately 17% of the time. Or, in the long run, we could expect to pull an ace of spades from a well-shuffled deck approximately 2% of the time. There are advantages to thinking of probabilities in terms of percentages, as we shall see when we apply probabilities to areas under the normal curve.

EMPIRICAL. The characteristic that differentiates empirical probability from classical is that empirical probabilities can be calculated only from previous observations. How would one calculate the probability that the next baby born at a given hospital will be a boy? If records for the past five years show 7223 babies born at this hospital, 3720 of them boys, our empirical probability that the next birth will be a boy would be $p = 3720/7223 = .515$. Or, again using the "long run" concept, we would say that 51.5% of the births would be boys.

In a similar fashion, the weather reporter's prediction of a 30% chance of rain is based on a number of past situations when meteorological conditions were identical to present conditions. Some precipitation occurred in the surrounding area on 30% of the occasions when these conditions were present, so she is predicting that $p = .30$ for some rain during the next 24 hours. (Not everyone comprehends this system, as illustrated by Figure 6.8. The same interpretation can be applied to the car insurance rates and the betting results for the boxer. Past results (amount of insurance claims or amount bet on the boxer to win vs. amount bet on him to lose) in each case determine the probability associated with each event.

It should be clear by now that the way to compute an empirical probability is simply to calculate the percentage of times that the event in which you are interested has occurred in past situations and convert that percentage to a proportion. So, if you find that there are 800 freshmen, 680 sophomores, 578 juniors, and 492 seniors at a college with a total enrollment of 2550, you would simply convert these to percentages to find that there are 31% freshmen, 27% sophomores, 23% juniors, and 19% seniors. Thus the probability that a name drawn at random from the student directory will be that of a senior would be $p = .19$, since 19% of the student body are seniors.

Reading Probabilities from the Normal Curve Table

We can go directly from the normal curve areas of Table B to statements of probability. This convenient arrangement is possible because probability in this case can be defined as

$$p = \frac{\text{area under a portion of the curve}}{\text{total area under the curve}}$$

For example, what is the probability of a z score being equal to or greater than 1.0? From Table B we note that 0.1587 of the area of the normal curve lies above a z of 1.0 (0.5000 − 0.3413), so

$$p = \frac{.1587}{1.0} = .1587$$

FIGURE 6.8. **Probability statements in weather forecasts are not universally understood.** (Reprinted by permission: Tribune Media Services.)

In other words, the proportions given in Table B *are* probabilities, since dividing by the total area of 1.0 always yields the numerator. So, in this example we would conclude that the probability of a z score being 1.0 or greater is $p = .1587$ or .16.

What is the probability of a z score being between −1.96 and 1.96? Since 95% of the area under the curve is between these two values, $p = .95$ that any z score would be between those values.

It should be obvious by now that in any of the previous examples where we calculated a percentage of the area under the curve we could easily make a probability statement. In Figure 6.6a, for example, we found that 3.01% of the population would have an IQ score of less than 70. Thus the probability that a person selected at random would have an IQ score below 70 is $p = .03$. Similarly, in Figure 6.6b and 6.6c, $p = .006$ that a person would have an IQ score above 140, and $p = .47$ that an individual would have a score between 90 and 110.

Odds and Ends

Any observer of the American sporting scene is aware of a unique way of stating the probability of success in a sporting event—in terms of the "odds for" or "odds against" a particular outcome. A football power might be a 2-to-1 favorite, a horse might be a 40-to-1 long shot, or the chances of filling out a spade flush in your friendly poker game might be 1 in 25. Of course, such statements are not limited to sporting events or games of chance but are often made about political elections, weather forecasts, and any other event where the outcome is something less than a certainty.

Undoubtedly, most of these statements are rather subjective, as shown by Figure 6.9, and are based upon hunches, past experiences, and the like. For example, in horse racing, the "morning line" of a newspaper contains the day's picks by one or more sportswriters, and statements like "Whirligig is a 100-to-1 long shot" simply imply that this creature may not even make it from the stable area to the starting gate. However, in some cases, actual data may be involved; thus the fact that a sample of voters has given two-thirds of their votes to

FIGURE 6.9. **An example of a subjective odds statement.**

one candidate may lead to a statement that "Senator Blank is 2-to-1 favorite."

It would be interesting to continue discussing horseflesh and gambling, but we will have to confine our treatment of the topic to classical probability and areas under the normal curve. We will then be able to use "odds for" or "odds against" as just another way of stating probability. A homely way of stating the *odds for* an event to occur would be

Frequency of occurrence to frequency of nonoccurrence.

Thus the *odds for* obtaining a "3" in the roll of a die would be 1 to 5, since out of six possible ways the die can land 1 is the frequency of the specified event and 5 is the frequency of the remaining events. In a similar fashion, the *odds for* drawing an ace of spades from a deck of cards would be 1 to 51, and the *odds for* drawing any ace would be 4 to 48. You might note that the *odds for* flipping a head, with an unbiased coin, would be 1 to 1; that is, the odds are even. In all cases note that the total frequency in the odds statement adds up to the total number of ways *any* event can occur. That is, 1 to 5 = 6 sides, 1 to 51 = 52 cards, and so on.

The *odds against* an event occurring are given simply by reversing the *odds for* statement. The *odds against* rolling the "3" are 5 to 1, and the *odds against* drawing the ace of spades are 51 to 1.

When odds statements are made concerning areas under the normal curve, we are able to go directly from probability to odds by using 100 as the base figure. For example, we noted that about 68% of the area under the curve lies between z's of -1.0 and 1.0. We then stated that the probability of a z score chosen at random being between -1.0 and 1.0 was $p = .68$. Thus the *odds for* a z score being between -1.0 and 1.0 are 68 to 32. Similarly, in Figure 6.6, the *odds for* an individual having an IQ of less than 70 would be 3 to 97, the *odds for* an individual having an IQ greater than 140 would be less than 1 to 99 (actually 6 to 994), and the *odds for* an individual having an IQ between 90 and 110 would be 47 to 53.

Some Concluding Remarks

As was mentioned earlier, the discussion of probability in this chapter is a very superficial treatment of the topic. We did, however, spend quite a bit of time with the normal curve, both examining its mathematical properties and applying these characteristics to some practical situations. But you still must be aware of the fact that we have just barely scratched the surface as far as the normal curve goes. By using this normal curve model, mathematical statisticians have been able to solve a number of theoretical issues that would otherwise be extremely tedious if not impossible. Many of the techniques to be discussed in later chapters are based on such a theoretical foundation. If this topic has whetted your appetite for a mathematical approach to statistics, by all means pursue the matter with your favorite mathematics department.

SAMPLE PROBLEM

A manufacturer of hockey equipment is designing a new set of protective gear for college hockey players, and the design engineers are basing their model on an average college male chest measurement of 39 inches with a standard deviation of 2.5 inches. We will assume that chest circumferences are normally distributed in the college male population.

What percentage of the population has a chest measurement of 44 inches or over?

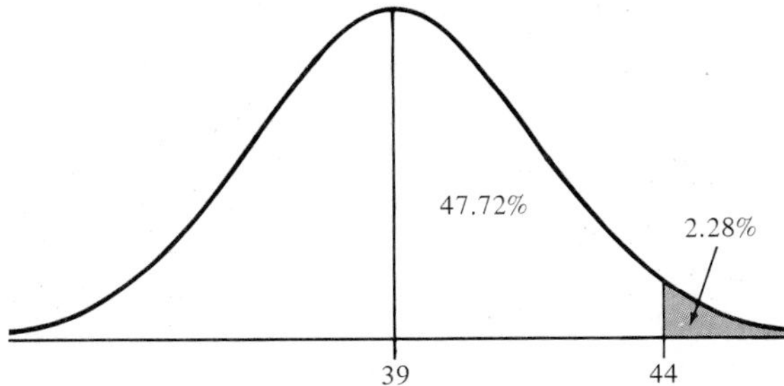

After drawing the curve, we calculate the z score for 44, which is $z = (44 - 39)/2.5 = 2.0$. In Table B of Appendix 4 we note that 47.72% of the area is between a z of 2.0 and the mean, leaving a total of 2.28% with a chest measurement of 44 inches or more.

If an engineer wants his design to fit the middle 99% of all players, what range of chest measurements will be covered?

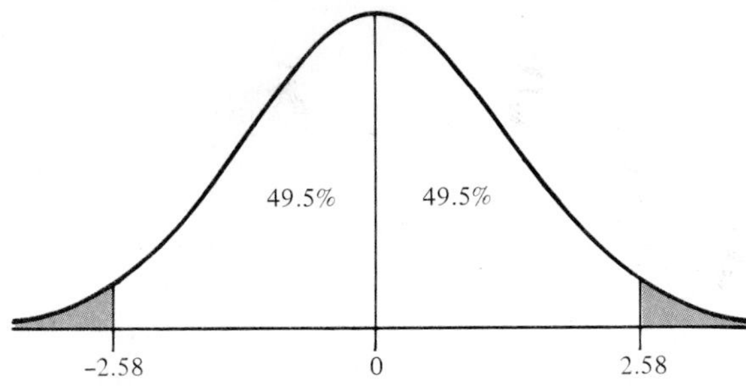

The middle 99% would be divided so that 49.5% of the distribution is on either side of the mean. In Table B we see that z scores of −2.58 and 2.58 are the boundaries for this 99%. Converting the z scores to chest measurements, we have

$$X = z\sigma + \mu = -2.58(2.5) + 39 = 32.6$$
$$X = 2.58(2.5) + 39 = 45.5$$

So we conclude that the design would have to be adjustable from 32.6 inches to 45.5 inches in order to fit the middle 99% of the players.

A husky defenseman claims that he is in the top 1% in chest expansion. What would his measurement have to be in order for him to make this claim?

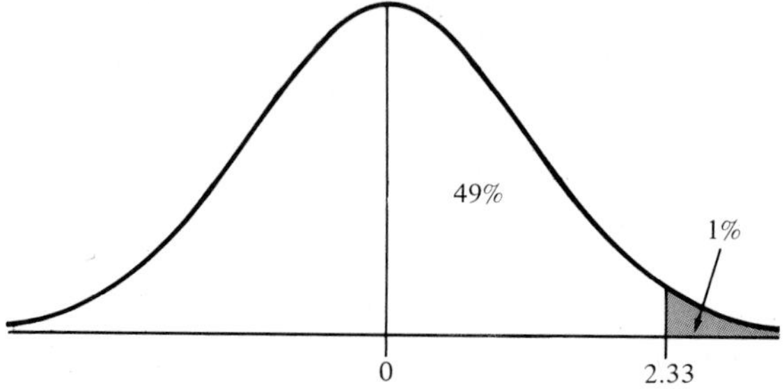

From the drawing and Table B we note that 49% of the area would be between the mean and the point needed, which would give a z score of 2.33. Converting the z score to inches, we would have

$$X = z\sigma + \mu = 2.33(2.5) + 39 = 44.8$$

His chest measurement would have to be 44.8 inches.

What is the probability that a hockey player, chosen at random, would have a chest measurement of less than 37 inches?

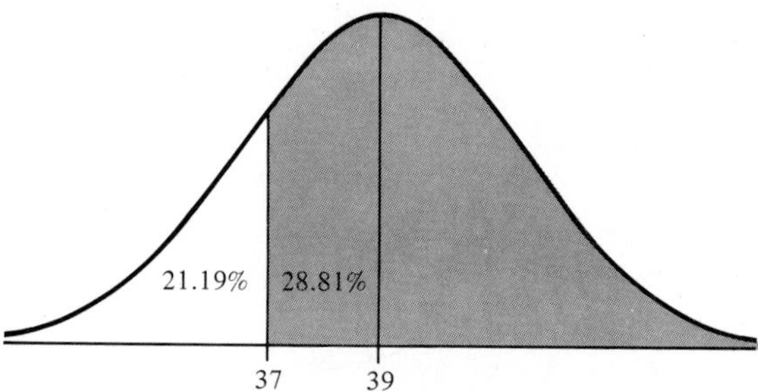

After drawing the curve, we calculate the z score for 37, which is $z = (37 - 39)/2.5 = -0.8$. In Table B we note that 28.81% of the area

SAMPLE PROBLEM (*cont.*)

is between the mean and a z of −0.8. This would leave 21.19% below this point, so the probability that any hockey player chosen at random would have a chest measurement of less than 37 inches would be $p = .21$.

STUDY QUESTIONS

1. Mathematical models are used for what two purposes?
2. How is the unit normal curve different from other normal curves?
3. A score equivalent to a z of −2.3 is located where in the normal curve?
4. How would you find the area under the normal curve between 2 negative z scores?
5. What is the difference between classical and empirical probability?
6. What is the relationship between a probability p and the "long run"?
7. How are odds statements used in classical probability? In empirical probability?
8. What is the difference between the statement "The odds are a thousand to 1 against my getting that job" and the statement "The odds are 3 to 1 for the mayor to win reelection?"

EXERCISES

1. What percentage of the area under the normal curve lies within the following boundaries?
   a. Between the mean and a z of 1.35.
   b. Between the mean and a z of −0.75.
   c. Between z scores of −2.31 and 1.43.
   d. Between z scores of −1.96 and 1.96.
   e. Above a z of 1.96.
   f. Below a z of −2.58.
   g. Above a z of −1.45.
   h. Below a z of 1.96.
2. What percent of the area under the normal curve lies
   a. between the mean and a z of −1.73?
   b. between the mean and a z of 1.96?
   c. between z scores of −1.82 and 0.90?
   d. between z scores of −2.58 and 2.58?
   e. below a z of 1.80?
   f. below a z of −1.96?

g. above a z of −1.00?
h. above a z of 2.25?

3. According to the National Center for Health Statistics the average male college student weighs 154 pounds and the distribution of weights has a standard deviation of 21.4 pounds. Assume that this distribution is normally distributed in the population to answer the following.
   a. What percent of college men weigh 200 pounds or more?
   b. Tami tells her friends that she will not date any man that weighs less than she does. Tami weighs 125. What percent of the college men need not bother to call?
   c. Bill states that he is heavier than 99 percent of the college male population. How much would Bill have to weigh to make this statement?
   d. A hot air balloonist wants to hire a male assistant who weighs 140 pounds or less. What percent of the college male population would be eligible for this job?
   e. The middle 95% of the college male population would be included between what two weights?
   f. What weight corresponds to the 90th percentile?

4. The National Center for Health Statistics also states that the population of college women has a mean height of 5 feet 4 inches with a standard deviation of 2.48 inches. Assume that this variable is normally distributed in the population to make the following predictions.
   a. What percent of college women are 5 feet or shorter?
   b. An eccentric restaurant owner wants his waitresses to be 5 feet 10 inches or taller. What percent of college women would be eligible for these jobs?
   c. Kim states that she is so short that she is in the bottom 10% of women's heights. What height would she have to be to make that statement?
   d. Pete says that he will not date any woman who is taller than he. Pete's height is 5 feet 6 inches. What percent of college women are missing this opportunity?
   e. The middle 99% of college women are between what two heights?
   f. What height corresponds to the 98th percentile?

5. A popular college entrance examination has a mean of 50 and a standard deviation of 10. Assume that this variable is normally distributed in the college-bound population.
   a. What percent of the population would score 30 or below?
   b. The Office of Admissions at one college decided to admit anyone who scored 35 or above. What percent of the population is eligible for admission?
   c. Maria learns that her test score is in the top 3% of the distribution. If this is true, how high must her test score be?
   d. A guidance counselor claims that no one scoring in the bottom 20% should consider going to college. This group would have scores below what value?

6. The National Heart, Lung, and Blood Institute completed a large-scale study of cholesterol and heart disease. They estimated that the average serum cholesterol reading in the United States was 210 milligrams per deciliter of blood (mg/dl) with a standard deviation of 33 mg/dl. Assume that this variable is normally distributed in the population to answer the following.
   a. One of the requirements for participation in the study was to have a cholesterol reading of 265 mg/dl or greater. What percent of the population would fulfill this requirement?
   b. Some physicians have recommended that anyone with a reading of 240 mg/dl or more should go on a low cholesterol diet and take cholesterol lowering drugs. What percent of the population would be in this category?
   c. Some authorities believe that even though the "average" cholesterol reading in the population is 210 mg/dl, it is too high and should be lowered to where the present 38th percentile now is. What average cholesterol level are they recommending?
   d. Sue is proud of her nutrition and exercise program and claims that her cholesterol reading is in the bottom 20% of the population. How low would her reading have to be to enable her to make that statement?

7. There are 60 cars in a parking lot: 14 Chevrolets, 12 Fords, 6 Plymouths, 5 Buicks, 4 Pontiacs, and 19 foreign cars. A women crosses the street and enters the lot to retrieve her car.
   a. What is the probability that she is the owner of a Ford?
   b. What is the probability that she owns a foreign car?
   c. What are the odds for her owning a foreign car? Odds against?

8. A leather bag contains 50 marbles of different colors: 17 red; 10 yellow; 9 blue; 5 green; and 3 each of white, brown, and black. You reach into the bag to get a marble.
   a. What is the probability it will be blue?
   b. What is the probability it will be white, brown, or black?
   c. Use the term "in the long run" to explain your answer in part (a).
   d. What are the odds for the marble being blue? Odds against?

9. In the distribution of college men's weights discussed in Exercise 3:
   a. What is the probability that a man chosen at random would weigh 200 pounds or more? What are the odds for this occurrence? Odds against?
   b. What is the probability that a man chosen at random would not meet Tami's requirements? Odds for? Odds against?

10. In the distribution of college women's heights discussed in Exercise 4:
   a. What is the probability that a college woman chosen at random would be 5 feet or shorter? What are the odds for this occurrence? Odds against?
   b. What is the probability that a woman chosen at random would be eligible for the restaurant job? Odds for? Odds against?

11. The distribution of scores for a popular college entrance exam was discussed in Exercise 5.
    a. What is the probability that a college-bound examinee chosen at random would score 30 or below?
    b. What is the probability that a college-bound examinee chosen at random would score 35 or above?

12. The distribution of cholesterol readings was discussed in Exercise 6.
    a. What is the probability that someone chosen at random would have a cholesterol reading of 265 mg/dl or greater?
    b. What is the probability that someone chosen at random would have a cholesterol reading of 240 mg/dl or greater?

CHAPTER 7

Sampling Theory

As the guard changes at Buckingham Palace, so must we change emphasis in our treatment of statistical topics. Earlier we noted the distinction between descriptive statistics and inferential statistics, and it is with this chapter that we begin to be concerned about making *inferences* about the *population* instead of being content merely to describe a particular distribution of data. To be sure, descriptive statistics are useful—even indispensable—for a teacher measuring the performance of a class of algebra students or a college administration calculating grade-point averages, but as Chapter 1 pointed out, educators and behavioral scientists often gather evidence from samples for the express purpose of generalizing their results to the population.

Descriptions Versus Inferences

If a social psychologist wants to know the relationship between TV violence and aggression in children, or if a guidance counselor wants to know if participation in extracurricular activities is an indicator of a healthy self-image, or if a personnel manager wants to see if performance on a manual dexterity test will predict job success on the assembly line, all three will undoubtedly select a group of individuals, gather data, and generalize the results from the small sample to a much larger population.

Note that the difference between descriptive statistics and inferential statistics is not in the *kind* of statistical measures employed but in the *purpose* for doing the measuring. If your intent is merely to describe a collection of data, you will construct a frequency distribution and frequency polygon in order to picture the distribution and you will calculate a mean, median, or mode to discover its central tendency and a standard deviation to find its variability. All of these procedures, of course, *describe* the distribution.

Now, if your purpose is to make *inferences* about a population, you will still calculate means and standard deviations, but your interest will focus on how well these statistics estimate a *population* mean and standard deviation. The rest of this chapter will be devoted to procedures for taking samples and for estimating the population values by using the sample statistics.

The terms *sample* and *population* were formally defined earlier in Chapter 3, but a review might be in order. Basically, a *population* is a group of elements that are alike on one or more characteristics as defined by the researcher, such as all fourth-graders in a state or all college students in the United States or all property owners in a county. Or in terms of the social psychologist, the guidance counselor, and the personnel manager mentioned earlier, the populations could be all the preschool children in the Midwest, all the junior high school students in a metropolitan school district, or all the welding machine operators in automotive assembly plants owned by General Motors. The important thing to remember is that the population is defined by a researcher *for a particular purpose* and *all* the elements satisfying the criteria are members of that population. Size alone does not determine whether or not a group of elements is a population, although in practice most populations turn out to be rather large.

A *sample*, on the other hand, is a group of elements that is selected from the population and is *smaller* in number than the size of the population. A sample may be chosen in a number of ways, and you can have random samples, stratified samples, or cluster samples. Each type involves a different procedure for determining how the data are to be gathered; the following section describes the different sampling procedures.

Types of Samples

Random Samples

The random sample is chosen in such a way that *each element in the population has an equal chance of being in the sample*. This is an important definition and one that must be strictly followed before the procedures outlined in this chapter can be applied. But how does one draw a random sample in practice?

As an example, let us say that we would like to interview a random sample of college students on a given college campus about their attitudes toward the college administration. It would be simple to stand in the entrance to the student union and pick the first 50 students that come by, but the result would *not* be a random sample. The fact that many polls are conducted in just this fashion does not make it right, and we must look for a more statistically defensible procedure. There are two generally acceptable ways of obtaining a random sample: a table of random numbers and a "counting-off" technique. The random number method is preferred, although the counting method will usually approximate a random sample.

TABLE OF RANDOM NUMBERS. Table L in Appendix 4 is a collection of random numbers, random in that any digit or any grouping of four digits bears no relationship to any other digit or grouping of digits in the table. In other words, in terms of our definition of a random sample, in any position in the table, each digit from 0 to 9 has an equal chance of appearing.

To obtain a random sample from a college population of 5000 students, we would first obtain a listing of all students and assign each one a number from 1 to 5000. We would then enter Table L at a point determined by chance (a colleague of mine closes his eyes and stabs at the page with his fingertip or a pencil). From this starting point, we would take as many numbers from the columns as we needed for our sample. If we ran across a duplicate number, we would ignore it and go on to the next one. If we wanted one-, two-, or three-digit numbers, we would use the same procedure but would take only the first one, two, or three digits of the four-digit numbers. The students with the numbers corresponding to the ones we selected would constitute our sample.

COUNTING-OFF PROCEDURE. The technique described above can be quite time-consuming when the population is large; assigning a number to all the names in a metropolitan telephone directory, for instance, would require great dedication. In such a case it may be more convenient to use a counting-off procedure and take every hundredth or every thousandth name in the list. For example, if we wanted a random sample of 200 from a population of 60,000, we would take every 300th name. A counting-off procedure is used frequently in consumer surveys, where a questionnaire may be inserted in every 30th box of the manufacturer's product or mailed along with a monthly billing to every 10th credit card customer.

Sometimes referred to as "systematic sampling," this procedure involves entering a directory or list of names by making the first draw a random one. For example, if we wanted a sample of 50 from a list of 500 names, we would need every 10th name on the list. The starting place for counting would be determined by a table of random numbers, so we would choose a number from 1 to 10 at random. If the number 6 came up, we would begin our drawing at the 6th name on the list and take every 10th name after that.

Stratified Random Sampling

When the results of the poll or experiment are likely to be affected by certain characteristics of the population, a stratified random sample is often used. For example, if a campus survey is conducted and the attitudes being evaluated would depend on sex, year in school, size of high school graduating class, and family income, it would be essential to have the sample contain the proportions that exist in the campus population. A number of subgroups would be set up (e.g., freshman women from a high school graduating class of 300 to 399 whose annual family income is between $40,000 and $50,000), and *random* samples would be drawn from each of the subgroups in such a way that the size of each subgroup in the sample was proportional to the size of the same subgroup in the total campus population. This, of course, would result in a sample that is a miniature population, with all relevant characteristics represented.

Although stratified sampling may require a colossal amount of work, it is essential in many instances where a *representative* sample is required. In Chapter 4 we noted that the sample of children used in constructing norms for an early IQ test was *stratified* according to the father's occupation, since it was known that IQ was related to socioeconomic status. Similarly, in voter polls prior to an election, stratified samples need to have proportionate representation of Democrats and Republicans, of men and women, and of relevant variables such as education, income, ethnic group, and the like.

Cluster Sampling

In practice it may be very difficult to stratify a population on a nationwide basis, so cluster sampling is often used. Whereas the members of a given stratum or subgroup of the stratified sample are alike (all men, or all Democrats, etc.), a cluster is itself composed of the different variables. An example might help clarify this distinction.

Let us say that we would like to construct a set of national norms for a test that we have developed and would like fourth-graders from all over the nation to be represented in the sample. Clusters could be any intact unit such as states, counties, or legislative districts, but we would probably choose to use school districts as our "population" of clusters. As you can see, each school district would have fourth-graders enrolled in one or more elementary schools, and all the variables affecting the test scores (e.g., socioeconomic level, sex, age) would be represented in the cluster.

After the clusters had been defined, a random sample of clusters would be chosen in the first stage of sampling, and additional random sampling would be used to select particular schools, classrooms, or even individuals within a class that would make up our final sample. Again, as with stratified sampling, our objective is to achieve a sample that is *representative* of the population.

One popular national polling service uses more than 300 voting precincts for its samples. These would be considered clusters, since they contain a certain proportion of each sex, political preference, ethnic group, socioeconomic level, and so on. Within these clusters random samples can be taken; for example, every fourth dwelling on the block might be polled.

Some Concluding Remarks on Samples

Before progressing to the mathematics of sampling theory, you should be aware of some terms and concepts that you may encounter in reference to sampling.

Size of Sample

All other things being equal, a sample is more likely to be accurate (i.e., to faithfully describe the population characteristics) as it increases in size. Size alone does not insure an accurate sample (as we will note when we take up the topic of *biased* samples), but it is safe to say that if we follow the rules in obtaining whatever type of sample we are working with, a larger sample will more accurately reflect the characteristics of the population. You will have to accept this statement on faith for stratified and cluster samples, but later in the chapter we will have occasion to demonstrate mathematically the effect of the size of a random sample on the accuracy of results.

Biased Samples

Any time that our samples contain a systematic error, they are said to be *biased*. A very common example of bias in sampling is the case in which copies of a questionnaire are mailed out to a number of people and only a fraction of the total are sent back. It is entirely possible that those returning the questionnaire, thereby volunteering to have their data examined, are different in some respect from those who do not return them. The classical study on sexual behavior in men by Kinsey was criticized by some on just this point; the critics felt that those men willing to divulge such information in explicit detail were somehow different from the ordinary male in the population.

Perhaps the most famous example of biased sampling occurred in a voting preference poll sponsored by the *Literary Digest* in 1936. On the basis of a huge sample of voters gleaned from automobile registration records and telephone directories, the magazine predicted that Alf Landon would easily defeat Franklin D. Roosevelt in the 1936 presidential election. The systematic error in this poll was that voters who did not own automobiles or have telephones—the people who were most likely to vote for FDR and his social reform policies—were not represented in the sample. They did vote, however, and Roosevelt was elected by an overwhelming majority.

Incidental Sampling

Much of the data constituting the foundation for theories in education and the behavioral sciences originates in studies using college students. It seems that introductory psychology courses are a very handy source of subjects for an infinite variety of experiments, and it is the rule rather than the exception to have each student participate in one or more experiments as a course requirement. Critics of this technique are quick to point out that college sophomores in the introductory psychology course may not be the representative of even college students in general, to say nothing of the population at large. Whenever such *incidental sampling* is employed (grade school classes, a college faculty, county welfare recipients, etc.), it is difficult, to say the least, to generalize to a population of all grade schools, college faculties, or welfare recipients. The particular sample at hand may be unique, so that the results of a study simply cannot be generalized to a larger population without serious error. Thus, if incidental sampling must be used because of cost considerations, great pains must be taken to examine the sample for any possible biases. Otherwise, it could be the *Literary Digest* poll all over again!

Sampling with Replacement

When random samples are drawn from a comparatively small population, it is usually necessary to replace the observation before taking the next one. If your population consists of a deck of cards, a bag of colored marbles, numbers in a hat, or the like, you should return the card or whatever to the group before making another draw. For example, let us say a bag contains 15 yellow, 3 blue, and 2 red marbles. The probability of drawing a blue marble would be $\frac{3}{20}$ or $p = .15$. However, if you didn't get a blue one on your first draw, the probability of getting a blue one on the second draw depends on whether or not you return the first marble to the bag. If you don't, the probability is now $\frac{3}{19}$ or $p = .16$. Similarly, the probability of pulling an ace from a well-shuffled deck is $\frac{4}{52}$ or $p = .077$. If you don't get an ace on the first draw and do not replace the card, the probability on the second draw is now $\frac{4}{51}$ or $p = .078$. The differences due to nonreplacement in the above examples are small, but nonreplacement could affect your calculations in a complex probability problem.

Parameters Versus Statistics

In the sections to come we will have occasion to talk about means, medians, standard deviations, and other characteristics of samples, as well as means and standard deviations of a population. We will attempt to avoid confusion by referring to these measures of central tendency and variability as *statistics* when they describe a *sample* and as *parameters* when they describe a *population*. In other words the mean and standard deviation of a sample are *statistics*, and their namesakes in the population are *parameters*. As we noted earlier in Chapters 3 and 5, we use the lowercase Greek letter μ (mu) for the population mean and the lowercase Greek letter σ (sigma) for the population standard deviation. As before, $\overline{X}$ and S are the sample mean and standard deviation, respectively.

Estimating the Population Mean

Much of the energy expended in inferential statistics goes toward the *estimation* of population parameters. This estimation process is necessary because in most cases it is virtually impossible to measure an entire population due to its size and ever-changing nature. If one wants to know the average IQ of high school seniors, manual dexterity of aerospace electronics technicians, reading readiness scores of 6-year-olds, or what have you, the respective populations are so large and inaccessible that the only reasonable procedure is to take a random sample and estimate the population values from this sample alone.

Although this procedure may seem questionable to you, it is the only practical approach to the problem. Statisticians have worked out the necessary formulas, tested mathematical models, and developed techniques to enable us to take a sample, calculate its mean, and estimate with a given degree of certainty how well this sample $\overline{X}$ estimates the "true" mean or population mean, μ. The following sections will describe the theory behind the techniques and the methods to be followed in estimating population parameters.

The Sampling Distribution

Let us suppose that we would like to know the average height of all adult males in a large midwestern university. We follow the rules listed earlier for obtaining a random sample and, using a sample size of 100, we come up with a mean of 69.2 inches and a standard deviation of 2 inches. The sample $\overline{X}$ of 69.2 is called *an unbiased estimate of the population mean*, μ. By *unbiased*, we mean that if we continued to take samples like the one we just did the *mean of all these sample means* would approach the value of the population mean.

After we have gathered our data and calculated the sample mean, we ask, "How good an estimate of the population mean is our sample mean of 69.2 inches?" "Is it too high?" "Too low?" "Just about right?" Since we rarely know the true mean (population mean), we can never answer these questions exactly, but we can make a *probability statement* about its most likely value.

In order to do this, we must first indulge in a mathematical fantasy. Let us perform a hypothetical exercise in which we continue to take samples of size 100 and calculate the mean of each sample. Keeping track of these means, we might get

$$\overline{X}_1 = 69.2$$
$$\overline{X}_2 = 68.7$$
$$\overline{X}_3 = 69.6$$
$$\overline{X}_4 = 68.5$$

and so on.

If we continued to do this for an *infinite* number of samples (you see why it is a *hypothetical* exercise), the following would occur:

1. A batch of means, called a *sampling distribution of means*, would result.
2. The mean of this sampling distribution, $\overline{X}_{\overline{X}}$ (say "mean of means"), would be equal to the population mean, μ.
3. The batch of sample means would be normally distributed around the mean of the distribution, $\overline{X}_{\overline{X}}$, with a standard deviation of $\sigma/\sqrt{N}$.

This last point summarizes the *central limit theorem*, and it is an important foundation for much of our work in statistical inference. Stated more precisely, the central limit theorem in this instance says that *if a population has a mean* μ *and a standard deviation* σ, *then the distribution of sample means drawn from this population approaches a normal distribution* as N *increases, with a mean* $\overline{X}_{\overline{X}}$ *and a standard deviation* $\sigma/\sqrt{N}$. Regardless of the shape of the population from which we draw our samples, the sampling distribution of means will be normal *if the sample size is sufficiently large*. What is a "sufficiently large" sample? There is no easy answer, because the required sample size depends on the shape of the population distribution. You will find some statistics texts specifying an *N* of 30, and others an *N* of 50; certainly an *N* of 100 would remove all doubt about the resultant shape of the sampling distribution. In any event, the central limit theorem enables us to solve sampling problems without worrying whether or not the population from which we are sampling is normal.

The three characteristics of a sampling distribution noted above are of primary importance, so let us look at each in greater detail.

SAMPLING DISTRIBUTION OF MEANS. We would note as we calculated mean after mean that most of them tended to cluster about some central point, just as our frequency distributions tended to do back in Chapter 2. We should not be surprised that all the samples do not have the same mean; we expect that because of *chance* fluctuations the means will tend to vary slightly. Just as we would not expect to get 5 heads and 5 tails every time we flipped a coin 10 times, we also do not expect that all sample means would be the same.

Let us assume that the mean of means, $X_{\overline{X}}$, in our hypothetical exercise on the height of adult males in a midwestern university turned out to be 69 inches. We would notice that many of the sample means would be quite close to $\overline{X}_{\overline{X}}$, say 68.7, 69.2, or 68.6. However, we would also notice that a few of the sample means would be scattered a little farther away from $\overline{X}_{\overline{X}}$, such as 66.2 or 70.1. An occasional sample mean would be as far away as 65.1 or 72.3. We expect the sample means to be distributed in this fashion, *and this scattering of* $\overline{X}$*'s about* $\overline{X}_{\overline{X}}$ *is due to sampling error.*

THE MEAN OF MEANS. The mean of the sampling distribution of means, $\overline{X}_{\overline{X}}$, is equal to the population mean, μ. What we are saying is that if we take all possible samples of a given size from a population and calculate the mean of each sample, the mean of all these sample means will be identical to the population mean. The mathematical proof for the relationship between $\overline{X}_{\overline{X}}$ and μ is beyond the scope of this book, but this is a handy relationship, as we shall soon see.

NOTE 7.1

CHANCE, "DUMB LUCK," AND SAMPLING ERROR

There will be a number of occasions in the remaining chapters when we will be talking about *chance* or *sampling error*. Decisions must be made to determine whether some result is due to chance (i.e., sampling error), or whether the result represents a real departure from purely chance fluctuations. If you regularly take a coffee break with a friend and flip a coin to see who pays, you expect in the long run to win about half the time. In other words, who winds up paying on a given day is strictly due to "chance." But, as you well know, there might be times when one of you will win, three, or even four days in a row. Such fluctuations can still be chance effects and would be called, simply "sampling error."

On the other hand, if your friend seems to win the toss 90% of the time, you would probably demand to examine her coin, since such a departure from what you expect to happen "just by chance" leads you to suspect something other than an unbiased coin toss. But where do you draw the line? How much of a deviation from a 50-50 split do you endure before you can say that there is something uncanny about your friend's coin-tossing ability? 60-40? 70-30? 80-20? At what point do you say that the deviation is no longer sampling error but is a *real* nonchance happening? Decisions such as these are typical of inferential statistics, and we will be spending considerable time on just how such decisions are made.

NORMAL SAMPLING DISTRIBUTION. Since the sampling distribution of means is normal, we can use the characteristics of the normal curve that we noted back in the last chapter. For example, we could say:

1. 68.26% of all the sample means would fall between -1 and 1 standard deviation from the mean, $\overline{X}_{\overline{X}}$.
2. 95% of all the sample means would fall between -1.96 and 1.96 standard deviations from the mean, $\overline{X}_{\overline{X}}$.
3. 5% of all the sample means would fall outside the same interval, -1.96 and 1.96 standard deviations from the mean, $\overline{X}_{\overline{X}}$.

Similarly, the probability statements we made concerning the normal curve would apply to this sampling distribution as well.

1. The probability that a sample mean falls between $\overline{X}_{\overline{X}} \pm 1$ standard deviation would be .68.
2. The probability that a sample mean falls between $\overline{X}_{\overline{X}} \pm 1.96$ standard deviations would be .95.
3. The probability that a sample mean falls *outside* the same interval, $\overline{X}_{\overline{X}} \pm 1.96$ standard deviations, would be .05.

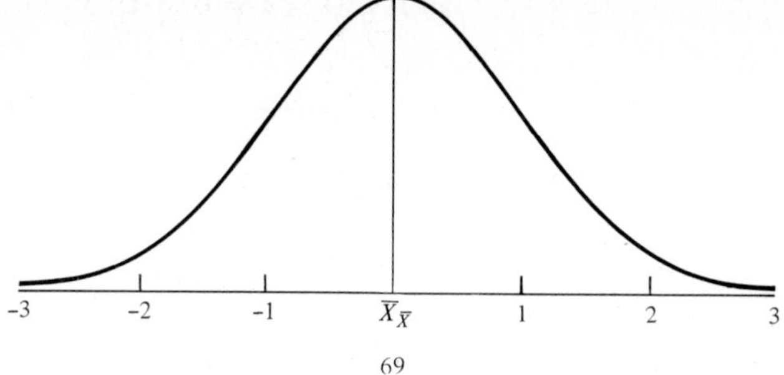

FIGURE 7.1. **Sampling distribution of means with $\overline{X}_{\overline{X}} = 69$.**

With these probability statements in mind, let us get back to our original question, "How well does our sample mean of 69.2 inches estimate the true mean?" In Figure 7.1, the hypothetical sampling distribution of means is shown with its mean, $\overline{X}_{\overline{X}}$, of 69. Basically, we are asking, "Where in this distribution of sample means does our mean of 69.2 fall?"

Standard Error of the Mean

As you can see from Figure 7.1, we could locate our sample mean of 69.2 and make probability statements about where it falls, *if we knew the standard deviation of this distribution.* If the standard deviation were 1 inch, 69.2 inches would be very close to $\overline{X}_{\overline{X}}$, but, if the standard deviation were 0.1 inch, the sample mean of 69.2 inches would be 2 standard deviations above $\overline{X}_{\overline{X}}$.

The standard deviation of this sampling distribution of means is called the *standard error of the mean*[†] and, as we noted earlier, is given by

$$\sigma_{\overline{X}} = \frac{\sigma}{\sqrt{N}} \tag{7.1}$$

where

$\sigma_{\overline{X}}$ is the standard error of the mean
σ is the standard deviation of the population from which the samples were drawn
N is the size of the sample

[†]The term *standard error of the mean* may sound awkward, but it is to be interpreted as any other standard deviation. We could just as well say "standard deviation of the mean," but convention is convention, and we will have to learn to live with the standard error of the mean. The important thing is that *standard error of the mean* means the standard deviation of the sampling distribution of means.

Note that this formula requires that we know the value of the population standard deviation, σ. For the situations where σ is *not* known, a different formula and procedure is required, and will be discussed in detail in Chapter 10.

Once we have calculated the standard error of the mean, $\sigma_{\overline{X}}$, we can make probability statements about where our sample mean falls in the sampling distribution and, by inference, just how good an estimate of the population mean our sample mean is. To illustrate the reasoning behind this procedure, let us use the example of our sample of the heights of 100 college men with $\overline{X} = 69.2$. We will assume that the standard deviation of the population of heights from which our sample came is $\sigma = 3.0$ inches.

To calculate the standard error of the mean, we substitute into the formula:

$$\sigma_{\overline{X}} = \frac{\sigma}{\sqrt{N}} = \frac{3}{\sqrt{100}}$$

$$\sigma_{\overline{X}} = 0.3$$

Some Trial Estimates of μ

Now, with our sample mean, $\overline{X}$, of 69.2 and our standard error, $\sigma_{\overline{X}}$, of 0.3, is it possible for the population mean, μ, to be 69? Figure 7.2a shows the sampling distribution with $\overline{X}_{\overline{X}}$ of 69 and its standard deviation ($\sigma_{\overline{X}}$) of 0.3. If the true mean (remember that $\overline{X}_{\overline{X}} = \mu$) is actually 69, where does our sample mean of 69.2 lie?

The z score is used to determine where in the distribution our mean of 69.2 is located. However, the usual z score formula, $z = (X - \overline{X})/S$, must be modified slightly in order for us to use it in a sampling distribution. Since the mean is $\overline{X}_{\overline{X}}$, the standard deviation is the standard error of the mean, $\sigma_{\overline{X}}$, and the "score" is our sample mean, $\overline{X}$, the z formula would now read

$$z = \frac{\overline{X} - \overline{X}_{\overline{X}}}{\sigma_{\overline{X}}} \tag{7.2}$$

It is still the same old z score, but the terms in the formula are changed to apply to a sampling distribution.

The z score for our sample mean of 69.2 would be $z = (69.2 - 69)/0.3 = 0.67$. In other words, it is approximately two-thirds of a standard deviation from the hypothesized true mean of 69. Since by sampling error it is just as easy to wind up with a sample mean 0.2 inches *below*

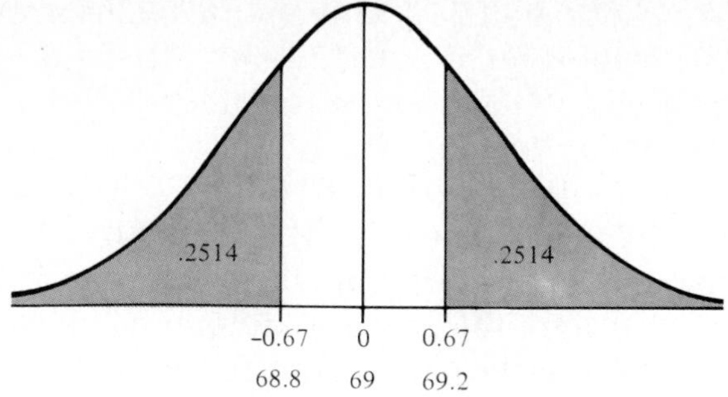

(a) $\overline{X}$ deviates 0.2 inches or more from $\overline{X}_{\overline{X}}$; $p = .2514 + .2514 = .5028$.

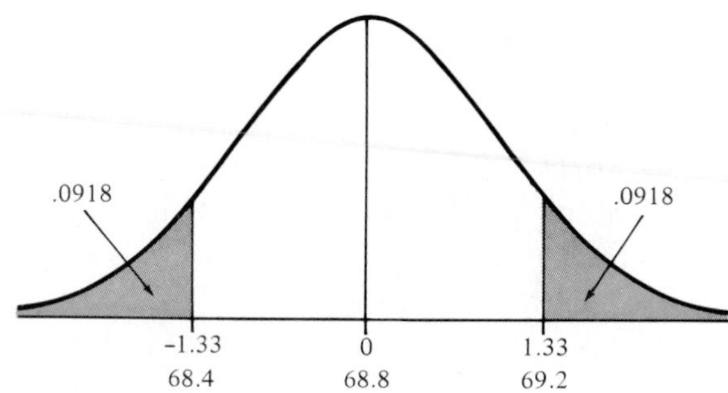

(b) $\overline{X}$ deviates 0.4 inches or more from $\overline{X}_{\overline{X}}$; $p = .0918 + .0918 = .1836$.

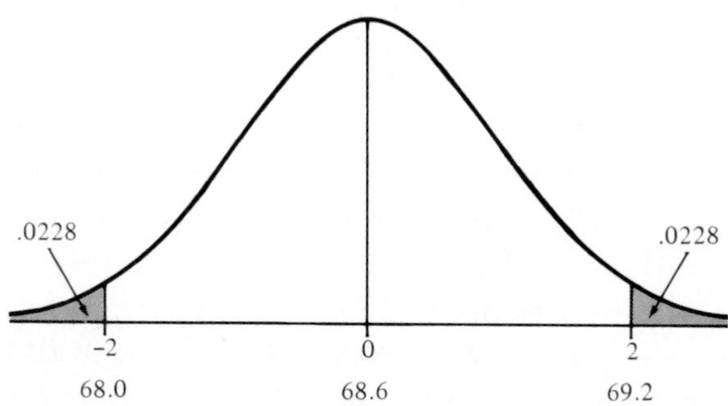

(c) $\overline{X}$ deviates 0.6 inches or more from $\overline{X}_{\overline{X}}$; $p = .0228 + .0228 = .0456$.

FIGURE 7.2. **Sample mean with different hypothesized true means.**

$X_{\overline{X}}$, we can rephrase our original question as follows: "What is the probability that any sample mean would deviate from the population mean by 0.2 inches or more?" Using Figure 7.2a and Table B of Appendix 4 we see that the sample means in the sampling distribution that deviate 0.2 inches or more are shown *below* a z score of -0.67 and above a z of 0.67, and that 50.28% of the normal curve area lies outside these two points. In other words, 50.28% of all sample means would fall outside these two z values, so *the probability that a sample mean, picked at random, would deviate 0.2 inches or more from this hypothesized true mean of 69 inches is .5028.*

From these results does it seem likely that the true mean is actually 69 inches? Remember that we do not know the exact value of the true mean but are simply hypothesizing a value and, on the basis of our sample, trying to make a decision about the most likely value of the true mean.

Suppose that we had hypothesized that the true mean was 68.8 inches. We then would have the situation shown in Figure 7.2b. With our sample mean again of 69.2, our question would now be "What is the probability that any sample mean would deviate from this hypothesized population mean by 0.4 inches or more?" If the true mean were in fact 68.8, the sampling distribution would look like Figure 7.2b, and we could calculate a z score for our sample mean of 69.2 as $z = (69.2 - 68.8)/0.3 = 1.33$. Since the deviation could be either plus or minus (again, remember that from a sampling point of view we are just as likely to get a sample mean that is 0.4 inches *below* the mean as one that is 0.4 inches above the mean), we can see from Figure 7.2b that this area is represented under the curve below a z of -1.33 and above a z of 1.33. We then conclude that if the true mean were 68.8, 18.36% of all the means in this sampling distribution would fall outside the two z scores of -1.33 and 1.33, or outside 68.4 and 69.2, so $p = .1836$ that any sample mean would fall outside these values.

With these results is it possible that the true mean could actually be 68.8? Is it possible with a sample mean of 69.2 to have a population mean that is only 68.8? Let us postpone an answer to these questions until we look at one more example.

Suppose that we had hypothesized that the population mean was 68.6 inches with the hypothetical frequency distribution shown in Figure 7.2c. With our sample mean again of 69.2, what is the probability that a sample mean, drawn at random, would deviate 0.6 inches or more from the true mean? The z score for our sample mean of 69.2 would now be $z = (69.2 - 68.6)/0.3 = 2.0$. Again, since this much of a deviation from the mean could be below the mean, we see in Figure 7.2c that this area is represented under the curve below a z of -2.0 and above a z of 2.0. We then conclude that if the true mean were actually 68.6, only 4.56% of all sample means in this sampling distribution would fall outside the two z scores of -2.0 and 2.0, or outside 68.0 and 69.2. The probability, then, of any sample mean chosen at random being outside these two values is $p = .0456$.

At last we have come to the point where we can make a decision about how good our sample mean is as an estimate of the population mean. (Again, keep in mind that we do not know the actual value of the population mean.) In Table 7.1 are the results of our three hypothetical estimates of the population mean, along with the probability of a sample mean deviating by as much as or more than our sample mean of 69.2 does. Examine this table carefully and, at the same time, inspect the three sampling distributions of Figure 7.2.

TABLE 7.1. **Hypothesized Population Means and Associated Probabilities of a Greater Deviation Than Our Sample Mean of 69.2 Inches**

Hypothesized Population Mean	Amount of Deviation from μ	Probability of Deviation
69 inches	0.2 inches or more	$p = .5028$
68.8 inches	0.4 inches or more	$p = .1836$
68.6 inches	0.6 inches or more	$p = .0456$

Note in Table 7.1 and Figure 7.2 that the farther away the hypothesized population mean is from our sample mean, *the smaller the probability that a given sample mean, drawn at random, would deviate by that amount of more.* On the basis of these three examples we note that:

1. When we hypothesized a population mean of 69 inches, we saw that more than 50% of all sample means would miss μ by as much as (or more than) our mean of 69.2 did. Since the probability of any sample mean, chosen at random (as ours was), deviating by as much as or more than that was .5028, we would conclude that it is entirely possible that the population mean is in fact 69, and our sample mean of 69.2 represents *sampling error* due to chance effects present in any sampling situation.
2. When we hypothesized that $\mu = 68.8$ inches, we saw that about 18% of all sample means would miss μ by as much as or more than 0.4 inches. The probability, then, of any sample mean deviating by that much or more than that was $p = .1836$, and we would probably conclude that it would be entirely possible for μ to be 68.8 and that our sample mean of 69.2 is due to sampling error. But, noting that fewer sample means (18% vs. 50%) deviate by this larger amount, we begin to feel a little uneasy about one of two possibilities. Either we doubt our hypothesis about the value of μ, or we wonder about the accuracy of our sample mean.

3. When we hypothesized that $\mu = 68.6$ inches, we noted that less than 5% of all the sample means would miss μ by as much as 0.6 inches or more. The probability of any sample mean, chosen at random, deviating by this much or more is $p = .0456$, and now we really begin to feel uneasy. Our sample mean deviates by so much that less than 5% of the sample means deviate by this much or more by *sampling error*. There are two possibilities, as indicated in statement 2.
   a. The population mean could really be 68.6 and our sample mean of 69.2 is a rare occurrence, one of the few that we expect would deviate this much by sampling error.
   b. Our hypothesis is dead wrong. We have made a big mistake in hypothesizing the value of the population mean.

NOTE 7.2

HOW LARGE A DIFFERENCE IS STILL DUE TO CHANCE?

An example may be helpful in illustrating the dilemma posed in statements 3a and 3b, where we were faced with the problem of determining if the discrepancy between our sample mean of 69.2 and a hypothesized population mean of 68.6 was due to sampling error.

Let us imagine that you have a friend who claims to have extrasensory perception (ESP). To test this ability, you take two slips of paper and write an *A* on one and a *B* on the second. You hand him the two slips of paper, *face down*, and ask him to put them in the proper order (i.e., *A* first, then *B*) without looking at the writing. Let us say that he performs this task successfully. Does this demonstrate ESP ability?

Before you decide, let us take a careful look at his task. We know that there are only two possible ways that he can respond—he can put them down on the table as *A*, then *B*; or he can lay them down as *B*, then *A*. Whether he has ESP or not, only one of the two orders is correct, and we say that he would be right 50% of the time *just by chance*. The probability that he would be correct just by chance would, of course, be $p = .50$. If he were correct on this simple task, we would be unimpressed, since anyone, ESP or no, would be right 50% of the time.

But let us make the task a little more difficult. Suppose we had decided to use three slips of paper, lettered *A*, *B*, and *C*, and required him to put them in order in the same manner. There are now *six* possible ways in which he could place them on the table: *ABC*, *ACB*, *BAC*, *BCA*, *CAB*, and *CBA*. Since only one of these orders is correct, we know that the probability of his being correct, *just by chance*, is $\frac{1}{6}$, or $p = .17$. If he were correct, we would probably still be unimpressed, since anyone would be right 17% of the time just by chance.

We could make the task still more difficult by requiring him to order correctly four, five, six, or more slips of paper. The probabilities are as follows.

Number of Slips	Possible Orders	Probability of Correct Order
2 *AB*	2	$p = \frac{1}{2} = .50$
3 *ABC*	6	$p = \frac{1}{6} = .17$
4 *ABCD*	24	$p = \frac{1}{24} = .042$
5 *ABCDE*	120	$p = \frac{1}{120} = .008$
6 *ABCDEF*	720	$p = \frac{1}{720} = .001$

The point of this whole discussion is this: There is *no time* that you can say that your friend's performance is not due to chance and is definitely due to ESP. If he picks up four slips of paper and places them in the correct order, we begin to suspect him of possessing some kind of strange power, but we definitely cannot say that it is not chance, since it is still possible (4 chances out of 100) that the result is simply due to chance. If the test is set up with five slips, the probability due to chance of his ordering them correctly is pretty slim (8 chances out of 1000), but it is still possibly a chance effect. The same reasoning applies to the test with six slips, but note that the correct order would happen just by chance only 1 time out of 1000 ($p = .001$).

So, you ask, when is the result definitely, positively, irrevocably due to ESP (or whatever is responsible for your friend's strange ability) and *not* due to chance? You *never* know for certain; you just have varying degrees of certainty that the result could have happened by chance alone. If the probability of a result happening by chance is .042, we are more certain that this result could have happened by chance than a result whose probability due to chance was .008. Clearly, what we need are some guidelines to help us make our decision as to whether chance is responsible for our results.

It turns out that we must rely on our *sample* mean, so, if the discrepancy between our sample mean and some hypothesized population mean is too large, *we must reject the hypothesized population mean as being untenable*. When there is a conflict between the value of the sample mean and the hypothesized population mean, we will go along with the sample mean.

Significance Levels

It should be obvious by now that the crux of the whole problem is where to draw the line. At what point do you say that your result is such a rare occurrence that it is highly unlikely to be sampling error? When do you start rejecting the notion that the discrepancy is due to sampling error? Fortunately, statisticans have rallied to our cause and have proposed the concept of *significance levels*. Simply stated, a significance level is the probability that a result is due to sampling error, *and, if this probability is small enough, we reject the notion that sampling error is the cause.* We then conclude that there is a real difference between our result and what would logically be expected by chance. Traditionally, these significance levels have been set at .05 and .01.

THE .05 SIGNIFICANCE LEVEL. If the probability that our result happened by chance is .05 or less, we say that our results are significant at the .05 level. If the discrepancy between our result and what would be expected by chance alone would happen 5% of the time or less, we reject the notion that it is sampling error.

THE .01 SIGNIFICANCE LEVEL. If the probability that our result happened by chance is .01 or less, we say that our results are significant at the .01 level. If the discrepancy between our result and what would be expected by chance alone would happen 1% of the time or less, we reject the notion that it is sampling error.

Statisticians will be the first to admit that these levels are set in an arbitrary manner, but at least they represent some guidelines to use in making a decision. Let us apply these guidelines to our earlier problem—that of determining how good our sample mean is as an estimate of the population mean, μ.

The Confidence Interval Approach to Estimating μ

In an earlier example we tested various hypothesized values of the population mean and observed the hypothetical sampling distributions for possible population means of 69, 68.8, and 68.6 inches. We noted that as the hypothesized values deviated more and more from our sample mean of 69.2, the probability that we would obtain a sample mean that deviated by as much as or more than ours grew smaller and smaller.

So why not hypothesize a population mean that is the *same* value as our sample mean? With our sample mean of 69.2 and a population mean of 69.2, we would not have to worry at all about sampling error. Whenever we wished to estimate μ from a sample $\overline{X}$, all we would have to do is say that the most likely value of the population mean is the value of our sample mean.

However practical this approach may sound, there is an error in this type of reasoning. The population mean is a *fixed* value, *and it is the sample means that deviate about this fixed value.* As you saw in Figure 7.2, for any given value of μ, the sample means were normally distributed about this point. So instead of talking about possible values that

μ may take, given our sample $\overline{X}$, we are better off in setting up a *confidence interval* in which the true mean probably lies.

As an example let us use the information presented in Figure 7.2c, where we hypothesized a population mean of 68.6 and noted that 95.44% of all sample means fell between this mean $\pm 2\sigma_{\overline{X}}$, or between 68.0 and 69.2. You will remember from the last chapter that in a normal curve the interval containing the mean ± 1.96 standard deviations contained 95% of the cases and the mean ± 2.58 standard deviations contained 99% of the cases. So, in a similar manner, we could calculate the interval about the hypothetical sampling distribution with its mean of 68.6. This interval would be $\overline{X}_{\overline{X}} \pm 1.96\sigma_{\overline{X}}$, or $68.6 \pm 1.96(0.3) = 68.6 \pm .59$, or 68.01 to 69.19. So we would say that 95% of the sample means would fall between 68.01 and 69.19 if the population mean were 68.6. Similarly, with the interval of $\overline{X}_{\overline{X}} \pm 2.58\sigma_{\overline{X}}$, we would have $68.6 \pm 2.58(0.3)$, or 67.83 to 69.37. In this case we can say that 99% of the sample means would fall between 67.83 and 69.37.

However, when we use the confidence interval approach, we set up an interval with the *sample mean as the center*. The proof for this bit of algebraic maneuvering is shown in Note 7.3 for the benefit of the curious. Whether you understand the proof or not, we now have a technique which allows us to feel reasonably certain that the population mean is included in an interval, without having to start out by hypothesizing a specific value for μ.

THE 95% CONFIDENCE INTERVAL. To determine the 95% confidence interval, we simply calculate the interval

$$\overline{X} \pm 1.96\sigma_{\overline{X}} \tag{7.3}$$

where $\overline{X}$ is the *sample mean* and $\sigma_{\overline{X}}$ is the standard error of the mean. So, for our sample mean of 69.2 and standard error of 0.3, we would have

$$\begin{aligned} \overline{X} \pm 1.96\sigma_{\overline{X}} &= 69.2 \pm 1.96(0.3) \\ &= 69.2 \pm .59 \\ &= 68.61 \text{ to } 69.79 \end{aligned}$$

Since the probability is .95 that the population mean is included in all intervals similarly constructed from all possible sample means, we are reasonably certain that the population mean is between 68.61 and 69.79.[†]

THE 99% CONFIDENCE INTERVAL. To determine the 99% confidence interval, we calculate the interval given by

$$\overline{X} \pm 2.58\sigma_{\overline{X}} \tag{7.4}$$

which gives

$$\begin{aligned}\overline{X} \pm 2.58\sigma_{\overline{X}} &= 69.2 \pm 2.58(0.3)\\ &= 69.2 \pm .77\\ &= 68.43 \text{ to } 69.97\end{aligned}$$

Since the probability is .99 that the population mean is included in all intervals similarly constructed from all possible sample means, we are even more certain (than with the 95% interval) that the population mean is between 68.43 and 69.97. Or, stated in still another way, if we continued to take more and more samples of the same size, we would find that 99 out of 100 times the population mean would be included in the interval.

COMPARISON OF THE TWO CONFIDENCE INTERVALS. In practice, which of the two intervals do you use? Note that the 99% confidence interval is *wider* (68.43 to 69.97) than the 95% confidence interval (68.61 to 69.79). This is to be expected, since in order to be more certain you must have a slightly wider interval. We have sacrificed some precision in order to be more certain, so, as a colleague of mine has stated, "You are more and more certain of less and less."

Let us use the confidence interval approach in several examples and see how it can help us in estimating the population mean, μ. Suppose that we administer a college entrance exam to a random sample of 200 high school seniors and calculate a mean of 102. From the Examiner's Manual accompanying the test, we learn that the standard deviation of all high school students tested is 12.0. So, in summary, we have $N = 200$, $\overline{X} = 102$, and $\sigma = 12$.

The 95% confidence interval, $\overline{X} \pm 1.96\sigma_{\overline{X}}$, requires us to calculate first the standard error of the mean, $\sigma_{\overline{X}}$, as follows:

[†]We might be tempted to say that the probability is .95 that μ is between 68.61 and 69.79. However, mathematicians remind us that meaningful probability statements are made only about variables (such as $\overline{X}$) and not fixed values (such as μ).

NOTE 7.3

PROVING THAT THE CONFIDENCE INTERVAL IS AROUND THE SAMPLE MEAN, $\overline{X}$

Remembering that in a sampling distribution 95% of the sample means (in terms of z scores) fall inside the interval,

$$-1.96 \leq z \leq 1.96$$

But $z = \dfrac{\overline{X} - \overline{X}_{\overline{X}}}{\sigma_{\overline{X}}}$, so we can write

$$-1.96 \leq \frac{\overline{X} - \overline{X}_{\overline{X}}}{\sigma_{\overline{X}}} \leq 1.96$$

Considering first the left side of the inequality and multiplying by $\sigma_{\overline{X}}$, we obtain

$$-1.96\sigma_{\overline{X}} \leq \overline{X} - \overline{X}_{\overline{X}}$$

We add $\overline{X}_{\overline{X}}$ to both sides of the inequality to obtain

$$\overline{X}_{\overline{X}} - 1.96\sigma_{\overline{X}} \leq \overline{X}$$

and add $1.96\sigma_{\overline{X}}$ to both sides to get

$$\overline{X}_{\overline{X}} \leq \overline{X} + 1.96\sigma_{\overline{X}}$$

We now go through the same procedure with the right side of the original inequality, multiplying first by $\sigma_{\overline{X}}$ and then by adding $\overline{X}_{\overline{X}}$ to both sides to obtain

$$\overline{X} \leq 1.96\sigma_{\overline{X}} + \overline{X}_{\overline{X}}$$

Subtracting $1.96\sigma_{\overline{X}}$ from both sides, we get

$$\overline{X} - 1.96\sigma_{\overline{X}} \leq \overline{X}_{\overline{X}}$$

We now put both sides together to obtain

$$\overline{X} - 1.96\sigma_{\overline{X}} \leq \overline{X}_{\overline{X}} \leq \overline{X} + 1.96\sigma_{\overline{X}}$$

Thus we can see that the interval $\overline{X} \pm 1.96\sigma_{\overline{X}}$ will contain (in the long run) $\overline{X}_{\overline{X}}$ or μ (remember that $\overline{X}_{\overline{X}} = \mu$) 95% of the time.

$$\sigma_{\overline{X}} = \frac{\sigma}{\sqrt{N}} = \frac{12}{\sqrt{200}} = \frac{12}{14.14} = 0.85$$

Substituting into the interval formula yields

$$\begin{aligned} \overline{X} \pm 1.96\sigma_{\overline{X}} &= 102 \pm 1.96(0.85) \\ &= 102 \pm 1.67 \\ &= 100.33 \text{ to } 103.67 \end{aligned}$$

and we can be reasonably certain that the population mean is in that interval.

To calculate the 99% confidence interval, we substitute to obtain

$$\begin{aligned}\overline{X} \pm 2.58(0.85) &= 102 \pm 2.19 \\ &= 99.81 \text{ to } 104.19\end{aligned}$$

and we are even more certain that this wider interval contains the population mean.

As a second example, let us say that a friend of yours feels that college women are much taller today than when he was in school 10 years ago. He claims the average height of college women today is 5 feet 6 inches. To check out his claim, you take a random sample of 49 college women and find a mean of 65 inches. Then from an insurance company's table of ideal heights and weights, you find that the standard deviation of women in this age group is 2.5 inches. To summarize, $N = 49$, $\overline{X} = 65$ inches and $\sigma = 2.5$ inches. Could your friend's claim of 66 inches be correct? With what degree of confidence?

Calculating the standard error of the mean and both the 95% and 99% confidence intervals would give

$$\sigma_{\overline{X}} = \frac{\sigma}{\sqrt{N}} = \frac{2.5}{\sqrt{49}} = \frac{2.5}{7} = 0.36$$

95% CI: $$\begin{aligned}\overline{X} \pm 1.96\sigma_{\overline{X}} &= 65 \pm 1.96(.36) \\ &= 64.29 \text{ to } 65.71\end{aligned}$$

99% CI: $$\begin{aligned}\overline{X} \pm 2.58\sigma_{\overline{X}} &= 65 \pm 2.58(.36) \\ &= 64.07 \text{ to } 65.93\end{aligned}$$

In this example we can be very confident that μ is included in the interval 64.07 to 65.93. We would have to deny our friend's claim that the average college woman is 66 inches tall, since we are reasonably certain that the population mean is in the 64.07 to 65.93 range and 66 is outside that interval.

Confidence Intervals and Significance Levels

A few paragraphs ago we had occasion to consider the concept of levels of significance, and we noted that this concept referred to the probability of a result being caused by sampling error. Let us consider this last example of women's heights to show how confidence intervals, significance levels, and sampling error are related.

We note that your friend's claim that the average college woman's height is 66 inches is really a hypothesized value for the mean of the population of college women. However, our random sample of 49 women yields a mean of 65. Our question is "If the population mean is in fact 66, could we through sampling error come up with a sample mean of 65?" Using the 99% confidence interval, we found that we are reasonably certain that the population mean is included in the 64.07 to 65.93 interval, *which does not include 66.*

So we are again faced with the dilemma of determining whether the discrepancy between the hypothesized population mean of 66 and our sample mean of 65 is one of sampling error or whether the hypothesized population mean is incorrect. If we use the 95% confidence interval, the probability that all possible intervals include the population mean is .95. But this also means that the probability is .05 that they do *not* include the population mean. Similarly, with the 99% confidence interval, the probability is .01 that these intervals do not include the population mean.

Since the discrepancy between a possible $\mu = 66$ and our sample $\overline{X} = 65$ is such a rare occurrence (happening less than 1% of the time just by sampling error), we conclude that the discrepancy is *not* sampling error but that the hypothesized μ is incorrect. But there is a slight chance (probability less than .01) that we are wrong and that the discrepancy *is* due to sampling error.

We can summarize by making two general statements. With the 95% confidence interval, in the long run we stand to be in error 5% of the time; that is, 5%of the time our intervals will *not* contain the population mean. If we use the 99% confidence interval, we could be wrong 1% of the time; that is, 1% of the time our intervals would *not* contain the population mean.

These points can best be illustrated in the graph of Figure 7.3, which shows a number of 95% confidence intervals calculated from different samples. The population from which the samples were drawn consisted of the heights of college women, and the population mean is shown as $\mu = 65.3$ inches. The confidence intervals have been constructed around the mean of each sample, with the dot in the center of each interval representing the mean, and the horizontal bars representing lower and upper limits of the interval. For example, the first confidence interval shown (Sample 1) was constructed from our sample described on the last page with a mean height of $\overline{X} = 65$ inches and the 95% CI limits of 64.29 inches to 65.71 inches.

Notice that most of the intervals include the population mean of 65.3 inches. However, Sample 4 has a confidence interval that does *not* include a μ of 65.3 inches. If we use the 95% confidence interval, we expect that 5% (or 5 out of every 100) of the samples will yield a con-

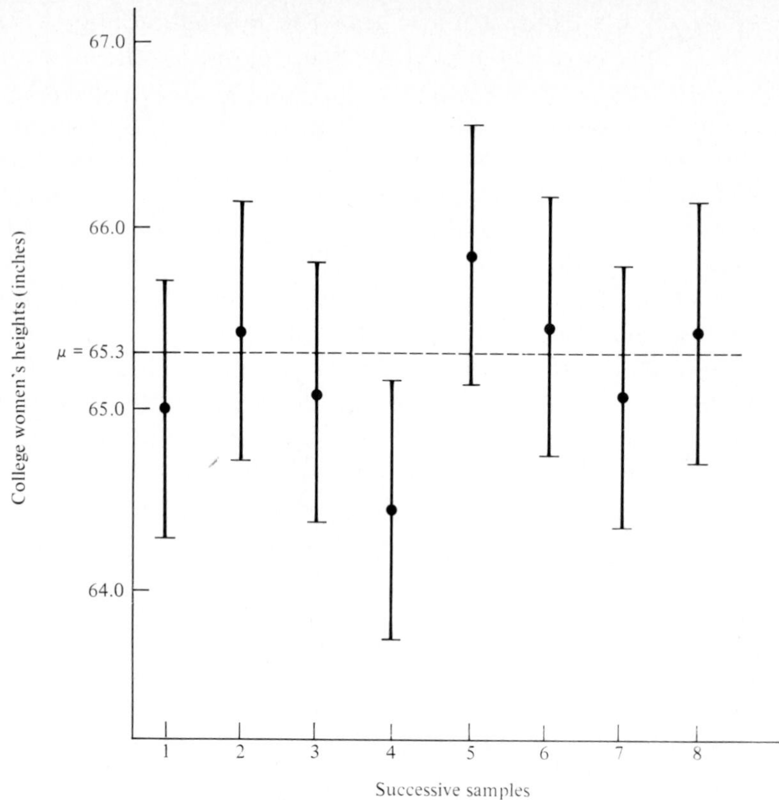

FIGURE 7.3. **Interval estimates of the population mean of college women's heights in successive samples.**

fidence interval that will not include μ and, like Sample 4, will miss μ on either the high end or low end. If we use the 99% confidence interval, we expect that 1% of the confidence intervals will not include μ.

Standard Error and Size of Sample

In previous sections we have looked at how we go about the business of estimating an unknown population mean on the basis of the data from a single sample. You have seen that we do not attempt to specify a *single* value for μ but rather an *interval* in which μ is most likely included. It should be obvious that we would like to have the confidence interval as *narrow* as possible, since we then have a better idea of what the population mean "really" is. Knowing that the IQ of the average college sophomore is between 110 and 130 is not every helpful; an interval of 118 to 123 would give us a much clearer picture of what is going on.

What factors are responsible for reducing the width of the interval (without causing a less stringent probability level, such as .90 or .80) so we can more accurately pinpoint the population mean? In examining the formula for the standard error of the mean:

$$\sigma_{\overline{X}} = \frac{\sigma}{\sqrt{N}}$$

you note that σ and N are the two variables affecting $\sigma_{\overline{X}}$ and consequently the confidence interval itself. We could reduce the size of $\sigma_{\overline{X}}$ either by *decreasing* the standard deviation of the population or by *increasing* the size of the sample. Since we obviously do not have any control over the amount of variability in the population, we must concentrate our efforts on increasing the sample N. Table 7.2 shows how the confidence interval is reduced as you go from a sample size of 10 to 20 to 50. It follows that we would like to have our sample size as large as possible.

TABLE 7.2. **Decreasing the 95% Confidence Interval by Increasing the Sample Size**

Sample $\overline{X}$ = 60 **Population σ = 5**

If N = 10, $\sigma_{\overline{X}} = \frac{\sigma}{\sqrt{N}} = \frac{5}{\sqrt{10}} = \frac{5}{3.1623} = 1.58$

$\overline{X} \pm 1.96\sigma_{\overline{X}} = 60 \pm 3.10 = 56.90$ to 63.10

If N = 20, $\sigma_{\overline{X}} = \frac{\sigma}{\sqrt{N}} = \frac{5}{\sqrt{20}} = \frac{5}{4.4721} = 1.12$

$\overline{X} \pm 1.96\sigma_{\overline{X}} = 60 \pm 2.20 = 57.80$ to 62.20

If N = 50, $\sigma_{\overline{X}} = \frac{\sigma}{\sqrt{N}} = \frac{5}{\sqrt{50}} = \frac{5}{7.0711} = 0.71$

$\overline{X} \pm 1.96\sigma_{\overline{X}} = 60 \pm 1.39 = 58.61$ to 61.39

But What If We Do Not Know the Population Standard Deviation, σ?

We noted earlier that the formula for the standard error of the mean, $\sigma_{\overline{X}} = \sigma/\sqrt{N}$, requires that we know the value of the population standard deviation, σ. If you will remember, in all of the preceding examples we knew the value of σ, and were able to calculate the standard error of the mean ($\sigma_{\overline{X}}$), use it in a confidence interval (e.g., $\overline{X} \pm 1.96\sigma_{\overline{X}}$), and make a statement regarding the possible value of the population mean.

But what do we do if we do not know the exact value of σ? Since *most* of the time we will *not* know what σ is, we will have to use a different formula than $\sigma_{\overline{X}} = \sigma/\sqrt{N}$ and use it in a slightly different procedure. But we will leave these new concepts until Chapter 10, and remember that the techniques described in this chapter require that we know the population standard deviation when we are trying to estimate the population mean.

Other Sampling Distributions

The bulk of this chapter has been spent on the sampling distribution of means, the standard error of the mean, and estimated confidence intervals for the population mean. We could go into depth for other statistics, such as the median, standard deviation, and proportion. However, in an introductory textbook it is just not possible to cover all these topics, and you are referred to any advanced statistics textbook for a treatment of these sampling distributions. But the method of approach is similar for each statistic. Each has its own sampling distribution with its standard deviation being, for example, the standard error of the median or standard error of the proportion.

In light of an earlier discussion we might briefly examine the formula for the standard error of the median, which is

$$\sigma_{med} = \frac{1.253\sigma}{\sqrt{N}}$$

Note that the standard error of the median is approximately 1.25 times as large as the standard error of the mean. In an earlier discussion we noted that the mean was a more *stable* measure of central tendency than was the median. This should now be obvious, since the confidence interval using the standard error of the mean about the population *mean* is narrower than the confidence interval using the larger standard error of the median about the population *median*. This is just another way of saying that *means* of repeated samples will vary less than will the *medians* of repeated samples.

Estimating the Population Variance

Back in Chapter 5 the concept of *variance* was introduced, and we saw that the variance of a distribution was simply the square of the standard deviation. In other words, variance is another method of showing the variability of a set of scores around the mean of the distribution.

However, it is much more than just another method of measuring variability, and we will have occasion to use this important concept in a number of applications in the remaining chapters of this book. For this reason let us carefully consider the development of the concept in the following sections, first demonstrating a *biased* estimate of the population variance for the data of Table 7.3 and then an *unbiased* estimate.

TABLE 7.3. **Biased and Unbiased Estimates of the Population Variance**

X	$(X - \overline{X})$	$(X - \overline{X})^2$
14	5	25
13	4	16
11	2	4
11	2	4
10	1	1
9	0	—
8	−1	1
7	−2	4
6	−3	9
1	−8	64
$\Sigma X = 90$	0	$\Sigma(X - \overline{X})^2 = 128$
$\overline{X} = 9$		

Biased Estimate: $S^2 = \dfrac{\Sigma(X - \overline{X})^2}{N} = \dfrac{128}{10} = 12.8$

Unbiased Estimate: $s^2 = \dfrac{\Sigma(X - \overline{X})^2}{N - 1} = \dfrac{128}{10 - 1} = 14.22$

Biased Estimate of Population Variance

In examining the scores of Table 7.3 we note that the procedure for calculating the variance, S^2, is identical to the technique we used in computing the standard deviation using the deviation formula in Chapter 5 except that we do not take the square root of $\dfrac{\Sigma(X - \overline{X})^2}{N}$. To calculate the variance, you subtract the mean from each score, square each deviation, add the squared deviations to obtain $\Sigma(X - \overline{X})^2$, and finally divide by N. In other words, the variance is simply the average of the squared deviations from the mean. In Table 7.3 the value $S^2 = 12.8$ represents how much, on the average, each score deviated from the mean in squared units. As it stands, the S^2 in Table 7.3 is a *descriptive statistic*, describing the variability of that distribution of scores.

Let us suppose, however, that the 10 scores are a random sample from some population and we are interested in estimating the population variance from this sample. If we do this, we must note that the variance, S^2, is a *biased* estimate of the population variance, σ^2. We noted in an earlier section that the sample mean, $\overline{X}$ is an *unbiased* estimate of the population mean, μ, because as more and more samples are drawn from the population, the mean of the sample means tends to approach the value of the population mean. Such is not the case with the biased variance estimate, S^2, since this variance (shown in Table 7.3) *tends to underestimate the population variance.* S^2 is, on the average, *too small* and obviously would give a misleading picture of the population distribution. (Note also that the *standard deviation* of a sample, S, is also a biased estimate of the population *standard deviation*, σ.)

Unbiased Estimate of Population Variance

The calculation of an *unbiased* estimate of the population variance involves only a minor change in the formula. For the unbiased estimate, $\Sigma(X - \overline{X})^2$ is divided by $N - 1$ instead of by N. The term $N - 1$ is called the *degrees of freedom*, a rather strange term that we will define later, in Chapter 10. For now, however, we say that an unbiased estimate of the population variance can be made from the sample through division of $\Sigma(X - \overline{X})^2$ by its degrees of freedom. Note that the symbol s^2 (lowercase s) is used instead of S^2 as the symbol indicating an *unbiased* estimate.

Again, if the 10 scores in Table 7.3 are a random sample from some population, the *unbiased* estimate of the population variance is calculated as $s^2 = 14.22$. Note that the effect of dividing $\Sigma(X - \overline{X})^2$ by $N - 1$ is to increase the size of the variance and to bring it in line with the actual population variance, σ^2. However, if we take the square root of s^2, we get $s = \sqrt{\Sigma(X - \overline{X})^2/N - 1}$, which is *still* a biased estimate of the population standard deviation, σ. We might note that $s = \sqrt{\Sigma(X - \overline{X})^2/N - 1}$ is slightly *less* biased than $S = \sqrt{\Sigma(X - \overline{X})^2/N}$, but it is still a biased estimate.†

In summary, then, we would say that

$$S = \sqrt{\frac{\Sigma(X - \overline{X})^2}{N}} \tag{5.2}$$

†Common sense might tell us that if s^2 is an unbiased estimator, then s should also be unbiased. However, mathematicians tell us that the square root of an unbiased estimator is not necessarily unbiased also. So much for common sense.

is a biased estimate of the population standard deviation, σ,

$$S^2 = \frac{\Sigma(X - \overline{X})^2}{N} \quad (7.5)$$

is a biased estimate of the population variance, σ^2, and

$$s^2 = \frac{\Sigma(X - \overline{X})^2}{N - 1} \quad (7.6)$$

is an unbiased estimate of the population variance, σ^2.

Random Versus Representative Samples

Let us close out this chapter by recalling an earlier discussion of the topic of a representative sample. We noted that a representative sample was a "miniature" population that contained all the relevant characteristics of the population. Now that we have covered a number of topics on random sampling, we should remember that a random sample may or may not be a representative one; that is, a random sample may not necessarily be an accurate representation of the population. Using again the example of a "population" of colored marbles, let us say that a bag contained 50 blue, 30 red, and 20 yellow marbles. A sample of 10 that contained 5 blue, 3 red, and 2 yellow marbles would certainly be a representative sample. On the other hand, it would be entirely possible for a *random* sample to contain 10 blue marbles (remember sampling error?) and obviously not be representative of the population at all.

So why don't we *always* use stratified or cluster sampling, where we are likely to obtain representative samples, and dispense with random sampling altogether? The reason is that we know the shape of the sampling distribution for random samples and can estimate its standard deviation in order to set up confidence intervals, but we do not always know the nature of the distribution for other kinds of samples. As a result we may freely use representative samples for opinion polls, consumer surveys, and the like, but when we need to set up confidence intervals for estimating population means, we resort to the old reliable random sample.

SAMPLE PROBLEM 1

A group of researchers was investigating the effect of early childhood intervention on the IQ tested later on in the upper elementary grades. A

SAMPLE PROBLEM 1 (cont.)

well-known IQ test with a mean of 100 and a standard deviation of 16 was used to assess intelligence (i.e., $\mu = 100$ and $\sigma = 16$). A group of 40 ten-year-olds from an economically disadvantaged area of a midwestern city was given this IQ test, and a mean of 82 was calculated.

1. Calculate the 95% CI for the true mean.
2. The Project Director felt that children that had received early intervention should have an average score of 90 when tested at age 10. Does this study support her belief?

Our first step is to calculate the standard error of the mean.

$$\sigma_{\overline{X}} = \frac{\sigma}{\sqrt{N}} = \frac{16}{\sqrt{40}} = \frac{16}{6.3246} = 2.53$$

1. The 95% CI, $\overline{X} \pm 1.96\sigma_{\overline{X}}$, would give

$$82 \pm 1.96(2.53)$$
$$82 \pm 4.96$$
$$77.04 \text{ to } 86.96$$

2. We would be reasonably certain that the true mean for the population from which this sample came would be between approximately 77 and 87. Since an IQ of 90 is not included in the interval, we would have to conclude that the Project Director's statement that this age population would have an average IQ of 90 is not supported.

SAMPLE PROBLEM 2

In The National Heart, Lung, and Blood Institute (NHLBI) study on cholesterol and heart disease mentioned in the last chapter, researchers reported that the national average for blood cholesterol level for 50-year-old males was 210 mg/dl with a standard deviation of 33. A total of 89 men with cholesterol readings in the average range (200 to 220) volunteered for a low cholesterol diet for 12 weeks. At the end of the dieting period their average cholesterol reading was 204 mg/dl.

1. Calculate the 95% CI for the true mean.
2. Is it possible that after 12 weeks of dieting the true mean could still be 210?

Calculating the standard error of the mean, we find

$$\sigma_{\overline{X}} = \frac{\sigma}{\sqrt{N}} = \frac{33}{\sqrt{89}} = \frac{33}{9.4340} = 3.50$$

1. The 95% CI, $\overline{X} \pm 1.96\sigma_{\overline{X}}$, would yield

$$204 \pm 1.96(3.50)$$
$$204 \pm 6.86$$
$$197.14 \text{ to } 210.86$$

2. We would be reasonably certain that the true mean for the population from which this sample came would be between approximately 197 and 211. Since the population mean of 210 is still within this interval, we would have no reason to believe that the low cholesterol diet was effective in lowering blood cholesterol readings. Or stated in statistical terms, with a population mean of $\mu = 210$, it is possible that we could have gotten a sample with a mean of $\overline{X} = 204$ just by sampling error.

STUDY QUESTIONS

1. How are the terms *descriptive* and *inferential* used with the terms *sample* and *population*?
2. Twenty thousand students are enrolled at an eastern university. Describe the procedure you would use to obtain a random sample of 500.
3. Why do you think stratified random sampling is sometimes called horizontal sampling while cluster sampling is called vertical sampling?
4. What is a sampling distribution of means? What do the symbols $\overline{X}_{\overline{X}}$ and $\sigma_{\overline{X}}$ represent in this distribution?
5. How does the central limit theorem apply to a sampling distribution of means?
6. Define the *standard error of the mean.*
7. What do the terms *sampling error* and *chance* mean?
8. How is the sampling error versus real difference dilemma "solved" by using arbitrary significance levels (e.g., .05 or .01)?
9. What effect does sample size have on the size of $\sigma_{\overline{X}}$? On the size of the confidence interval?
10. What is meant by an *unbiased* estimate? Why is S^2 a *biased* estimate of the population variance?

EXERCISES

1. Differentiate among the terms *sample, sampling distribution,* and *population.*

2. Sketch a sample distribution, a sampling distribution of means, and a population distribution. Indicate $\overline{X}$ and S, $\overline{X}_{\overline{X}}$ and $\sigma_{\overline{X}}$, and μ and σ on these distributions and note what they tell us about each distribution.

3. Draw a sampling distribution of means and mark off $-1.96\sigma_{\overline{X}}$ and $1.96\sigma_{\overline{X}}$. What percentage of the sample means will fall in the region outside $\pm 1.96\sigma_{\overline{X}}$?

4. Using the same sketch you drew in Exercise 3, darken the region outside $\pm 1.96\sigma_{\overline{X}}$. What is the probability that a sample mean will deviate by this much or more just by sampling error?

5. You are in charge of testing a random sample of 121 college-bound seniors. You administer the American College Testing Program (ACT) college entrance exam and calculate the means of the following four subtests.

English	Mathematics	Social Studies	Natural Sciences
$\overline{X} = 17.3$	$\overline{X} = 18.5$	$\overline{X} = 18.6$	$\overline{X} = 21.6$

   a. According to the test publisher the standard deviation of the population of scores on the *Natural Sciences* subtest is $\sigma = 6.4$. Calculate the 95% CI for this subtest.
   b. A friend of yours says that the national average on the Natural Sciences subtest is around 20. In light of the confidence interval you calculated in part (a), is it likely that $\mu = 20$? Why or why not?

6. Again, let us use the ACT data from Exercise 5.
   a. According to the test publisher the standard deviation of the population of scores on the *English* subtest is $\sigma = 5.6$. Calculate the 95% CI for this subtest.
   b. Is it likely that the population mean would be 18? Explain your answer.

7. A random sample of 80 vegetarians has a mean cholesterol level of 201 mg/dl. The NHLBI study described in Sample Problem 2 at the end of this chapter reported a population standard deviation of 33 mg/dl.
   a. Calculate the 95% CI for this data.
   b. Is it likely that the true mean is 210?

8. What conclusions would you have reached in Exercise 7 if you had obtained the same results but had a sample size of 36? What would you conclude concerning the relationship between sample size and estimating μ?

9. A sample of 49 fourth-graders is given a popular intelligence test and a mean of 104 is calculated. Sample Problem 1 at the end of this chapter noted that according to this test the standard deviation of IQ scores in the population was $\sigma = 16$.
   a. Calculate the 95% CI.
   b. A friend of yours is looking over your shoulder and remarks that the average IQ should be 100. Given the confidence interval that you calculated in part (a), is this likely?

10. What conclusions would you have reached in the last exercise if you had gotten the same results but had a sample size of 81? What does this say about the relationship between sample size and estimating μ?

11. The number of pounds lost by the 17 women in the National Institutes of Health study described earlier in Chapter 5 are shown below. Calculate both a *biased* and an *unbiased* estimate of the population variance. Explain what these values represent.

30	18	23	17	10
18	19	20	26	5
18	22	15	13	20
15	17			

12. The heights (in inches) of a men's basketball team at the NCAA Division 2 playoffs first mentioned in Chapter 3 are shown again here. Calculate both a *biased* and an *unbiased* estimate of the population variance. What are these values estimating?

78	77
81	78
76	75
72	76
73	79
77	76

CHAPTER 8

Correlation

If we looked back over the contents of several earlier chapters we would recall that all discussions to date have concerned a *single* distribution. Frequency distributions, measures of central tendency, measures of variability—all have given us information about a single distribution of scores. At this point it is necessary to shift our emphasis to cases involving *two* distributions, and in this chapter and the next we will be concerned with measuring the amount of *relationship* between *two* distributions of scores.

Although the phrase "relationship between two distributions" may sound unfamiliar to you, the concept of relationships between variables is a part of our everyday language. We commonly hear relational statements such as "I just know it's going to rain—it's so humid out," or "She is really articulate; I'll bet she's either in politics or journalism," or even "Of course, he's got a temper—all redheads are that way." In these statements, relationships are implied between the probability of rain and relative humidity, verbal skills and success in a profession and temper and hair color.

The above examples probably belong more properly in the category of folklore than statistics, but let us consider some other common examples of relationships between variables. Haven't we all, at some time or another, heard questions such as "What is the relationship between high school grades and college success?" or "What is the relationship between your IQ and your ability to remember?" or "What is the relationship between college entrance exam scores and college performance?"

Questions such as these about relationships can be answered by a statistical technique called *correlation*. If there is a relationship between two variables, such as high school grades and college success, we say that they are *correlated*. The statistical techniques to be developed in Chapters 8 and 9 will demonstrate two major functions of correlation.

First, we would like to develop techniques that indicate the *strength* or *amount* of the relationship so that a single value will tell us at a glance how two variables are related. Second, we would like to be able to *predict* scores on one variable from knowledge of another variable. For example, if there is a relationship between high school grades and college success, we would like to be able to actually predict your college grades on the basis of your high school grades.

Before beginning such an ambitious undertaking, let us first consider the meaning of correlation by looking at two approaches to this new concept. The first is an intuitive approach, and the second is a graphical method, called a *scattergram*.

An Intuitive Approach

Let us use a rather homely example of correlation and assume that we would like to see if there is a relationship between athletic ability and scholastic ability. Are good athletes also good students? Or are the best athletes poor students, while the poorer athletes are good students? To test our hypotheses, we choose five college football players and have the coaches rank them on their football ability. (Obviously, we would want more than five subjects in an actual experiment, but such a small number will make our illustration easier.) After the coaches have ranked the five on football ability, we consult the Registrar's Office and obtain the players' grade-point averages (GPAs), which are a measure of their academic success. We then rank the five on their scholastic ability and get the results shown in Table 8.1.

TABLE 8.1. **Ranks on Football Ability and GPA: Positive Correlation**

Player	Football Ability	GPA
Al	1	1
Bob	2	2
Carl	3	3
Don	4	4
Ed	5	5

Note that Al was the best football player and also the best student of this small sample. Bob was second best at both, and so on down to poor Ed, who brings up the rear as the poorest football player and the poorest student. This is an illustration of *perfect positive correlation*. It is *per-*

fect because there are no reversals or changes from the 1-1, 2-2, 3-3, 4-4, 5-5 pairs of ranks, and it is *positive* because both variables *increase together*. If you are high on one variable, you are high on the other, and, if you are low on one, you are low on the other. *Perfect positive correlation is denoted by a coefficient of + 1.00.* (We will come back to the meaning of a coefficient later, but for now regard it as a descriptive number.)

But let us suppose that our little experiment had turned out just the opposite way and the rankings on football ability and GPA resulted in the data shown in Table 8.2.

TABLE 8.2. **Ranks on Football Ability and GPA: Negative Correlation**

Player	Football Ability	GPA
Al	1	5
Bob	2	4
Carl	3	3
Don	4	2
Ed	5	1

As you can see, there is a definite relationship here, but in just the opposite direction. Al, who is the *best* in football, has the *worst* GPA, the second *best* football player has the second *worst* GPA, and so on down to Ed, who is the worst football player but has the highest GPA. This is an illustration of *perfect negative correlation*. It is *perfect* because there are no changes or reversals from the best-worst, second best-second worst, third best-third worst, and so on, pairs of ranks, and it is *negative* because as one variable *increases* the other *decreases*. The better one is at football, the poorer he is at getting grades. *Perfect negative correlation is denoted by a coefficient of −1.00.*

Of course, there is the possibility that we would find no relationship at all between football ability and GPA. Al might be the best at football and the third best student, while Carl might be the third best at football and the second best student. In other words, there might be no pattern of relationship shown in the data. Thus there would be no correlation, and the coefficient would be simply 0, indicating no relationship.

The first two examples illustrate the extreme cases, where the correlation was either perfect positive or perfect negative, that is, 1.00 or −1.00. In practice we find that correlation coefficients may take any value between −1.00 and 1.00, such as −.80, .43, or .60. These three hypothetical examples are shown on a continuum in Figure 8.1 and might illustrate the correlation between juvenile delinquency and socioeconomic level (−.80), manual dexterity and assembly line production (.43), and height and weight (.60).

Let us ponder further the correlation between height and weight. What does a correlation coefficient of .60 tell us about the data? Any coefficient less than perfect means that there have been some reversals or changes in the relative ranking. In Table 8.1, for example, suppose that Carl had a GPA ranking of 4 while Don's GPA ranking was 3. This reversal would result in a coefficient that was less than 1.00, but it would still be quite high and would still be positive, maybe around .90. We would still say that the relationship between football ability and GPA was "high" and "positive." We would say that the better players "tended to be" the better students while the poorer players "in general" were poorer students.

As there get to be more and more reversals in the relative ranks, we find that the correlation gets lower and lower. In our height-weight example with a coefficient of .60, we would note that, in general, taller people are heavier and shorter people are lighter—but there are an ample number of reversals, with some tall, skinny, light people and some short, stocky, heavy people.

There are two important characteristics of a correlation coefficient to keep in mind when evaluating a relationship: first its *sign* and then its *size*. If the sign is positive, we know that as one variable increases so does the other. So, if the relationship between height and weight or between manual dexterity and assembly line performance is positive, we know that tall people tend to be heavier than short people, or that factory workers with high dexterity scores will produce more units than their lower-scoring fellow workers. If, on the other hand, the sign is negative, we know that as one variable increases, the other decreases. Thus, if the correlation between coordination and age is negative for people over age 40, we know that as age increases, coordination de-

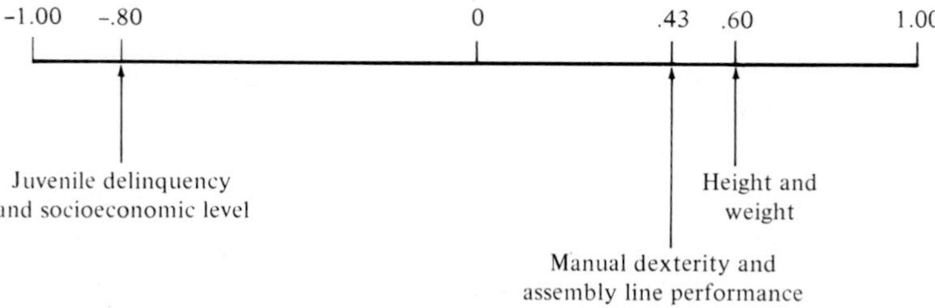

FIGURE 8.1. **Examples of correlation coefficients on a continuum from −1.00 to 1.00.**

creases. Also, in the example shown in Figure 8.1, the negative relationship between juvenile delinquency and socioeconomic level indicates that as socioeconomic level increases, the incidence of juvenile delinquency decreases.

The *size* of the coefficient, as we noted earlier, indicates the *amount* of relationship. In Tables 8.1 and 8.2, there were no reversals in relative ranking, and the resultant coefficients were a perfect 1.00. As there get to be more and more changes in the relative rankings, the coefficient becomes lower and lower until it finally reaches 0, indicating no relationship between the variables. The correlation of .60 between height and weight, as we saw earlier, indicated that "in general" tall people tended to be heavier and short people lighter, but there would be a number of exceptions. The correlation of .43 between manual dexterity and assembly line performance shows a considerably poorer relationship than that between height and weight. We would still say that there is a tendency for good manual dexterity to be matched with good assembly line production, but there are a lot of exceptions.

The Graphical Approach

There is always something appealing about an intuitive approach to an unfamiliar topic, but, unfortunately, we cannot stop with the preceding section but must push on to a graphical and a mathematical interpretation of correlation. The intuitive approach was presented to provide a frame of reference for discussing correlation in general terms, and many of the concepts to be discussed in future sections will be a little more familiar because of that earlier treatment.

The graphical approach to correlation uses a *scattergram*. A scattergram is simply a graph showing the plotted pairs of values of the two variables being measured. In keeping with mathematical convention, we designate the vertical axis (the ordinate) as Y and the horizontal axis (the abscissa) as X, and plot each X,Y pair for all the pairs in our data. As an illustration let us say we would like to see if there is a relationship between height and weight among college males. We select a random sample of 20 college men, measure their height and weight, and enter the paired scores (X = height, Y = weight) on a data sheet, as in Table 8.3.

These X, Y pairs are plotted in the method described in Chapter 2 for graphing a functional relationship. The resulting *scattergram* is shown in Figure 8.2. Examining this scattergram you would expect the correlation to be positive (since those with greater heights tend to weigh more) and quite high. However, the correlation certainly is not perfect,

TABLE 8.3. **Heights and Weights of 20 College Men**

Student	X Height (inches)	Y Weight (pounds)	Student	X Height (inches)	Y Weight (pounds)
A	70	177	K	64	147
B	69	174	L	70	162
C	72	190	M	70	177
D	70	174	N	65	147
E	72	177	O	72	180
F	67	162	P	69	153
G	71	186	Q	68	168
H	67	165	R	68	150
I	66	159	S	71	168
J	70	171	T	69	159

since you can find a number of reversals. Student P, for example, is 3 inches taller than student I but weighs 6 pounds less. If you will find those two points on the scattergram (Student I is at [66,159] and Student P is at [69,153]), you will see that they contribute to the "scattering" of the points away from a straight line. Anyway, there are not too many of these reversals in the height–weight data, and the actual correlation coefficient is .81, which is quite high.

Scattergrams and Size of Correlation

The height–weight data of Figure 8.2, with its correlation coefficient of .81, illustrated a certain degree of "scattering" of the plotted points. Let

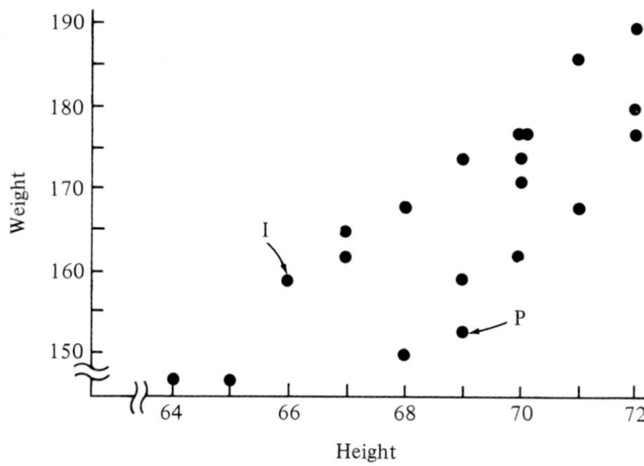

FIGURE 8.2. **Scattergram of heights and weights.**

us now take a look at a number of scattergrams with different patterns of plotted points and see how they are related to the size of the correlation coefficient. Four sets of data and their associated scattergrams are shown in Figure 8.3.

FIGURE 8.3a. Note that with perfect positive correlation the plotted points lie on a *straight line* going from the lower left-hand corner to the upper right hand corner. There are no reversals in X and Y; each increase in X is accompanied by a corresponding increase in Y.

FIGURE 8.3b. As the correlation coefficient decreases to .83, note that there is a scattering away from the straight line of graph A. The pattern of points is elliptical, but its major axis is still from lower left to upper right, indicating positive correlation. In general, low scores in X are paired with low scores in Y, and high values of X are paired with high values of Y. You can see that exceptions to this general statement contribute to the scattering away from the major axis of the ellipse.

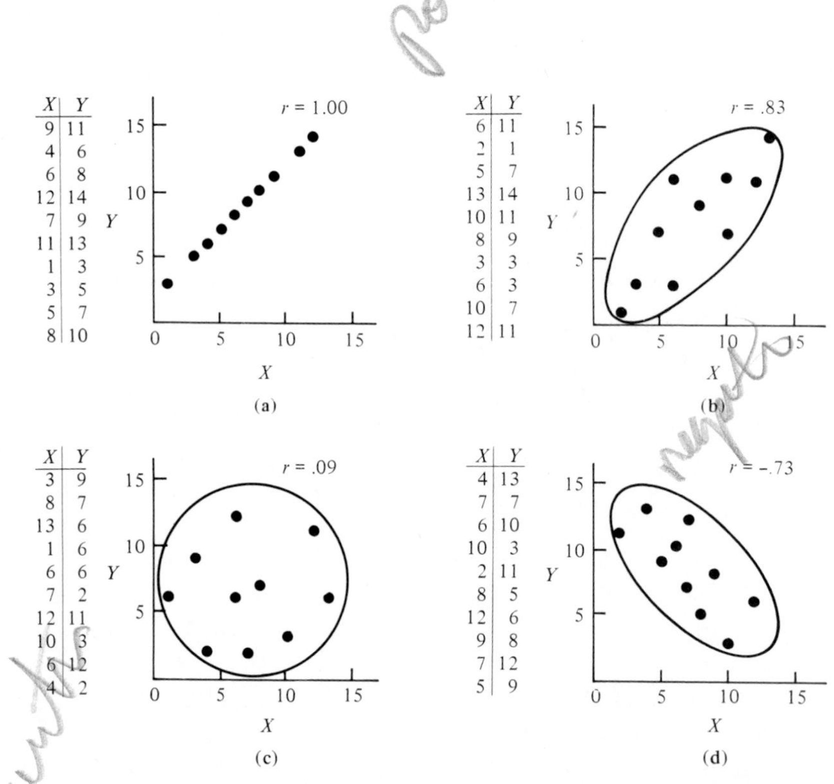

FIGURE 8.3. **Scattergrams and size of correlation coefficients.**

FIGURE 8.3c. By the time the coefficient drops to near 0 (.09 in this example), the pattern of points is almost circular and it is difficult to tell whether the relationship is positive or negative. Note that some high values of X are accompanied by *high* values of Y while other high values of X are paired with *low* values of Y. The same can be said of lower values of X. There apparently is little or no relationship between X and Y, and this is reflected in the coefficient of nearly 0.

FIGURE 8.3d. When the relationship between X and Y is *negative*, you can see that the points lie in an ellipse whose major axis goes from the upper left-hand corner to the lower right-hand corner. This is, of course, due to the fact that *low* values of X are paired with *high* values of Y and *high* values of X with *low* values of Y. And since a correlation of −.73 is quite high, we do not expect much scattering away from a straight diagonal line but a rather tight ellipse, as shown.

Calculating the Pearson *r* (*z* Score Method)

We have finally reached the point where we are ready to actually calculate a correlation coefficient. There are a number of techniques used to calculate a measure of correlation, but one stands alone as far as popularity and universality go. This coefficient is called the Pearson r (after Karl Pearson [1857–1936], an English statistician), and it is derived from the z scores of the two distributions to be correlated. The Pearson r may be computed for interval or ratio data.

As an example, let us say that we would like to see if there is any relationship between proficiency in algebra and proficiency in geometry. An algebra and a geometry test are given to 10 high school sophomores, with the result shown in Table 8.4.

In Table 8.4 are 10 pairs of scores, each individual having a score on X (the algebra test) and a score on Y (the geometry test). In addition to these raw scores, a z score has been calculated for both X and Y for each individual, by the simple division of each score's deviation from its mean by the standard deviation for that distribution. For example, individual A has an X score of 14, which is 1 unit above the mean of 13. That $(X - \overline{X})$ is then divided by the standard deviation of the X distribution, 2.5, for a resultant z score of 0.4. Similarly, A's z score on Y would be $z = (Y - \overline{Y})/S_Y = (24 - 21)/3 = 1.00$.

The next step in calculating the Pearson r is to obtain the *products of the pairs of z scores for each individual.* Each individual's z score on X is multiplied by his or her z score on Y. For individual A, $z_X = 0.4$ and $z_Y = 1.00$, so A's z_Xz_Y product would be 0.4×1.00, or 0.400.

TABLE 8.4. **The *z* Score Method for Calculating the Pearson *r***

	Algebra			Geometry			
Individual	X	$X - \overline{X}$	z_X	Y	$Y - \overline{Y}$	z_Y	$z_X z_Y$
A	14	1	0.4	24	3	1.00	0.400
B	16	3	1.2	23	2	0.67	0.804
C	17	4	1.6	25	4	1.33	2.128
D	13	—	—	21	—	—	—
E	13	—	—	20	−1	−0.33	—
F	14	1	0.4	23	2	0.67	0.268
G	14	1	0.4	23	2	0.67	0.268
H	10	−3	−1.2	18	−3	−1.00	1.200
I	9	−4	−1.6	17	−4	−1.33	2.128
J	10	−3	−1.2	16	−5	−1.67	2.004
	$\overline{X} = 13$			$\overline{Y} = 21$			$\Sigma z_X z_Y = 9.2$
	$S_X = 2.5$			$S_Y = 3.0$			

$$r = \frac{\Sigma z_X z_Y}{N} = \frac{9.2}{10} = .92$$

Similarly, individual B would have a $z_X z_Y = 1.2 \times 0.67 = 0.804$. After $z_X z_Y$ products are obtained for all individuals, the $z_X z_Y$ column is summed and $\Sigma z_X z_Y$ is obtained.

The final step in the calculation of the Pearson r is simply to find the *mean* of these summed products of z_X and z_Y by dividing by N, the number of pairs.

$$r = \frac{\Sigma z_X z_Y}{N} \tag{8.1}$$

For the data of Table 8.4,

$$r = \frac{9.2}{10} = .92$$

With r = .92 we would conclude that there is a high positive correlation between X and Y in the data of Table 8.4. If a person has a high level of ability in algebra, he or she also excels at geometry.

Let us stop a moment and consider just what these $z_X z_Y$ products are telling us about the data and how they indicate the amount of relationship between two distributions. Consider three examples: high positive, low positive, and high negative correlations.

HIGH POSITIVE VALUES OF *r*. If the correlation between two variables is high positive, we expect high scores in one variable to be matched with high scores in the other variable. In terms of z scores, if a person's z_X is high his or her z_Y should also be high. For example, in Table 8.4, individual C has a z_X of 1.6 and a z_Y of 1.33. This, of course, indicates that C is well above the mean in both the X and Y distributions. We also expect in a situation where r is high and positive that an individual who is low in X will also score low in Y. This point is illustrated in Table 8.4 by individual J, who had a z_X of -1.2 and a z_Y of -1.67, well below the mean in both distributions.

Note that in either example the $z_X z_Y$ products are both large and *positive*. If both of a pair of z scores are positive, the resultant $z_X z_Y$ product is, of course, positive. And, if both of a pair of z scores are negative, the resultant $z_X z_Y$ product is *still positive*. Thus the sum of the products $\Sigma z_X z_Y$, will be large and positive, yielding a high value of r.

LOW POSITIVE VALUES OF *r*. In light of the preceding discussion, can you see what factors are responsible for decreasing the value of r? Just what happens as the scores become more and more scattered, as shown earlier, in Figure 8.3? Instead of finding each positive z_X paired with a positive z_Y or a negative z_X paired with a negative z_Y, we now begin to see instances where some individuals scoring high on X score lower on Y. This means that some large positive z scores on X are paired with small positive z scores on Y, which, of course, eventually means a smaller $\Sigma z_X z_Y$ and a smaller r. As the strength of the relationship decreases still further, some individuals with a positive z_X may score *below* the mean on Y, which results in a negative z_Y. The resultant $z_X z_Y$ product for such individuals would be negative, and this would reduce the size of $\Sigma z_X z_Y$, and of r.

In the case where r is near 0, we would note that there would be about as many positive $z_X z_Y$ products as there were negative $z_X z_Y$ products, making $\Sigma z_X z_Y$ and r about 0.

HIGH NEGATIVE VALUES OF *r*. We noted in the introductory material to the topic of correlation that if there were a high negative correlation between two variables, high scores on X would be paired with low values on Y, and low values on X would be matched with high values on Y. In terms of z scores, note that this means that a large *positive* z score on X is paired with a large *negative* z score on Y, and a large *negative* z_X is paired with a large *positive* z_Y. As a result the $z_X z_Y$ product for most individuals will be negative. This, of course, means that the sum of the cross products, $\Sigma z_X z_Y$, will also be negative and will yield a negative r.

Calculating the Pearson r (Raw Score Methods)

The z score method, just described, for calculating r is extremely tedious and time-consuming, since computing a $z_X z_Y$ for each individual involves a number of different steps. Much more convenient techniques make use of *raw score methods,* where we deal with pairs of raw scores instead of the derived z scores. We have two formulas using raw scores—one that requires the means and standard deviations of the two distributions and one that requires only the sums of the various columns on a correlation worksheet.

Pearson r. Mean and S Formula

The formula for the Pearson r using the means and standard deviations of X and Y is

$$r = \frac{\Sigma XY/N - \overline{X}\overline{Y}}{S_X S_Y} \tag{8.2}$$

where

N is the number of *pairs* of scores
ΣXY is the sum of the products of each *pair* of scores
$\overline{X}$ is the mean of the X distribution
$\overline{Y}$ is the mean of the Y distribution
S_X is the standard deviation of the X distribution
S_Y is the standard deviation of the Y distribution

All of the terms in the above formula are familiar except ΣXY, which is the sum of the products of each person's pair of scores. For example, in Table 8.5, the XY product for individual A is $14 \times 24 = 336$; for individual B it is $16 \times 23 = 368$, and so on. The sum of this column is $\Sigma XY = 2799$. Let us repeat the data from Table 8.4 for the algebra and geometry scores and use formula 8.2 to calculate r.

Table 8.5 shows the calculated r to be .93, almost the same as the r calculated by the z score method shown in Table 8.4. The z score formula and formula 8.2 are algebraically identical, and the difference between .93 and .92 is due simply to rounding error.

Pearson r. Machine Formula

If the means and standard deviations of X and Y are not immediately available, the calculation of the Pearson r is simplified greatly with a pocket calculator and the following machine formula.

TABLE 8.5. **Calculating the Pearson r Using Means and Standard Deviations**

Individual	Algebra X	Geometry Y	X^2	Y^2	XY
A	14	24	196	576	336
B	16	23	256	529	368
C	17	25	289	625	425
D	13	21	169	441	273
E	13	20	169	400	260
F	14	23	196	529	322
G	14	23	196	529	322
H	10	18	100	324	180
I	9	17	81	289	153
J	10	16	100	256	160
	$\Sigma X = 130$	$\Sigma Y = 210$	$\Sigma X^2 = 1752$	$\Sigma Y^2 = 4498$	$\Sigma XY = 2799$

Means:

$$\overline{X} = \frac{\Sigma X}{N} = \frac{130}{10} = 13 \qquad \overline{Y} = \frac{\Sigma Y}{N} = \frac{210}{10} = 21$$

Standard Deviations:

$$S_X = \sqrt{\frac{\Sigma X^2}{N} - \overline{X}^2} = \sqrt{\frac{1752}{10} - 169} \qquad S_Y = \sqrt{\frac{\Sigma Y^2}{N} - \overline{Y}^2} = \sqrt{\frac{4498}{10} - 441}$$

$$S_X = \sqrt{6.2} = 2.49 \qquad S_Y = \sqrt{8.8} = 2.97$$

Pearson r:

$$r = \frac{\frac{\Sigma XY}{N} - \overline{X}\,\overline{Y}}{S_X S_Y} = \frac{\frac{2{,}799}{10} - (13)\,(21)}{(2.49)\,(2.97)} = \frac{279.9 - 273}{7.40}$$

$$r = \frac{6.9}{7.40} = .93$$

$$r = \frac{N\Sigma XY - \Sigma X \Sigma Y}{\sqrt{N\Sigma X^2 - (\Sigma X)^2}\,\sqrt{N\Sigma Y^2 - (\Sigma Y)^2}} \tag{8.3}$$

where N is again the number of pairs and the rest of the terms are simply the sums of the X, X^2, Y, Y^2, and XY columns. Table 8.6 shows the calculation of the Pearson r for the algebra-geometry data using this machine formula. We again calculate $r = .93$, since the machine formula is algebraically equivalent to formula 8.2.

Although the machine formula may look somewhat imposing, this formula and a pocket calculator will simplify computations considerably. Some helpful hints on how to use a pocket calculator to compute r are given in Note 8.1. It should be emphasized that, of the three for-

TABLE 8.6. **Calculating the Pearson *r* Using the Machine Formula**

Algebra	**Geometry**
$\Sigma X = 130$	$\Sigma Y = 210$
$\Sigma X^2 = 1{,}752$	$\Sigma Y^2 = 4{,}498$

$$\Sigma XY = 2{,}799$$

$$r = \frac{N\Sigma XY - \Sigma X \Sigma Y}{\sqrt{N\Sigma X^2 - (\Sigma X)^2}\sqrt{N\Sigma Y^2 - (\Sigma Y)^2}}$$

$$= \frac{10(2{,}799) - (130)(210)}{\sqrt{10(1{,}752) - (130)^2}\sqrt{10(4{,}498) - (210)^2}}$$

$$= \frac{27{,}990 - 27{,}300}{\sqrt{17{,}520 - 16{,}900}\sqrt{44{,}980 - 44{,}100}}$$

$$= \frac{690}{\sqrt{620}\sqrt{880}} = \frac{690}{(24.90)(29.66)}$$

$$r = \frac{690}{738.53} = .93$$

mulas, the machine formula (8.3) will result in the most accurate calculation of r, since it is affected the least by rounding error.

The raw score formulas are obviously preferred, since the z score method, as was mentioned earlier, requires a number of tedious steps with each score value. However, the z score formula was included to show just what the Pearson r is measuring, since with the raw score methods it is not possible to see what the individual pairs of scores are doing.

Testing *r* for Significance

Up to this point we have been treating r simply as a descriptive statistic, meaning that it describes the mathematical relationship between pairs of scores. This is useful, to be sure, and there are many times when we may wish to know the amount of relationship in a set of data without needing to make inferences concerning a population. However, if we wish to use correlational techniques to *predict* one variable from another, or to make inferences regarding the amount of relationship between two variables in the *population*, we find that correlation is indeed a very powerful tool in statistical analysis.

In order to be able to use correlation in this way, we must observe some rules of sampling, just as we did for the mean in the last chapter. These rules, or assumptions, are stated explicitly in Chapter 9, but for

NOTE 8.1

CALCULATING THE PEARSON *r* WITH A POCKET CALCULATOR

Even the most inexpensive pocket calculator can make quick work of calculating the Pearson *r* if it has memory and square root functions. Before proceeding, you might review Note 5.1 in Chapter 5, which gave explicit directions on calculating ΣX and ΣX^2. The same directions, of course, would apply to the calculations of ΣY and ΣY^2. To calculate ΣXY, it would be necessary to multiply each X by its paired Y, and enter each product in M^+ storage. For example, in the Algebra–Geometry data of Table 8.5, you would multiply the first number, 14, by 24 and enter the product, 336, by pressing M^+. The second, X, Y pair, 16 and 23, would then be multiplied and entered in M^+, and so on, until all pairs had been entered. Recalling from memory would give $\Sigma XY = 2799$.

Calculations are simplified greatly if N, ΣX, ΣX^2, ΣY, ΣY^2 and ΣXY are summarized in one place—let us use Table 8.6 as a model. Attacking the numerator first, to obtain $N\Sigma XY$ we would multiply 2799 by 10 and enter in M^+. We would then obtain $\Sigma X\Sigma Y$ by multiplying 130 by 210 and entering in M^- which performs the necessary subtraction. We then recall memory to obtain the result, 690. Making sure we have cleared memory, we now begin the denominator by first obtaining $N\Sigma X^2$ by multiplying 1752 by 10 and entering the product in M^+. We then subtract $(\Sigma X)^2$ by multiplying 130 by 130 and entering the result in M^-. Recalling memory, we obtain 620 and extract the square root to obtain 24.90. After clearing memory, we compute the right-hand term in the denominator, obtaining $N\Sigma Y^2$ by multiplying 4498 by 10 and entering the result in M^+. We then square 210 to obtain $(\Sigma Y)^2$ and enter the result in M^-. We recall memory to obtain 880 and take the square root to yield 29.66. Finally, multiplying 29.66 by 24.90 and dividing the result into 690 yields our Pearson *r* of .93.

now we are concerned primarily with just one—that of a random sample from the population.

If we wanted to know, for example, the relationship between intelligence and school grades for children in the sixth grade, we would want to conduct our study with a *random* sample from a population of sixth-graders. We would then be able to generalize the results from our sample of 100 or 500 or 1000 to the population of sixth-graders everywhere.

But we face a problem similar to that mentioned in the last chapter: Does our calculated correlation coefficient represent the actual situation out there in the real world, or *is it due to sampling error?* Is it possible that in the population from which we drew our sample the true r is 0 and the r calculated from our sample is due simply to sampling error? We can answer this question, at least on a probability basis, by taking the next step in a correlational analysis—testing our obtained r for significance. Stated more formally, after we have calculated the r for the sample, we must determine if our r could have arisen by chance alone from a sampling distribution of r's whose mean is 0.

CHANCE CORRELATION. The last statement above might be confusing at first, so let us use an example to help clarify some of the important concepts implicit in that statement. Imagine yourself in a room with 39 other people, taking part in a little demonstration. Forty slips of paper are numbered from 1 to 40, dropped in a hat, and mixed thoroughly. The hat is then passed around, and each of you draws a number without looking into the hat. After this is done, another 40 slips of paper, numbered from 1 to 40, are placed in the hat, and the procedure is repeated. Each of you now has two slips of paper, and we ask, "What is the relationship between the first number you drew and the second number?" Since both were random draws, you would undoubtedly say that there is no relationship at all. However, if we calculated r for the 40 pairs of numbers we would very likely not get an r that is exactly 0. It would probably be very close to 0, such as .07 or −.02, but it certainly does not surprise us that we do not calculate an r that is exactly equal to 0. This discrepancy, similar to that noted between a sample mean ($\overline{X}$) and the population mean (μ) in the last chapter, is due to sampling error.

So, we certainly would not be very concerned in this hypothetical demonstration if we actually obtained an r that is .07 or −.02 or .11 instead of exactly 0. What *does* concern us is the reverse situation, where we run a correlational study, compute an r, and then wonder whether the "true" r in the population is really 0 and our r is due to sampling error.

Fortunately, we can use the confidence interval approach that was introduced in the last chapter to help us out of our dilemma. But, instead of taking an infinite number of sample *means* from some population and constructing a hypothetical sampling distribution of sample means, let us construct a sampling distribution of sample r's.

SAMPLING DISTRIBUTION OF *r*. We first assume that the true r in the population from which we draw our samples is 0. If we then draw an infinite number of random samples from this population and calculate an r between the two variables we are interested in, we will have a *sampling distribution* of r. This sampling distribution of r's would be normal (if the size of each of our samples was greater than 30), its mean would be 0, and its standard deviation would be called the standard error of r, s_r.

As with the sampling distribution of means, we could then make such statements as:

1. 95% of all sample r's would fall between $0 \pm 1.96s_r$.
2. 99% of all sample r's would fall between $0 \pm 2.58s_r$.

The formula for the standard error of r, which is an estimate of the standard deviation of this sampling distribution of r's, is

$$s_r = \frac{1}{\sqrt{N - 1}} \tag{8.4}$$

where N is the size of the sample.

For an example, let us say that we have tested a random sample of 50 high school juniors to find the correlation between their college entrance exam scores and their scores on a current events test and found r = .45. Calculating s_r, we find

$$s_r = \frac{1}{\sqrt{N - 1}} = \frac{1}{\sqrt{50 - 1}} = \frac{1}{\sqrt{49}} = \frac{1}{7} = 0.143$$

We can now set up a confidence interval and say that if the population r is 0, we would expect 95% of the sample r's to fall between $0 \pm 1.96s_r$. The equation would be as follows:

$$0 \pm 1.96(0.143) = 0 \pm .28 = -.28 \text{ to } .28$$

Similarly, 99% of the sample r's would fall between $0 \pm 2.58s_r$:

$$0 \pm 2.58(0.143) = 0 \pm .37 = -.37 \text{ to } .37$$

Figure 8.4 shows this hypothetical sampling distribution with the 99% confidence interval about the hypothesized mean of 0.

But where does our sample r = .45 enter the picture? Note from Figure 8.4 that our r = .45 is *outside* this interval and, given the above assumptions, would be branded as a very rare occurrence. Specifically, if the true r in the population were indeed 0, we would obtain an r as large as or larger than .45 due to sampling error less than 1% of the time. Again, following the logic of the last chapter, we must now make a decision. Is the true r really 0 and our sample r one of those rare occurrences caused by sampling error? Or is it not sampling error at all, and is the true r some other value than 0? Since by sampling error

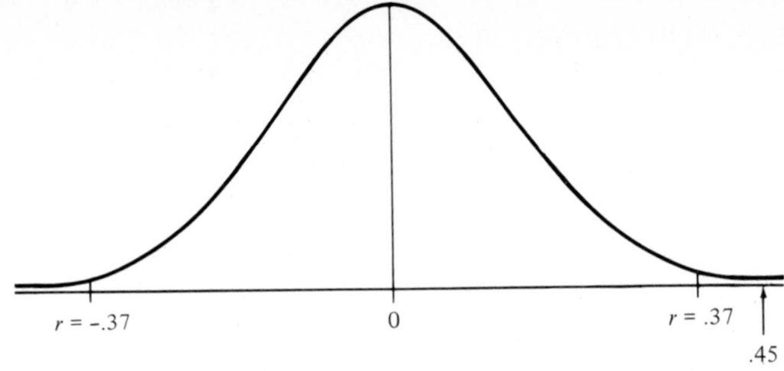

FIGURE 8.4. **Sampling distribution of *r* when hypothesized population *r* is 0.**

alone, r's as large as or larger than .37 happen less than 1% of the time, we conclude that our r of .45 is not sampling error and that the true r is something larger than 0. In terms of our example, the correlation between college entrance exam scores and knowledge of current events *in the population* is not 0. In effect, we have shown that a relationship does exist on the basis of our sample.

USING TABLE C TO TEST *r* FOR SIGNIFICANCE. The confidence interval approach outlined above is not only rather time-consuming but should be avoided if the size of your sample is less than 30. The shape of the sampling distribution of r when N is less than 30 is *nonnormal,* and the normal curve values of 1.96 and 2.58 for the 95% and 99% confidence intervals would be in error. For this reason a table of r's necessary for r to be significant at the .05 and .01 levels is printed in Table C in Appendix 4. The tabled values are to be interpreted in the same manner as in the confidence interval approach.

Before using Table C in an example, you must note that the left-hand column contains the term *degrees of freedom* (*df*). We had occasion to call attention to this term in Chapter 7, where we noted that the degrees of freedom was used in calculating an unbiased estimate of the population variance and was equal to N − 1, where N was the size of the sample. For purposes of using Table C, we must remember that the degrees of freedom in calculating r is equal to N − 2, where N is the number of pairs of scores. We will have more to say later about the degrees of freedom concept, but for now just remember when you use Table C that the degrees of freedom is equal to N − 2.

As an example of the use of Table C, let us say that we found r = .42 for 25 students given a manual dexterity test (*X*) and a hand steadiness test (*Y*). Is this r significantly different from 0? We turn to Table C and note that for 23 degrees of freedom (*df* = 25 − 2 = 23), r at the .05 level is .396 and at the .01 level is .505. These values are to be interpreted in the same way as when we constructed a confidence interval; that is, 95% of the sample r's would fall between −.396 and .396 if the true population r were 0. Similarly, 99% of the sample r's would fall between −.505 and .505 if the true r in the population were 0.

With our sample value of $r = .42$, we note that this is a rather rare occurrence (since r's outside the −.396 to .396 interval occur less than 5% of the time by random sampling from a distribution whose mean r is 0), so we conclude that the true r is not 0. We can say that the obtained r of .42 is *significantly different from 0 at the .05 level.*

Note that with this size sample you would have to obtain an r of at least .505 to be able to say it was significantly different from 0 *at the .01 level.* We would conclude that the probability that our r of .42 arose through sampling error is between .01 and .05.

If we would examine Table C carefully, we would note that with very large sample sizes (an N of 500 or even 1000) our correlation coefficient can be quite small, and still be significantly different from zero. This could result in declaring a coefficient of .10 or .20 as "very significant" whereas, in fact, it is so low as to be of little or no practical significance. So, it is always necessary to examine the actual size of correlation coefficients, instead of relying on a statement that "the correlation between such and such was highly significant." We will treat this topic again under the "How High Is High?" section at the end of the chapter.

Establishing Confidence Intervals for the True *r*

After one has concluded that the obtained r is significantly different from 0, the logical next step would be to set up a confidence interval for the true r, in much the same way that we set up confidence intervals for the true mean in the last chapter. It would then be possible to make a statement that, for example, $p = .95$ that the population r is included in the interval .47 to .61. However, determining confidence intervals for the true r is a rather complicated affair, and you are referred to any advanced statistics text book for this involved operation. The reason that it is not a simple, straightforward technique is that the sampling distribution is nonnormal for population r's that are not 0. The sampling distribution of r becomes more and more skewed as the population r gets larger. An excellent treatment of establishing confidence intervals for the true r can be found in Guilford and Fruchter (1978).

Restrictions on Using the Pearson *r*

There are two restrictions on the Pearson r (other than the fact that it requires interval or ratio data) that must be kept in mind. First, the relationship between X and Y must be linear. Second, the technique requires pairs of values; that is, for every observation, you must have a value for X and a value for Y. This second restriction is self-explanatory, since you need X, Y pairs to plot a scattergram, but the linearity restriction deserves further clarification.

A linear relationship between X and Y means that the plotted points in a scattergram ascend (or descend, if r is negative) in a regular fashion such as shown in the scattergrams of Figure 8.3 (a, b, and d).

A nonlinear relationship between X and Y is shown in Figure 8.5, which plots coordination scores as a function of age. There is no doubt that there is a high degree of relationship between age and coordination. As you can see in Figure 8.5, coordination scores increase as the child gets older, up to about age 15. From 15 years of age until about 40, coordination scores stay the same, and they decline after age 40. The scattergram indicates a high degree of relationship since the points are tightly clustered, but $r = 0$ for these data, since the relationship of X and Y is curvilinear (the points are clustered about a curved line). Mathematically speaking, the Pearson r measures the amount of *linear* relationship present in two distributions, and the interpretation of r will be in error if the relationship between X and Y is nonlinear. It is a good practice to plot a scattergram to check for nonlinearity when using the correlational method.

The Spearman Rank-Difference Method

There is a convenient shortcut for calculating the correlation between two variables if N is relatively small, that is, less than 30 or so. The

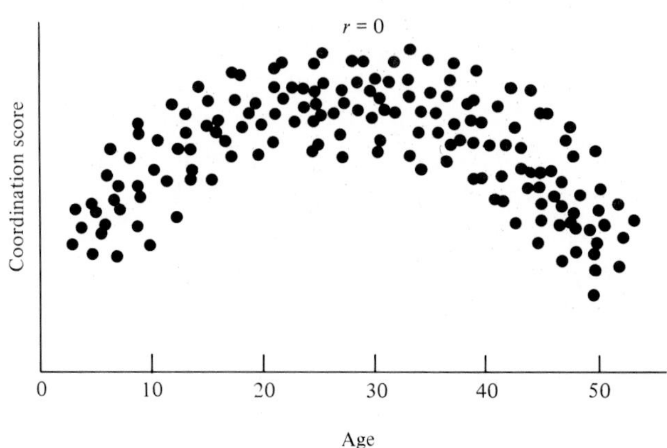

FIGURE 8.5. **A curvilinear relationship: coordination as a function of age.**

procedure is called the Spearman rank-difference method, and it uses the ranks of the scores instead of the scores themselves. The formula for the Spearman coefficient, r_s, is

$$r_s = 1 - \frac{6\Sigma D^2}{N(N^2 - 1)} \tag{8.5}$$

where

r_s is the coefficient (some older textbooks use the Greek symbol ρ)
ΣD^2 is the sum of the squared differences between ranks
N is the number of pairs of ranks

Calculating the Spearman r_s

The Spearman r_s can be calculated for data that are already in the form of ranks or that can be converted to ranks. Let us first consider the example of Table 8.7, where a teacher has ranked 10 children on "social responsiveness" and "conversational skills." Is there any relationship between these two variables?

TABLE 8.7. **The Relationship Between Teacher Rankings on Social Responsiveness and Conversational Skills**

Child	Social Rank	Skills Rank	D	D^2
A	2	1	1	1
B	4	3	1	1
C	1	2	1	1
D	8	8	0	—
E	3	6	3	9
F	6	4	2	4
G	9	10	1	1
H	5	5	0	—
I	10	9	1	1
J	7	7	0	—
			$\Sigma D^2 =$	18

$$r_s = 1 - \frac{6\Sigma D^2}{N(N^2 - 1)} = 1 - \frac{6(18)}{10(100 - 1)}$$

$$= 1 - \frac{108}{990} = 1 - 0.11$$

$$r_s = .89$$

Note that you obtain the quantity ΣD^2 by first listing the pairs of ranks for each individual. Then you subtract the rank on one variable from the rank on the other variable to obtain the difference (D) for each pair. You square these differences to get D^2 and sum the D^2 column to obtain ΣD^2. Note that after you substitute $\Sigma D^2 = 18$ and $N = 10$ into the formula, you subtract the resulting fraction, 0.11, from 1 to obtain $r_s = .89$. A coefficient of .89 would indicate a strong relationship between social responsiveness and conversational skills as judged by this teacher.

Converting Existing Data to Ranks

When one or both of the variables to be correlated are measurements of one sort or another, the scores must be ranked; the highest score receives a rank of 1, the next highest a rank of 2, and so on. As an illustration of this method, let us consider the results of an international study on highway safety. Fifteen nations participated in the survey, and the speed limits for highways other than superhighways (converted to miles per hour) and fatalities per 100 million miles driven are listed in Table 8.8. Let us use the Spearman rank-difference method to determine the correlation between speed limits and fatality rates.

Note that when there are tied values, *each observation is* assigned the average rank for the tied positions. For example, Hungary, Spain, and West Germany have the same speed limit, 62. Since these values occupy the rank positions of 2, 3, and 4, all three are assigned a rank of 3 in the first rank (R_1) column. Similarly, Belgium, France, Portugal, and Turkey all have a limit of 56, and these values are in rank positions of 6, 7, 8, and 9, so each is assigned the average rank of 7.5.

Is r_s Significantly Different from 0?

We have already seen that it is possible to obtain a correlation coefficient that is due to chance or sampling error, so we must test our obtained r_s to see if it is significantly different from 0. Table D in Appendix 4 lists the values of r_s that are necessary for r_s to be significantly different from 0 at the .05 and .01 levels of significance. For example, with an N of 20, the tabled value of r_s at the .05 level is .450. As before, this means that an obtained r_s of .450 or greater would happen by sampling error alone less than 5% of the time.

In our example in Table 8.8, we found $r_s = .31$. Is this significantly different from 0? Entering Table D with an N of 15, we find the tabled values of r_s at the .05 and .01 levels to be .525 and .689, respectively. Since our r_s of .31 *is less than* .525, we must conclude that a correlation coefficient of this size could have happened just through sampling error, and we would state that it is not significantly different from 0. In terms of the study, we would conclude that no relationship has been dem-

TABLE 8.8. **Correlation Between Speed Limits and Highway Fatalities for 15 Countries**

Country	Speed Limit (miles per hour)	Fatalities per 100 million miles	R_1	R_2	D	D^2
Belgium	56	10.5	7.5	6	1.5	2.25
Britain	60	4.0	5	14	9	81
Denmark	50	4.8	12	11	1	1
France	56	8.0	7.5	7	.5	.25
Greece	37	12.9	14	4	10	100
Hungary	62	14.5	3	3	—	—
Italy	68	6.4	1	9	8	64
Japan	31	4.7	15	12	3	9
Netherlands	50	6.0	12	10	2	4
Norway	50	4.2	12	13	1	1
Portugal	56	22.5	7.5	2	5.5	30.25
Spain	62	12.4	3	5	2	4
Turkey	56	32.2	7.5	1	6.5	42.25
United States	55	3.3	10	15	5	25
West Germany	62	7.9	3	8	5	25
					$\Sigma D^2 =$	389.00

$$r_s = 1 - \frac{6\Sigma D^2}{N(N^2 - 1)} = 1 - \frac{6(389)}{15(225 - 1)}$$

$$= 1 - \frac{2334}{3360} = 1 - .69$$

$$r_s = .31$$

onstrated between speed limits and fatality rate, at least for this small sample of countries.

The Meaning of r_s

The rationale for interpreting the rank-difference method is a simple one. Note that when there is perfect positive correlation between two variables the pairs of ranks for each individual would be identical. This means that all the differences (D) would be 0, the differences squared would all be 0, and the fraction $6(\Sigma D^2)/N(N^2 - 1)$ would be 0, leaving $r_s = 1 - 0 = 1$.

As the relationship drops, the differences and ΣD^2 increase and r_s, of course, gets smaller. Finally, when there is a negative correlation the differences and ΣD^2 are very large indeed so that the fraction to be subtracted from 1 is greater than 1, resulting in a negative r_s.

The Spearman r_s can be computed on pairs of measurements where one or both of the variables are expressed in an ordinal scale. The Spearman coefficient is nothing more than the Pearson r applied to *untied* ranks, and it would be interpreted in the same manner as the Pearson r. However, when some of the ranks are *tied*, the formula for r_s does not yield the same result as the formula for r and is only an approximation of r. The difference is negligible if there are not too many tied ranks. Also, since r_s is computed on ordinal data (remember that ranking yields an ordinal scale), it cannot take the place of the Pearson r in the regression equation or the standard error of estimate—two concepts to be discussed in the next chapter. Given these shortcomings of r_s, we are still likely to see the Spearman method used occasionally because of its computational simplicity.

Some Concluding Remarks on Correlation

If it should turn out that your calculated r indicates that two variables are correlated, this does not necessarily indicate a *cause-and-effect* relationship. After all, the Pearson r merely tells the strength of a *mathematical* relationship between X and Y: It is left to the researcher to determine the reason for the correlation. One may find that there is a correlation between socioeconomic level and school performance, or between a father's salary and his child's IQ, but we cannot say that one causes the other unless we have additional information about the variables involved. The calculation of r is only the first step in a correlational study, and it indicates the *degree* to which two variables are related. The *why* of the relationship is a nonmathematical matter left up to the ingenuity of the researcher.

A very common result noted in correlational studies is that the relationship between two variables is caused by a third variable. For example, a study of the mental ages of elementary school pupils and their height in inches would indicate a substantial correlation. Such a spurious correlation, obviously, does not mean that there is any meaningful relationship between mental age and height, because the relationship is due to a third variable: the chronological age of the children. Both mental age and height increase with chronological age, and thus the relationship between mental age and height is a statistical artifact. When you read the results of correlational studies, it is a good idea to consider the possibility of other variables contributing to an obtained correlation. There are statistical techniques designed to handle this problem, and you are referred to one of the advanced textbooks listed in the References for information on these techniques. (See also Table 8.10.)

Restriction of Range

The correlation coefficient is highly sensitive to the range of scores on which it is calculated. As the range becomes more and more restricted, the size of the coefficient decreases. For example, a college entrance exam, given to incoming freshmen during the first week on campus, might correlate .60 with college GPA at the end of the freshman year. However, if we continue to calculate the correlation coefficient at the end of the sophomore, junior, and senior years, we would find a marked reduction in the size of r. One of the reasons for a declining r is that many of those with lower entrance exam scores would drop out of school after an unsuccessful freshman year, a few more after the sophomore year, and so on, and the range of the scores would decrease as these lower scores drop out of the picture.

Similarly, the Graduate Record Exam (GRE), used for screening applicants for admission to graduate school, correlates quite highly with success in postgraduate study. However, if we select a given graduate program (e.g., a psychology department at a large university) and attempt to correlate GRE with course grades, we would be lucky indeed to find an r as large as .35 or .40. (See Table 8.9.) These low coefficients are due to the fact that the range of GRE scores has been severely restricted, since only those with very high GRE scores were initially admitted to this graduate program. There are several techniques for dealing with the problem of a restricted range, but they are beyond the scope of this book; you are referred to one of the advanced textbooks listed in the References.

How High Is High?

It is a very common practice to describe the strength of a correlation by such descriptive adjectives as high, low, moderate, strong, weak, and the like. And some textbooks encourage this practice by stating that correlation coefficients can be described according to the following scheme:

Very high r = .80 or above

Strong r = .60 to .80

Moderate r = .40 to .60

Low r = .20 to .40

Very low r = .20 or less

TABLE 8.9. **Some Examples of Correlation Studies**[†]

Variables	*r*
Consummate confidence (bet the mortgage)	
IQ test reliability	.90s
Standardized School Achievement Test reliability	.90s
IQ and school achievement—grade 1	.85–.90
IQ and school achievement—college from high school	.50–.55
GRE and graduate school grade point average	.00–.40
IQ of identical twins reared apart	.75
IQ of identical twins reared together	.85
IQ of fraternal twins and/or siblings	.50
IQ and memory (higher with age into adulthood)	.50–.70
Height and weight	.55–.60
The ubiquitous .35 correlation	
School achievement (cognitive) and affective	.35
School achievement and socioeconomic status (SES)	.35
School achievement and self-concept	.35
School achievement and motivation	.35
School achievement and student ratings of teacher effectiveness	.44
IQ and self-concept	.35
IQ and creativity	.35
Considerable confidence (bet the rent)	
Traffic fatalities and indices of progress in third world countries	−.70
IQ and age of extremely deprived children	−.70
IQ of parents and children	.50
IQ between spouses	.50
Physical similarity between spouses (believe it or not)	~.40
Different creativity tests	.35
Aversive maternal behavior and aversive child behavior	.55
Reading achievement and television viewing	−.05
Some confidence (don't bet, but say that you did)	
IQ (school achievement) and sibsize	−.30

[†]From John Follman (1984), "Cornucopia of Correlations," *American Psychologist*, 39, 701–702, by permission of the author.

Although such descriptors may be convenient for summarizing a series of research studies ("The correlation between cigarette consumption and heart disease is very high"), it makes much more sense to use the actual value of the correlation coefficient itself. When asked how tall a friend of yours is, you don't answer "very tall" if you know he is exactly 6 feet 4 inches. You simply say that he is 6 feet 4 inches tall. So, in the interest of precision, you are encouraged to use the exact value of r (or range of values, if more than one study is being cited) rather than the vague and somewhat ambiguous descriptors above.

Aside from scientific precision, there is another reason for avoiding descriptive terms for the strength of a correlation. Whether a correlation coefficient is high, moderate, or low depends, to a certain extent, on what variables are being correlated. For example, the correlation between two forms of an intelligence test would be considered low if $r = 80$, while the correlation between college entrance exam scores and college success would be exceedingly high with the same value of r. Again, the scientist can avoid confusion by stating the exact value of r and letting the readers make their own value judgment.

As we noted earlier, we need to be aware of the actual size of the correlation coefficient, as well as whether or not it is significantly different from zero, before we can draw conclusions from a correlational study. A third component could well be added to these two—that of repeatability. In other words, if a correlational study is repeated, would we get similar results? In fact, the more often a study is repeated with similar results, the more confidence we have in any single study.

Follman (1984) called attention to the importance of repeatability by classifying a number of correlational studies into three types as shown in Table 8.9. He notes that the correlations reported in the "Consummate Confidence" category have been reported many times, but those in the other two categories have been found less frequently, and he suggests that they be treated accordingly.

Other Correlational Techniques

In our introduction to the concept of correlation, we have just barely scratched the surface of correlational topics, and the inclusion of the Pearson r and Spearman r_s in this chapter was dictated mainly by the popularity of these methods. There are a wide variety of other techniques, however, developed for rather specific applications. Some are modifications of the Pearson r, while others are based on probability functions. Some of the more popular techniques and possible applications are shown in Table 8.10.

TABLE 8.10. **Some Other Correlational Methods**

Method	Application
Point-biserial r:	One dichotomous variable (yes/no; male/female) and one interval or ratio variable
Biserial r:	One variable forced into a dichotomy (grade distribution dichotomized to "pass" and "fail") and one interval or ratio variable

TABLE 8.10. (continued)

Phi coefficient: Both variables are dichotomous on a nominal scale (male/female vs. high school graduate/dropout)

Tetrachoric *r*: Both variables are dichotomous with underlying normal distributions (pass/fail on a test vs. tall/short in height)

Correlation ratio: There is a curvilinear rather than linear relationship between the variables (also called the eta coefficient)

Partial correlation: The relationship between two variables is influenced by a third variable (e.g., a correlation between mental age and height, which is strongly influenced by chronological age)

Multiple *R*: The maximum correlation between a dependent variable and a combination of independent variables (a college freshman's GPA as predicted by her high school grades in English, biology, government, and algebra)

Table 8.10 is not intended to be an exhaustive list of possible correlational methods but a sampling of the possible ways in which correlational research can be used. The serious student is referred to one of the advanced textbooks listed in the References.

SAMPLE PROBLEM 1

The Drug Enforcement Administration (DEA) provides federal funds to the states for marijuana eradication. Listed below is a random sample of 30 states showing the amount spent (in thousands of dollars) by each state last year and the number (in thousands) of marijuana plants destroyed. Use both raw score formulas for the Pearson r to see if there was a relationship between the amount of money spent and the number of plants reported destroyed. For example, the first state mentioned reported spending $60,000 to destroy 110,000 plants.

State	Spent (X)	Plants (Y)	X^2	Y^2	XY
1	60	110	3,600	12,100	6,600
2	40	6	1,600	36	240
3	80	88	6,400	7,744	7,040
4	40	23	1,600	529	920
5	75	107	5,625	11,449	8,025
6	30	5	900	25	150
7	55	69	3,025	4,761	3,795
8	15	13	225	169	195
9	20	5	400	25	100
10	15	1	225	1	15
11	22	89	484	7,921	1,958

12	20	4	400	16	80
13	60	107	3,600	11,449	6,420
14	50	62	2,500	3,844	3,100
15	40	2	1,600	4	80
16	15	4	225	16	60
17	10	3	100	9	30
18	23	4	529	16	92
19	15	3	225	9	45
20	70	101	4,900	10,201	7,070
21	21	22	441	484	462
22	30	24	900	576	720
23	20	25	400	625	500
24	30	11	900	121	330
25	25	4	625	16	100
26	10	7	100	49	70
27	60	26	3,600	676	1,560
28	87	32	7,569	1,024	2,784
29	50	16	2,500	256	800
30	25	65	625	4,225	1,625
Sums	1,113	1,038	55,823	78,376	54,966

Means:

$$\overline{X} = \frac{\Sigma X}{N} = \frac{1{,}113}{30} = 37.1 \qquad \overline{Y} = \frac{\Sigma Y}{N} = \frac{1{,}038}{30} = 34.6$$

Standard Deviations:

$$S_X = \frac{1}{N}\sqrt{N\Sigma X^2 - (\Sigma X)^2} = \frac{1}{30}\sqrt{30(55{,}823) - (1{,}113)^2}$$

$$= \frac{1}{30}\sqrt{435{,}921} = \frac{660.2431}{30}$$

$$S_X = 22.01$$

$$S_Y = \frac{1}{N}\sqrt{N\Sigma Y^2 - (\Sigma Y)^2} = \frac{1}{30}\sqrt{30(78{,}376) - (1{,}038)^2}$$

$$= \frac{1}{30}\sqrt{1{,}273{,}836} = \frac{1{,}128.6434}{30}$$

$$S_Y = 37.62$$

Pearson r (Formula 8.2):

$$r = \frac{\dfrac{\Sigma XY}{N} - \overline{X}\overline{Y}}{S_X S_Y} = \frac{\dfrac{54{,}966}{30} - (37.1)(34.6)}{(22.01)(37.62)}$$

$$= \frac{1{,}832.2 - 1{,}283.66}{828.0162} = \frac{548.54}{828.0162} = .66$$

Pearson r (Formula 8.3):

$$r = \frac{N\Sigma XY - \Sigma X \Sigma Y}{\sqrt{N\Sigma X^2 - (\Sigma X)^2}\sqrt{N\Sigma Y^2 - (\Sigma Y)^2}}$$

$$= \frac{30(54{,}966) - (1{,}113)(1{,}038)}{\sqrt{30(55{,}823) - (1{,}113)^2}\sqrt{30(78{,}376) - (1{,}038)^2}}$$

$$= \frac{1{,}648{,}980 - 1{,}155{,}294}{\sqrt{1{,}674{,}690 - 1{,}238{,}769}\sqrt{2{,}351{,}280 - 1{,}077{,}444}}$$

$$= \frac{493{,}686}{\sqrt{435{,}921}\sqrt{1{,}273{,}836}} = \frac{493{,}686}{(660.24)(1{,}128.64)}$$

$$= \frac{493{,}686}{745{,}173.27} = .66$$

From Table C in Appendix 4 we note that with 28 degrees of freedom ($N - 2 = 28$), r at the .01 level is .463. Since our value of .66 exceeds this tabled value, we conclude that $r = .66$ is significantly different from 0 at the .01 level. There are a number of exceptions in the data shown, but *in general* greater numbers of plants eradicated are associated with larger expenditures of funds.

SAMPLE PROBLEM 2

In an article on the changing role of women in different countries, Tifft, O'Reilly, and Vollers (1985) cited several descriptive statistics from selected countries concerning literacy, life expectancy, and so on. The results are shown in the following table. Use the Spearman rank-difference method to see if there is any relationship between life expectancy and the number of estimated births per woman in these 10 countries.

Women Worldwide

				% enrolled in school				
	Female Population in Millions	% of Women in Labor Force	% of Women Who Are Literate	Elementary and Secondary	University	Female Life Expectancy	Estimated Births per Woman	Females as % of National Legislature
Australia	7.6	33	99	74	25	78	1.9	10
Brazil	68.2	23	76	64	18	67	3.7	1
China	519.0	37	55	38	1	72	2.3	21
Denmark	2.6	38	99	81	29	78	1.6	27
Ghana	7.0	41	43	44	*	55	6.5	*
Hungary	5.6	43	98	77	13	75	2.0	31
Italy	29.4	29	95	72	23	77	1.7	7
Morocco	11.9	16	22	36	4	62	5.7	0
U.S.	120.1	39	99	84	61	78	2.1	5
U.S.S.R.	147.1	49	98	69	22	75	2.4	33

Source: Selected countries from *Women . . . A World Survey* by Ruth Leger Sivard.
*Not available.

Ranking the two columns, we get the following data.

Life Expectancy	Births	R_1	R_2	D	D^2
78	1.9	2	8	6	36
67	3.7	8	3	5	25
72	2.3	7	5	2	4
78	1.6	2	10	8	64
55	6.5	10	1	9	81
75	2.0	5.5	7	1.5	2.25
77	1.7	4	9	5	25
62	5.7	9	2	7	49
78	2.1	2	6	4	16
75	2.4	5.5	4	1.5	2.25
				$\Sigma D^2 =$	304.5

$$r_s = 1 - \frac{6\Sigma D^2}{N(N^2 - 1)} = 1 - \frac{6(304.5)}{10(100 - 1)}$$

$$= 1 - \frac{1,827}{990} = 1 - 1.85$$

$$r_s = -.85$$

In Table D we see that with $N = 10$, r_S at the .05 level is .648 and at the .01 level is .818. Since our value of $-.85$ is larger than the tabled values, we conclude that there is a significant negative relationship between female life expectancy and number of births per woman in the countries listed. In other words, high birth rates are associated with lower life expectancy.

STUDY QUESTIONS

1. Show how you would use the intuitive approach to explain to a friend the difference between a positive and a negative correlation.
2. Also use the intuitive approach to explain the difference between correlation coefficients of .85 and .62.
3. Use the graphical approach to explain the situations found in Questions 1 and 2.
4. How would the scattergram plotted for data with $r = .08$ differ from another scattergram where $r = -.86$?
5. During the calculation of a Pearson r using the z score formula, how can you tell from glancing at the $z_X z_Y$ column whether r will be positive or negative? Large or small?

6. In the sampling distribution used for testing r for significance, what does s_r represent?
7. What is the mean of the sampling distribution described in Study Question 6?
8. Under what condition would the Pearson r and the Spearman r_s for a set of data be identical?
9. What part does *repeatability* play in the interpretation of a correlation coefficient?

EXERCISES

1. For the variables listed, indicate whether you would expect a positive, negative, or zero correlation.
   a. College GPA and age.
   b. IQ and head size.
   c. Annual income and educational level.
   d. Miles per gallon and weight of automobile.
   e. Blood pressure and incidence of stroke.
   f. Occupational noise level and degree of hearing loss.
   g. Educational level and unemployment rate.
2. What type of correlation (positive, negative or zero) would you be likely to find between the following variables?
   a. Thickness of wall insulation and intensity of sound conducted.
   b. Motor coordination and age.
   c. Job seniority and salary.
   d. Scholastic Aptitude Test (SAT) scores and manual dexterity.
   e. Outdoor temperature and restaurant sales of hot chocolate.
   f. Speed of automobile and braking distance.
   g. Shoe size and glove size.
3. One of the purposes of the study of Type A behavior of college students mentioned in the Sample Problem in Chapter 5 was to see if there was any difference between Type A and Type B students in the amount of time spent studying per week. The Jenkins Activity Survey (JAS) was used to measure Type A behavior, with higher JAS scores indicative of greater Type A behavior. A summary of the JAS scores and number of hours studied per week is shown below for a sample of 50 students. Calculate the Pearson r using the mean and standard deviation formula (formula 8.2).

JAS (*X*)	Hours Studied (*Y*)
$\overline{X} = 17.0$	$\overline{Y} = 7.14$
$S_X = 8.68$	$S_Y = 3.16$

$N = 50$

$\Sigma XY = 6661$

4. In the College Life Questionnaire first discussed in the Sample Problem in Chapter 2, college students reported their GPA and estimated the number of hours they studied each week. In a random sample of 30 students the GPAs and hours studied were as shown below. Calculate the Pearson r using
   a. The mean and standard deviation formula (formula 8.2).
   b. The machine formula (formula 8.3).

GPA	Hours Studied	GPA	Hours Studied
2.3	18	3.8	13
2.2	10	2.7	15
3.9	16	3.7	24
2.7	13	1.7	10
2.5	10	3.6	15
3.7	7	2.0	12
3.1	11	3.6	19
3.3	15	3.6	16
3.4	10	3.1	10
3.6	10	2.6	10
3.0	15	3.1	18
3.2	30	2.9	11
2.9	10	3.3	12
1.7	21	3.5	25
3.1	15	2.6	15

5. A well-known college entrance exam is administered to a sample of 50 entering freshmen at a small midwestern college. The data for the exam scores and for the GPAs at the end of the freshman year are summarized below. Use the machine formula (formula 8.3) to calculate the Pearson r for these data.

Entrance Exam (X)	GPA (Y)
$\Sigma X = 3{,}019$	$\Sigma Y = 132.3$
$\Sigma X^2 = 185{,}037$	$\Sigma Y^2 = 365.59$

$N = 50$
$\Sigma XY = 8{,}098.3$

6. An instructor in an educational psychology course was interested in seeing if her midterm exam was a good measure of the students' performance. She reasoned that if it were a good exam, the exam scores should be related to the students' course grades at the end of the term. She converted the grades in the course to numbers (A = 12, A− = 11, B+ = 10, etc.). The data for the midterm exam and the course grades are summarized below. Calculate the Pearson r using the machine formula to see if there was a relationship between scores on the midterm exam and the students' course grades in educational psychology.

Midterm Exam (X)	Course Grades (Y)
$\Sigma X = 1{,}569$	$\Sigma Y = 347$
$\Sigma X^2 = 62{,}451$	$\Sigma Y^2 = 3{,}205$
$N = 40$	
$\Sigma XY = 13{,}921$	

7. A sample of the heights and weights of 20 students from the College Life Questionnaire follow. Calculate the Pearson r for these data.

Height	Weight	Height	Weight
68	145	72	155
69	150	68	110
72	145	62	120
74	148	73	155
70	130	65	115
63	116	68	138
67	123	74	160
66	120	68	112
69	150	72	195
72	142	68	125

8. Heart rates (HR) at a university health service were taken by a student nurse when each patient first entered the examining room. The heart rate was again taken by the examining physician a few minutes later. Both the first and second heart rates are given for a sample of 25 students. Use the Pearson r to determine if there was a relationship between the two measurements of heart rate.

First HR	Second HR	First HR	Second HR
58	58	54	56
84	84	82	80
58	56	62	66
78	76	70	70
92	94	72	74
70	66	60	62
61	70	68	66
76	76	70	62
66	76	68	66
62	68	68	68
90	88	76	76
62	62	82	84
72	70		

9. Depth perception (the ability to visually align two vertical rods from a distance of 10 feet) was measured in 20 subjects. In one condition they were allowed to turn their heads from side to side as they lined up the rods, while in the other condition they held their heads steady. The amount of error in millimeters was measured for each subject under both conditions. Use the Pearson r to see if there was a relationship between their depth perception in the moving head condition and in the fixed head condition.

Fixed	Moving	Fixed	Moving
45	18	19	7
17	9	24	12
17	9	19	14
20	15	50	22
10	9	36	11
36	15	34	16
40	20	40	18
34	13	22	10
12	7	18	8
10	8	31	14

10. Student drivers at a driving school were given a two-choice and a three-choice visual reaction time test. On the two-choice test, they pushed a lever forward if the red light came on and backward if the green light appeared. On the three-choice test the lever was pushed forward for the red light, to the side for amber and backward for green. The results for 17 student drivers follow. Use the Pearson r to see if there was a relationship between their two-choice reaction time and their three-choice reaction time.

Two-Choice	Three-Choice	Two-Choice	Three-Choice
35	34	38	38
32	37	33	34
34	36	42	40
33	39	30	30
31	32	40	45
31	31	39	37
38	36	26	30
29	34	42	45
34	43		

11. A high school counselor found a correlation of .43 between the mathematics and verbal subtests of the Scholastic Aptitude Test (SAT) with a sample of 25 graduating seniors. Was this correlation significantly different from zero? (Use Table C.)

12. Another researcher found a correlation of .22 between the mathematics subtest of the SAT described in Exercise 11 and GPA after the freshman year in college. Her study was based on 72 college freshmen. Use Table C to see if this correlation was significantly different from zero.

13. In Sample Problem 2 a table of descriptive statistics concerning women in a number of countries was shown. Calculate the Spearman r_S to see if there is a relationship between the percent of women who are literate and female life expectancy. Is this correlation significantly different from zero? (Use Table D.)

14. Using the same table described in Exercise 13, calculate the Spearman r_s for the percent of women enrolled in school (elementary and secondary) and estimated births per woman. Use Table D to see if this correlation is significantly different from zero.

15. Shown below are the populations of 10 midwestern communities, the number of full-time firefighters employed, and the number of firefighters responding to each house fire call. Use the Spearman r_s to see if there is a relationship between the population size and the number of firefighters responding to each call. Is this correlation significantly different from zero? (Use Table D.)

Population	Firefighters	Number Responding
61,308	72	10
29,998	27	7
43,765	60	9
44,485	40	6
32,843	42	9
66,798	84	8
81,343	135	12
57,855	86	16
42,566	48	11
91,811	146	13

16. In the firefighter staffing issue discussed in Exercise 15, what is the relationship between the number of firefighters employed and the number of firefighters responding to each call? (Use the Spearman r_s.) Use Table D to see if this correlation is significantly different from zero.

CHAPTER 9

Prediction and Regression

Prediction is a fascinating topic, as evidenced by the rapt attention we give to astrologers, medical doctors, and Super Bowl oddsmakers. And, few of us can resist a quick peek at the 10-item "Are You a Good Mate?" or "AreYou a Dangerous Driver?" tests in the Sunday supplement. From childhood to old age, we listen to predictions of our school success, athletic prowess, job satisfaction, and life expectancy from teachers, parents, coaches, and physicians. Some of these predictions, fortunately, do not come true.

Most of the above examples are predictions based on a multitude of factors; for example, your doctor's diagnosis may be based on 10 or 20 physiological indicators. For the purposes of our discussion in this chapter, however, we will restrict our topic to predictions made from a single variable. In the last chapter we noted that the two purposes of correlation were (1) to indicate the amount of relationship between two variables, and (2) to enable us to predict one variable from the knowledge of another variable. It is this kind of prediction that we will be concerned with here—the prediction of an individual's score on one variable on the basis of that person's score on another variable.

The Use of r in Prediction

After a significant correlation has been obtained between two variables, X and Y, we would like to be able to take a score on X and *predict* the associated score on Y. For example, if a correlation has been established between high school grade-point average (GPA) and college GPA, we would like to develop a method that would allow us to select a high school senior, calculate his or her GPA, and actually predict what the college GPA will be (within limits) after the freshman year. This procedure, which at first glance may appear to border on the impossible,

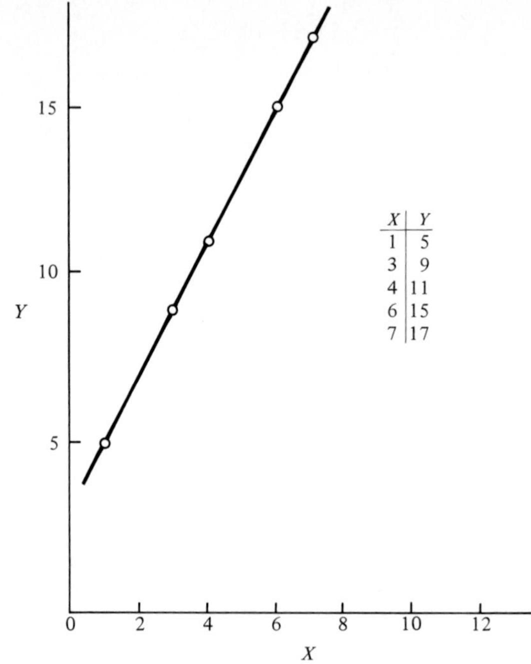

FIGURE 9.1. **Graph of the straight line $Y = 2X + 3$.**

is really rather simple if one progresses step by step to the different concepts involved. Let us begin with some very basic algebra.

Equation for a Straight Line

Somewhere back in ninth-grade algebra you probably learned that the equation of a straight line was of the general form $Y = mX + b$, and you plotted various values of X and Y to get a graph similar to that of Figure 9.1. The equation of the line in Figure 9.1 is $Y = 2X + 3$, and five pairs of X,Y values are plotted. In the general equation $Y = mX + b$, the quantity b is called the *Y intercept* and m is the *slope* of the line. For $Y = 2X + 3$, in Figure 9.1, the Y intercept is 3, since the line will cross the Y axis at $Y = 3$. The slope of the line is 2, indicating that for an increase of one unit in X, there is an increase of two units in Y.[†]

This graph of $Y = 2X + 3$ shows a *functional relationship* between X and Y, with X as the independent variable and Y as the dependent variable. (It might be helpful to review the topic on functional relation-

[†]There is an easy way to remember these two concepts. Go one unit to the right of your line and the *slope* is the distance you need to go up (or down) to get back on the line. The *Y intercept* is the value of Y when X is 0.

ships in Chapter 2.) That is, we can insert any value of X and *predict* the value of Y which is paired with that value of X. For example, in Figure 9.1, we can specify an X of 5 and "predict" that a Y of 13 will satisfy the equation and fall on the same straight line. Note that a similar straight line has occurred in Figure 8.3a, where *perfect correlation* is shown between X and Y. When the correlation between X and Y is *perfect* (either positive or negative), we are able to predict Y from X without error. Or in terms of the example of several paragraphs ago, if $r = 1.00$ between high school GPA and college GPA, we would be able to predict accurately your college GPA from your high school record alone!

The Prediction Equation

But back to the world of reality. How can you predict Y from X when r is not 1.00 and the points do not all fall on the straight line? The answer is simple: We calculate a straight line that "best fits" the points and from X predict the most likely value of Y. In other words, we can take a group of paired scores, regardless of the value of r (as long as it is significantly different from 0), calculate the best-fitting straight line, and for any given value of X predict the most likely value of Y.

Three sets of data are shown in the scattergrams of Figure 9.2, with the best-fitting straight lines for each set. Note that when r is high (.83) most of the points are fairly close to the line and when r is low (.09) the points are scattered well away from this line. This means that the higher r is, the better will be our predictions of Y from a given X. We will have more to say about this later on.

The equation that will result in the best-fitting straight line for a set of paired scores is

$$Y' = \left(\frac{rS_Y}{S_X}\right) X - \left(\frac{rS_Y}{S_X}\right) \overline{X} + \overline{Y} \tag{9.1}$$

where

Y' is the predicted† value of Y
r is the correlation between X and Y
S_Y is the standard deviation of the Y distribution
S_X is the standard deviation of the X distribution
X is the variable in the formula $Y = mX + b$

†We will use Y' to denote any value that falls on the regression line and satisfies the equation, while Y will refer to any of the *observed* values of that variable in the distribution.

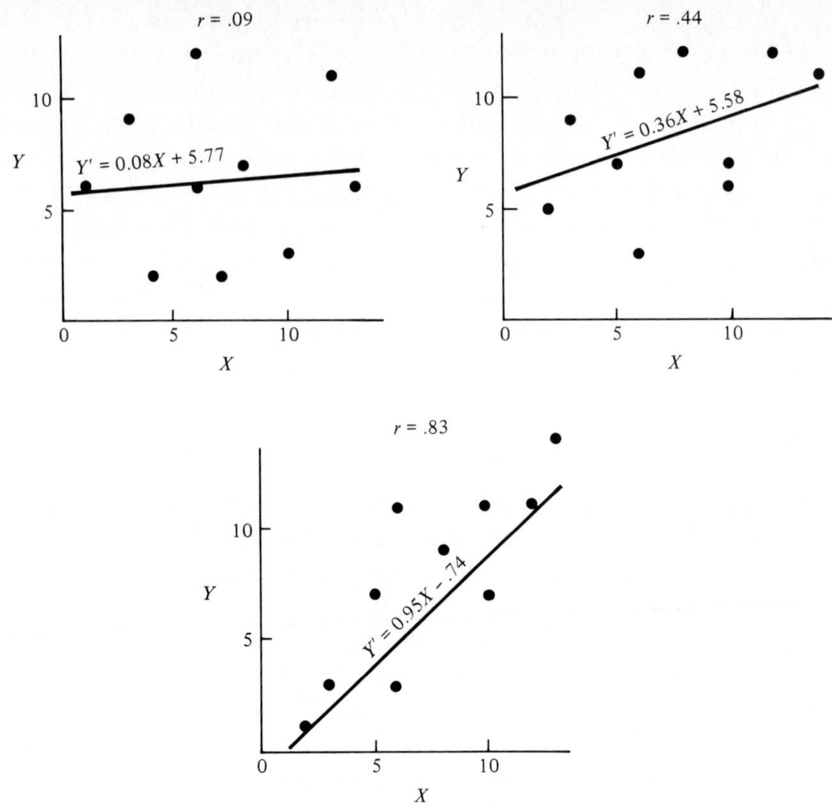

FIGURE 9.2. **Scattergrams and best-fitting straight lines.**

$\overline{X}$ is the mean of the X distribution
$\overline{Y}$ is the mean of the Y distribution

This formula is admittedly an imposing one, but after all the calculations have been performed, it reduces to the familiar form of $Y = mX + b$. Let us use the data of the algebra and the geometry scores (summarized in Table 9.1) and calculate the best-fitting straight line that would predict an individual's geometry score (Y) from his or her algebra score (X).

TABLE 9.1. **Algebra–Geometry Scores and Summary Statistics**

Individual	**Algebra** X	**Geometry** Y
A	14	24
B	16	23
C	17	25
D	13	21

TABLE 9.1. **(continued)**

Individual	Algebra X	Geometry Y
E	13	20
F	14	23
G	14	23
H	10	18
I	9	17
J	10	16

Algebra (X)	**Geometry (Y)**
$\overline{X} = 13$	$\overline{Y} = 21$
$S_X = 2.49$	$S_Y = 2.97$

$r = .93$

Substituting into the formula, we get

$$
\begin{aligned}
Y' &= \left(\frac{rS_Y}{S_X}\right) X - \left(\frac{rS_Y}{S_X}\right) \overline{X} + \overline{Y} \\
&= \frac{.93(2.97)}{2.49} X - \frac{.93(2.97)}{2.49}(13) + 21 \\
&= \frac{2.76}{2.49} X - \frac{2.76}{2.49}(13) + 21 \\
&= 1.11X - 1.11(13) + 21 \\
&= 1.11X - 14.43 + 21 \\
Y' &= 1.11X + 6.57
\end{aligned}
$$

This equation, sometimes called a *regression equation*, yields the best-fitting straight line for the algebra–geometry data. We can now plot a scattergram as shown in Figure 9.3 from the algebra–geometry data, and locate the line on this scattergram.

To draw this best-fitting straight line, also called the *regression line*, on a scattergram, choose a *low* value of X and a *high* value of X and calculate their corresponding Y' by using the regression equation above. Looking at the scattergram of Figure 9.3, we see that X values of 10 and 16 would do nicely, so calculating Y' by substituting for X in the regression equation would give

$$Y' = 1.11X + 6.57 = 1.11(10) + 6.57 = 11.1 + 6.57 = 17.67$$
$$Y' = 1.11X + 6.57 = 1.11(16) + 6.57 = 17.76 + 6.57 = 24.33$$

These two values of Y' are located on the scattergram above X values of 10 and 16, respectively, and a straight line connects these two points.

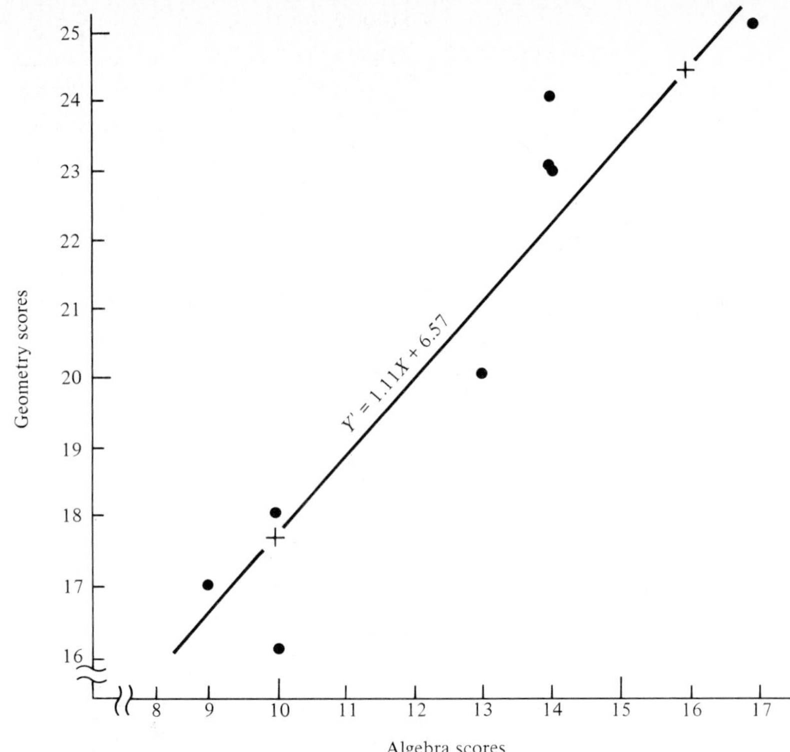

FIGURE 9.3. **Scattergram of algebra scores and geometry scores with regression line.**

This regression line is the best-fitting straight line for this set of data, and the equation for this line is $Y' = 1.11X + 6.57$.[†]

The regression equation can now be used to predict the most likely value of Y for any given value of X. For the algebra–geometry data, we can predict the most likely geometry score (Y) from an individual's algebra score alone. For example, let us suppose that some other individual takes the algebra test and scores 15. What would be her predicted geometry score (Y')? Substituting for X in the prediction equation, we would get

[†]"Best-fitting straight line" means that the regression line is so located that the sum of the squared discrepancies between each actual Y and its corresponding predicted Y' is as small as possible (i.e., $\Sigma(Y - Y')^2$ is a minimum). This approach is called the "least squares criterion."

$$
\begin{aligned}
Y' &= 1.11X + 6.57 \\
&= 1.11(15) + 6.57 \\
&= 16.65 + 6.57 \\
Y' &= 23.2
\end{aligned}
$$

and we conclude that for the individual scoring 15 on the algebra test our best prediction of her geometry score would be 23.2. This score, of course, falls on the regression line of Figure 9.3.

NOTE 9.1

ORIGIN OF THE REGRESSION CONCEPT

Sir Francis Galton (1822–1911) undertook a series of studies of inheritance that tested some of the hypotheses of his cousin, Charles Darwin. In studying the relation between the heights of parents and the heights of their offspring, Galton noted that the heights of offspring tended to *regress* toward the mean of the general population. In general, tall parents had children who were above average in height but who were not as tall as the parents. Short parents had children who were below average in height but who generally were taller than their parents. This "dropping back" toward the general mean was often referred to as the *law of filial regression*. The term *regression* came to be used whenever the relationship between two variables was studied.

Errors in Predicting *Y* from *X*

In the previous sections we have been discussing the prediction of the "most likely value of Y" without ever really defining what is meant by "most likely." You have probably gathered by now that the accuracy of a predicted value of Y is somehow related to the scattering of the points about the regression line. In Figure 9.3, for example, most of the points lie *near*, but not *on*, the regression line. If you will look back at Figure 9.2, you will note that, as r gets larger, the points are closer to the regression line.

Errors of Estimate

In fact, we can define the *error of estimate*, e, as the distance between the predicted value of Y and the observed score of Y at a given value of X. Stated algebraically, we would have

$$e = Y - Y'$$

where

e is the error of estimate
Y is the observed value for a given value of X
Y' is the predicted value of Y for that value of X

The algebra–geometry data are shown again in Table 9.2 with the predicted geometry score (Y') and the error of estimate ($Y - Y'$) shown for each individual.

TABLE 9.2. **Predicted Geometry Scores and Errors of Estimate for Algebra–Geometry Data**

Individual	Algebra X	Geometry Y	Predicted Y'	$Y - Y'$	$(Y - Y')^2$
A	14	24	22.11	1.89	3.5721
B	16	23	24.33	−1.33	1.7689
C	17	25	25.44	− .44	.1936
D	13	21	21.00	0.0	0.0
E	13	20	21.00	−1.00	1.0000
F	14	23	22.11	.89	.7921
G	14	23	22.11	.89	.7921
H	10	18	17.67	.33	.1089
I	9	17	16.56	.44	.1936
J	10	16	17.67	−1.67	2.7889
					11.2102

Summary Statistics (from Table 8.5):

$$\overline{X} = 13,\ S_X = 2.49,\ \overline{Y} = 21,\ S_Y = 2.97,\ r = .93$$

Standard error of estimate (deviation formula):

$$s_E = \sqrt{\frac{\Sigma(Y - Y')^2}{N}} = \sqrt{\frac{11.2102}{10}} = \sqrt{1.12102} = 1.06$$

Standard error of estimate (computational formula):

$$s_E = S_Y\sqrt{1 - r^2} = 2.97\sqrt{1 - (.93)^2} = 2.97\sqrt{.1351} = 1.10$$

Let us repeat part of the scattergram of the algebra–geometry data in Figure 9.4 and demonstrate this error of estimate for 2 students, A and J. As you can see, student A has an algebra score (X) of 14 and a geometry score (Y) of 24. However, the predicted score Y' for an X of 14 would be

$$Y' = 1.11X + 6.57 = 1.11(14) + 6.57 = 22.11$$

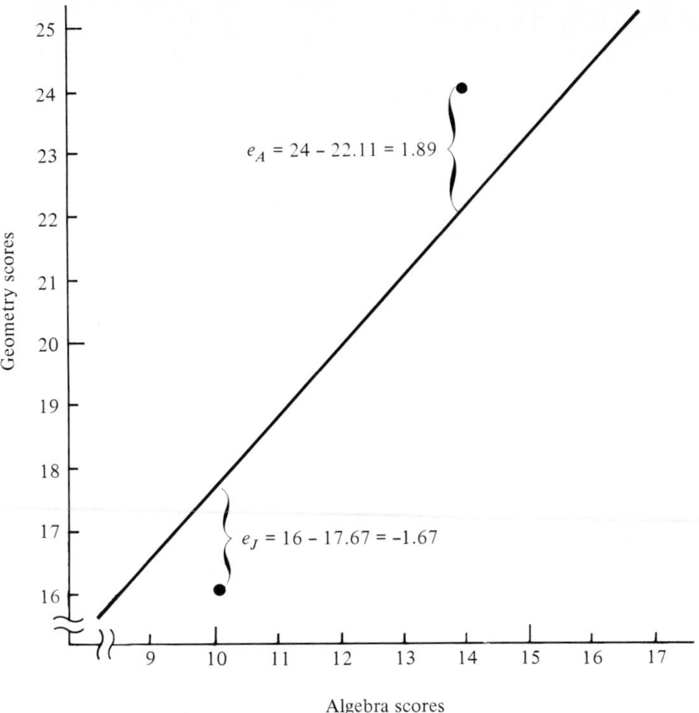

FIGURE 9.4. Errors of estimate for students A and J.

so the error of estimate for student A would be

$$e_A = Y - Y' = 24 - 22.11 = 1.89$$

which indicates that the actual score of 24 is 1.89 score units *above* the predicted Y' of 22.11 on the regression line. For student J, who has an algebra score of 10 and a geometry score of 16, the predicted value of Y is 17.67, which yields an error of estimate of

$$e_J = Y - Y' = 16 - 17.67 = -1.67$$

which indicates that the actual score of 16 is 1.67 units *below* the predicted Y' of 17.67 on the regression line. These values, 1.89 and -1.67, clearly indicate the amount of error in predicting Y from X in terms of deviations from the regression line.

Table 9.2 shows the errors of estimate for the other 8 individuals. It should be obvious that if, *on the average*, the errors of estimate are quite small, our predictions of Y from a knowledge of X will be relatively accurate. Also, as these errors get larger, our prediction of Y would become less accurate.

Standard Error of Estimate, s_E

These errors of estimate, $Y - Y'$, can be used to develop a statistic that will help us tell how accurate our predictions are. We first take each error, $Y - Y'$, square it to obtain $(Y - Y')^2$, and form an additional column shown in Table 9.2. The sum of these squared deviations, $\Sigma(Y - Y')^2$, is then divided by N and the square root is calculated. This statistic is called the *standard error of estimate,* and is given by

$$s_E = \sqrt{\frac{\Sigma(Y - Y')^2}{N}}$$

where

s_E is the standard error of estimate in predicting Y from X
Y is an observed value of the Y variable
Y' is the predicted value associated with each observed value
N is the number of pairs of observations

As you can see in Table 9.2, using this formula we calculate s_E to be 1.06. Before we discuss the meaning of s_E, we should note that this formula for s_E is the *deviation* formula and is a nuisance to calculate. So, as usual, we prefer to use the computational formula:

$$s_E = S_Y\sqrt{1 - r^2} \tag{9.2}$$

where

s_E is the standard error of estimate in predicting Y from X
S_Y is the standard deviation of the Y distribution
r is the Pearson correlation coefficient between X and Y

Let us use this formula for s_E to calculate the standard error of estimate for the algebra–geometry data. Substituting in the formula for S_Y and r, we have

$$\begin{aligned} s_E &= s_Y\sqrt{1 - r^2} \\ &= 2.97\sqrt{1 - (.93)^2} \\ &= 2.97\sqrt{1 - .86} \\ &= 2.97\sqrt{0.14} \\ &= 2.97(0.37) \\ s_E &= 1.10 \end{aligned}$$

We notice that $s_E = 1.10$ is slightly different from the 1.06 calculated with the deviation formula, but this is merely rounding error.

So, what does this standard error of estimate, s_E, represent? We noted both in Chapters 7 and 8 that a standard error has the properties of a standard deviation. In this case, s_E is a type of standard deviation about the regression line. The smaller the size of the standard error, the tighter the points will be clustered around the regression line.

The Confidence Interval About *Y′*

In fact, we can use the properties of the standard error of estimate, s_E, discussed above to construct a confidence interval about a predicted score, Y' if we can assume that X and Y are normally distributed in the population. In our example, this would mean that ability in algebra and proficiency in geometry are normally distributed in the population, which is a rather reasonable assumption. The formula for constructing the 68% confidence interval is

$$68\%\ \text{CI:} \qquad Y' \pm s_E \sqrt{1 + \frac{1}{N} + \frac{(X - \overline{X})^2}{NS_x^2}} \qquad (9.3)$$

where

Y' is the individual's predicted score from the regression equation
s_E is the standard error of estimate
X is the individual's score on X
$\overline{X}$ is the mean of the X distribution
S_x is the standard deviation of the X distribution
N is the number of pairs of observations

Let us again use the algebra–geometry scores and the summary statistics from Table 9.1 and our calculated standard error of estimate, $s_E = 1.10$. In the example of the woman who scored 15 on the algebra test, we used the regression equation to predict her geometry score would be $Y' = 23.2$. Calculating the 68% CI using formula 9.3, we have

$$68\%\ \text{CI:} \qquad Y' \pm s_E \sqrt{1 + \frac{1}{N} + \frac{(X - \overline{X})^2}{NS_x^2}}$$

$$23.2 \pm 1.10 \sqrt{1 + \frac{1}{10} + \frac{(15 - 13)^2}{10(2.49)^2}}$$

$$23.2 \pm 1.10\sqrt{1.1645} = 23.2 \pm 1.10(1.0791)$$

$$23.2 \pm 1.19 \text{ or } 22.01 \text{ to } 24.39$$

In other words, we are 68% certain ($p = .68$) that an individual who scores 15 on X will have a Y score between 22.01 and 24.39. We have accomplished what we set out to do: We can now make a probability statement about the interval in which a given Y score is likely to fall.

Prediction and the Size of *r*

For prediction purposes we would like to have the confidence interval as narrow as possible. It would not be very helpful, for example, in the algebra–geometry data if our algebra score of 15 resulted in a predicted geometry score of, say, 10 to 33 instead of our 21.92 to 24.48. Since we want the confidence interval as small as possible, we must have s_E as small as possible. A glance at the formula for s_E shows that this is a function of the size of r. When $r = 1.00$, the quantity $S_Y\sqrt{1 - r^2}$ reduces to 0, and s_E is 0. In other words, if r is 1.00, there is no error in predicting Y from X. This is obvious, because all the points on a scattergram would fall on the regression line. As r gets smaller, the quantity $S_Y\sqrt{1 - r^2}$ gets larger, of course; when there is no correlation ($r = 0$), s_E is the same size as S_Y, the standard deviation of the Y distribution. So we would have to conclude that the larger the value of r, the more informative will be our predictions.

What About Predicting *X* from *Y*?

All of the preceding sections have been devoted to topics concerning the prediction of Y from X. What about the reverse situation—predicting X from Y? It must be emphasized that we *cannot* take the formulas for the regression equation and the standard error of estimate, and predict X from Y by simply substituting X where Y should be, and vice versa. Because the formulas, as well as their interpretation, are different, you are advised to consult a more advanced statistics textbook given in the References.

As a practical matter, of course, there is no problem. We simply call whatever variable we wish to predict the Y variable, plot our scattergram accordingly, and calculate the usual regression equation.

The Coefficient of Determination, r^2

By squaring the Pearson r, we obtain r^2, a statistic named the *coefficient of determination*. It tells us how much of the variation in Y is due to changes in X. An example may help to clarify this definition.

Again, let us use the example of the relationship between height and weight, with height the X variable and weight the Y variable. Let us say that the correlation was $r = .60$. If we examined the variation of the Y values (the weights), we would see considerable variation of weights about the mean, $\overline{Y}$. What is the source of this variation? We notice that some of the variation in weight must be due to the heights of the individuals involved, since we did find a relationship between height and weight. After all, longer people should weigh more than shorter people.

But since the relationship is not perfect ($r = .60$, not 1.00), we might find two 6 foot 2 inch males—one weighing 160 pounds and the other 190 pounds. Clearly, factors other than height, such as bone density and muscle mass, cause variations in the weights about the mean, $\overline{Y}$.

The coefficient of determination, r^2, tells us the proportion of the variation in Y that is associated with changes in X. If the correlation between height and weight is .60, $r^2 = (.60)^2 = 0.36$, or 36% of the variation in the weights is caused by changes in heights. That means that 64% is caused by other factors.

Example of Correlational Research

The preceding sections have covered a number of topics, and it may be helpful at this point to examine an actual research study to see how the various concepts fit together. A midwestern college administered a nationally standardized entrance examination to a sample of 484 freshmen during their first week on campus. At the end of the freshman year the grade-point average (GPA) for these students was calculated. Since the administration was interested in eventually predicting college GPA from the exam scores taken the first week, the GPA was the dependent variable (Y) and the entrance exam was the independent variable (X). The means, standard deviations, and ΣXY for the 484 pairs of scores are as follows. The scattergram for the data is plotted in Figure 9.5.

Exam Scores (X)	GPA (Y)
$\overline{X} = 22.21$	$\overline{Y} = 2.53$
$S_X = 4.00$	$S_Y = 0.67$

$$N = 484$$
$$\Sigma XY = 27{,}884.06$$

The main interest, of course, is in the possible relationship between the exam scores and the GPAs, so the very first step is the calculation of r. Since the means and standard deviations are given, let us use formula 8.2 for the Pearson r.

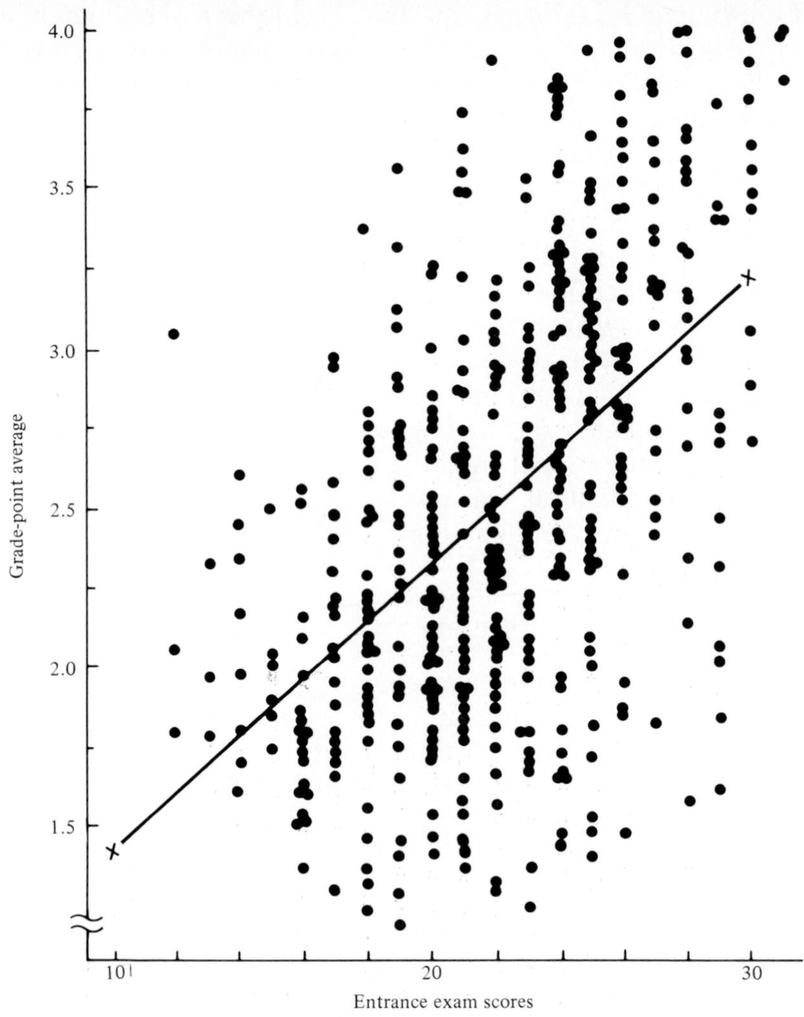

FIGURE 9.5. **Scattergram and regression line for predicting GPA from entrance exam scores.**

$$r = \frac{\frac{\Sigma XY}{N} - \overline{X}\overline{Y}}{S_X S_Y} = \frac{\frac{27{,}884.06}{484} - (22.21)(2.53)}{(4.00)(0.67)}$$

$$= \frac{57.6117 - 56.1913}{2.68} = \frac{1.42}{2.68} = .53$$

After r has been calculated, the next step is to see if it is significantly different from 0. From Table C in Appendix 4 we note that for 400 degrees of freedom (we actually have 484 − 2 = 482 degrees of freedom, but since Table C does not list that value, we take the next *lowest* value listed). An r must be at least .128 to be significantly different from 0 at the .01 level. Since an r of .53 is outside the $-.128$ to .128 interval, we reject the notion that the true r is 0. There is definitely a relationship between entrance exam scores and college GPA.

The next step is the calculation of the regression line. This is calculated to be

$$
\begin{aligned}
Y' &= \left(\frac{rS_Y}{S_X}\right) X - \left(\frac{rS_Y}{S_X}\right) \overline{X} + \overline{Y} \\
&= \frac{.53(0.67)}{4} X - \frac{.53(0.67)}{4} (22.21) + 2.53 \\
&= 0.089X - .089(22.21) + 2.53 \\
&= 0.089X - 1.98 + 2.53 \\
Y' &= 0.089X + 0.55
\end{aligned}
$$

We now position the regression line on the scattergram of Figure 9.5 by substituting any two values of X and solving the equation for Y'. Let us use X values of 10 and 30 and solve for the corresponding Y' values. This would give

$$
\begin{aligned}
Y' &= 0.089X + 0.55 = 0.089(10) + 0.55 = 0.89 + 0.55 = 1.44 \\
Y' &= 0.089X + 0.55 = 0.089(30) + 0.55 = 2.67 + 0.55 = 3.22
\end{aligned}
$$

These Y' values are plotted above their respective X values on the scattergram, a straight line is drawn between the two points, and this regression line, $Y' = 0.089X + 0.55$, is the best-fitting straight line for this data.

The last step before we can make predictions about an individual's GPA is to calculate the standard error of estimate, s_E. This is calculated by

$$
\begin{aligned}
s_E &= S_Y\sqrt{1 - r^2} = 0.67\sqrt{1 - (.53)^2} \\
&= 0.67\sqrt{1 - .28} = 0.67\sqrt{.72} \\
&= 0.67(0.85) = 0.57
\end{aligned}
$$

On the basis of the above research, the college administration is now ready to use the entrance examination for predicting college GPA for *future* students. If John Jones enrolls as a freshman the following year and scores 24 on the entrance examination, what would be the best prediction for his college GPA at the end of his freshman year? Using the regression equation, we find

$$Y' = 0.089X + 0.55 = 0.089(24) + 0.55 = 2.14 + 0.55 = 2.69$$

and setting up the 68% confidence interval about Y':

$$68\%\ \text{CI:} \qquad Y' \pm s_E \sqrt{1 + \frac{1}{N} + \frac{(X - \overline{X})^2}{NS_x^2}}$$

$$2.69 \pm 0.57 \sqrt{1 + \frac{1}{484} + \frac{(24 - 22.21)^2}{484(4)^2}}$$

$$2.69 \pm 0.57(1.0012) = 2.69 \pm 0.57$$

$$2.12 \text{ to } 3.26$$

We would conclude that $p = .68$ that John Jones' actual GPA at the end of his freshman year would be between 2.12 and 3.26.

Assumptions for the Pearson *r* in Prediction

We noted in the last chapter that we had to make two assumptions before we could meaningfully apply the Pearson method: linearity of X and Y and paired X, Y data. In a similar way, assumptions must be made before the Pearson r can be used for purposes of prediction. They are

1. The regression of Y on X is linear (linearity).
2. X and Y are normally distributed in the population (normality).
3. The standard deviation of the Y values about Y' for a given value of X is about the same for all values of Y' (homoscedasticity).

Linearity

We emphasized the importance of linearity in the last chapter, since r measures the degree of *linear* relationship between X and Y. Figure 9.6 shows the scattergram of the coordination scores as a function of age that we examined in the last chapter. But note that the "best-fitting" straight line is a horizontal line at the mean of the Y scores.† It is obvious that there is a high correlation between coordination scores and age, but the Pearson r is 0, because of the lack of linearity in the data. With a regression line that is horizontal at the mean of the Y scores, our prediction of Y' for any value of X would be the same, $Y' = \overline{Y}$! As we noted in the last chapter, we certainly would want to draw a scattergram, in order to examine the data for possible nonlinearity.

†You can demonstrate this for yourself very easily by substituting $r = 0$ in the regression equation and obtaining $Y' = \overline{Y}$.

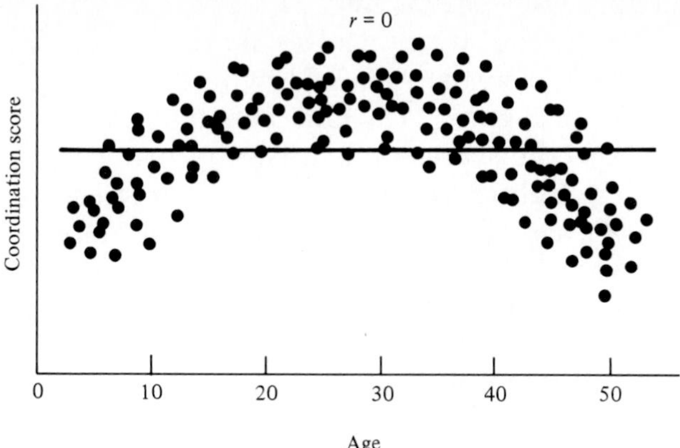

FIGURE 9.6. **A curvilinear relationship: coordination as a function of age.**

Normality

In order to use the standard error of estimate to establish the accuracy of a predicted Y', it is necessary that X and Y be normally distributed in the population. Since the confidence interval about a predicted Y' uses the traditional normal curve values of $\pm 1.96s_E$ and $\pm\ 2.58s_E$, the assumption of a normal population distribution of X and Y is a logical one.

Homoscedasticity

In order for us to use s_E for any predicted Y' on the regression line, it is necessary that there be homoscedasticity in the values of Y about Y'. Homoscedasticity (this tongue-twisting term can be roughly translated as "equal spread") means that the variance of the Y values around their mean of Y' for a given value of X should be about the same for all distributions of Y values about their respective Y' values. In other words, if we were to calculate the variance (or standard deviation) of all of the Y values for a given value of X, we would expect this variance to be approximately equal to the variance of another group of Y values for some other value of X.

To better understand this concept, let us refer back to Figure 9.5, where the scattergram of the entrance exam scores and college GPA is plotted. Let us arbitrarily choose an X value of 20. The Y values for this value of X are distributed about the mean, $Y' = 2.34$, which is on the regression line. If there is homoscedasticity, the variance of this distribution should be about the same as the variance of any of the other columns of Y values.

Strictly speaking, homoscedasticity is a property possessed by samples that are very large, but we can get a very rough idea from the shape of the scattergram. The scattergram of Figure 9.5 is roughly elliptical, but the scattergram of Figure 9.7 is not. This figure shows the relationship between IQ and the scores on a creativity test. Note that the creativity scores (Y) are tightly clustered about the regression line for the lower IQ scores (X), while the creativity scores are more dispersed for the higher IQ scores. This would tell us that those with low IQ scores tend to score lower in creativity, while those with the higher IQ scores may be low, average, or high in creativity. In other words, if you have a lower IQ, you probably are not a very creative person, but if you have a high IQ, you may or may not be very creative. This peculiar state of affairs results in an unequal spread of Y values about the regression line, and these data would not meet the assumption of homoscedasticity.

The Importance of Correlation and Regression

In two chapters we have barely scratched the surface of this extremely important tool in education and the behavioral sciences. Because of the

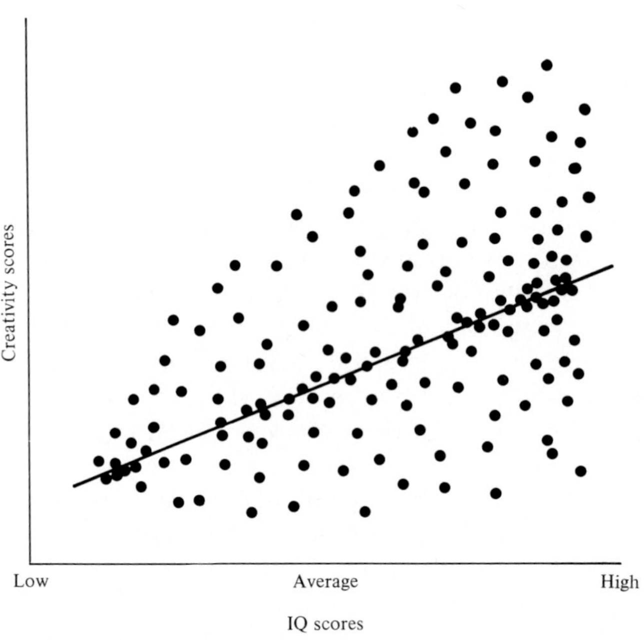

FIGURE 9.7. **Scattergram not meeting the homoscedasticity assumption.**

introductory nature of this book and space limitations on what is included, only a brief initiation to some selected topics in correlation and regression is possible. If you are interested in a further treatment of these topics, any of the more advanced textbooks in the References is recommended, especially the work by McCall (1986).

SAMPLE PROBLEM

A personnel director at a large factory would like to be able to predict the performance of work inspectors on an assembly line. He develops a visual search test that requires the examinee to identify target letters in a mass of letters and numbers on 9 inch × 12 inch cards. The score is the number of errors made on 20 cards. The assembly line performance requires the inspector to reject an assembly if it has a broken connection. An error is counted if the assembly gets by the inspector before he or she can see the faulty part and push a reject switch. Each inspector's score is the number of errors made in a 2-hour work period.

The search test (X) is given to 15 inspectors; later, their assembly line errors (Y) are noted during a 2-hour work period. How well does the search test predict assembly line performance?

Worker	Search Test (X)	Assembly Errors (Y)	X^2	Y^2	XY
A	16	9	256	81	144
B	17	12	289	144	204
C	17	10	289	100	170
D	15	8	225	64	120
E	14	8	196	64	112
F	11	6	121	36	66
G	11	5	121	25	55
H	12	5	144	25	60
I	13	6	169	36	78
J	14	5	196	25	70
K	4	1	16	1	4
L	7	4	49	16	28
M	12	7	144	49	84
N	7	1	49	1	7
O	10	3	100	9	30
	$\Sigma X = 180$	$\Sigma Y = 90$	$\Sigma X^2 = 2364$	$\Sigma Y^2 = 676$	$\Sigma XY = 1232$

Means:

$$\overline{X} = \frac{\Sigma X}{N} = \frac{180}{15} = 12 \qquad \overline{Y} = \frac{\Sigma Y}{N} = \frac{90}{15} = 6$$

Standard Deviations:

$$S_X = \frac{1}{N}\sqrt{N\Sigma X^2 - (\Sigma X)^2} = \frac{1}{15}\sqrt{15(2364) - (180)^2}$$

$$= \frac{1}{15}\sqrt{35{,}460 - 32{,}400} = \frac{\sqrt{3060}}{15}$$

$$S_X = \frac{55.32}{15} = 3.69$$

$$S_Y = \frac{1}{N}\sqrt{N\Sigma Y^2 - (\Sigma Y)^2} = \frac{1}{15}\sqrt{15(676) - (90)^2}$$

$$= \frac{1}{15}\sqrt{10{,}140 - 8100} = \frac{\sqrt{2040}}{15}$$

$$S_Y = \frac{45.17}{15} = 3.01$$

Pearson r:

$$r = \frac{N\Sigma XY - \Sigma X\Sigma Y}{\sqrt{N\Sigma X^2 - (\Sigma X)^2}\sqrt{N\Sigma Y^2 - (\Sigma Y)^2}}$$

$$= \frac{15(1232) - (180)(90)}{\sqrt{15(2364) - (180)^2}\sqrt{15(676) - (90)^2}}$$

$$= \frac{2280}{(55.32)(45.17)} = \frac{2280}{2498.8044}$$

$$r = .91$$

Is r significantly different from 0?

From Table C in Appendix 4, we note that with 13 degrees of freedom ($N - 2 = 13$), r at the .01 level is .641. Since our value of .91 exceeds this tabled value, we conclude that $r = .91$ is significantly different from 0 at the .01 level.

Calculate scattergram and regression equation.

Figure 9.8 shows the scattergram for the data. Calculating our regression equation, we get

$$Y' = \left(\frac{rS_Y}{S_X}\right)X - \left(\frac{rS_Y}{S_X}\right)\overline{X} + \overline{Y}$$

$$= \frac{.91(3.01)}{3.69}X - \frac{.91(3.01)}{3.69}(12) + 6$$

$$= 0.74X - 0.74(12) + 6$$

$$= 0.74X - 8.88 + 6$$

$$Y' = 0.74X - 2.88$$

To plot the regression equation on the scattergram of Figure 9.8, we need to choose a small and a large value of X and solve for Y'. Using X values of 5 and 15, we get

$$Y' = 0.74(5) - 2.88 = 3.70 - 2.88 = 0.82$$
$$Y' = 0.74(15) - 2.88 = 11.10 - 2.88 = 8.22$$

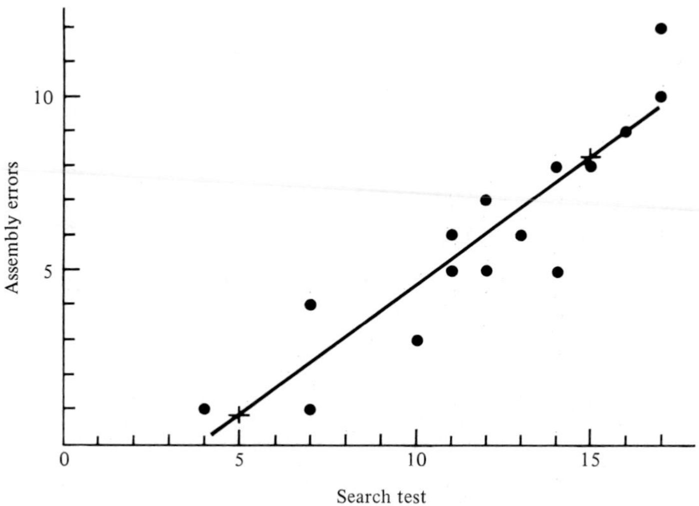

FIGURE 9.8. **Scattergram and regression line for assembly line performance.**

We then locate the points (5, 0.82) and (15, 8.22) on the graph and draw a straight line between them to obtain the regression line.

John Jones applies for a job as inspector and makes 9 errors on the search test. What would be our prediction of his performance on the assembly line?

$$Y' = 0.74(9) - 2.88 = 6.66 - 2.88 = 3.78$$

How accurate is our predicted score? To set up a confidence interval around a predicted Y' score of 3.78, we calculate the standard error of estimate:

$$s_E = S_Y\sqrt{1 - r^2} = 3.01\sqrt{1 - (.91)^2}$$
$$= 3.01\sqrt{1 - .83} = 3.01\sqrt{.17}$$
$$= 3.01(0.41) = 1.23$$

The confidence interval would be calculated as follows:

68% CI: $$Y' \pm s_E\sqrt{1 + \frac{1}{N} + \frac{(X - \overline{X})^2}{NS_x^2}}$$

$$3.78 \pm 1.23\sqrt{1 + \frac{1}{15} + \frac{(9 - 12)^2}{15(3.69)^2}}$$

$$3.78 \pm 1.23(1.054) = 3.78 \pm 1.30$$

$$2.48 \text{ to } 5.08$$

and we would conclude that $p = .68$ that his actual error performance would be between 2.48 and 5.08.

STUDY QUESTIONS

1. What is meant by the *slope* of a line? The *Y intercept*?
2. How do you predict Y from X when all the points on the scattergram are not on a straight line?
3. What is the difference between Y, Y', and $\overline{Y}$?
4. Which statistics are needed to predict Y from X?
5. What is meant by the term *best-fitting straight line*?
6. Explain the expression $e = Y - Y'$.
7. In looking at scattergrams and their regression lines, such as those of Figure 9.2, how can you tell which would have the larger standard errors of estimate and which would have the smaller?
8. The Pearson r tells us the strength of the linear relationship between two variables. What does r^2, the coefficient of determination, tell us about the two variables?
9. List the assumptions for the Pearson r when used for prediction and tell what each means.

EXERCISES[†]

1. On a sheet of graph paper, plot the equation for the line $Y = 3X + 5$. What is the slope of this line? What is its Y intercept?

[†]The answers you calculate for the Chapter 9 exercises may differ slightly from those given in the Answers due to the fact that rounding errors in means, standard deviations, and the Pearson r are magnified when all appear in the regression formula. This, of course, will also affect your calculation of Y'.

2. Plot the equation for the line $Y = -2X + 10$ and determine its slope and Y intercept.

3. In Sample Problem 1 of Chapter 8 we saw a list of 30 states showing how much money each had spent on marijuana eradication and the number of plants that were destroyed. The relevant statistics (in thousands) are repeated below.
   a. Calculate the regression equation for these data.
   b. Another state is planning to spend $60,000 on marijuana eradication. On the basis of this study what would be our prediction for the number of plants destroyed?

Money Spent *X*	**Plants Destroyed** *Y*
$\overline{X} = 37.1$	$\overline{Y} = 34.6$
$S_X = 22.01$	$S_Y = 37.62$
$r = .66$	

4. The College Life Questionnaire surveyed a sample of 30 students, asking the number of hours per week that they spent studying and their current GPA. The relevant statistics are reported below.
   a. Calculate the regression equation to predict GPA from the number of hours studied.
   b. Jane Doe says that she studies 20 hours per week. What would be our prediction for her GPA?

Hours Studied *X*	**GPA** *Y*
$\overline{X} = 14.53$	$\overline{Y} = 3.01$
$S_X = 5.14$	$S_Y = 0.60$
$r = .17$	

5. Calculate the coefficient of determination, r^2 for Exercise 3. What does this mean in terms of the relationship between funds expended and number of marijuana plants destroyed?

6. In Exercise 4, the correlation between the number of hours studied per week by a sample of college students and their GPA is given. Calculate r^2, the coefficient of determination, and tell what this means in terms of predicting GPA from the number of hours studied.

7. In the Chapter 8 exercises, we noted a study where the heart rate of 25 patients at a university health service was first taken by a student nurse and later by the examining physician. The correlation between the first and second measurements of heart rate was found to be $r = .93$. Other relevant statistics were: first heart rate $\overline{X} = 70.44$ and $S_X = 9.90$; second heart rate $\overline{Y} = 70.96$ and $S_Y = 9.65$. The data are shown below.
   a. Use graph paper to plot a scattergram of these data.
   b. Calculate the regression equation for predicting Y from X. Draw the regression line on the scattergram.
   c. Susan's heart rate is 80 when taken by the student nurse. Use the regression equation to predict what her heart rate would be when it is taken later by the physician.
   d. Calculate the 68% CI for Susan's predicted heart rate and describe what this interval means.

First HR	Second HR	First HR	Second HR
58	58	54	56
84	84	82	80
58	56	62	66
78	76	70	70
92	94	72	74
70	66	60	62
61	70	68	66
76	76	70	62
66	76	68	66
62	68	68	68
90	88	76	76
62	62	82	84
72	70		

8. An experiment on depth perception was described earlier in the Chapter 8 exercises where subjects tried to align two vertical rods from a distance of 10 feet. In one condition they held their heads steady (Fixed) and in the other condition, they were allowed to move their heads from side to side (Moving). The amount of error in millimeters was measured for each subject under both conditions, and the error scores are shown below. The

correlation between the two conditions was r = .87. Other relevant data: fixed condition $\overline{X} = 26.7$ and $S_X = 11.75$; moving condition $\overline{Y} = 12.75$ and $S_Y = 4.37$.

a. Plot a scattergram of these data on a sheet of graph paper.

b. Calculate the regression equation for predicting Y from X and draw the regression line on the scattergram.

c. John had an error of 10 millimeters in the fixed condition. Use the regression equation to predict what his error score would be in the moving head condition.

d. Calculate the 68% CI for John's predicted error score. What does this interval represent?

Fixed	Moving	Fixed	Moving
45	18	19	7
17	9	24	12
17	9	19	14
20	15	50	22
10	9	36	11
36	15	34	16
40	20	40	18
34	13	22	10
12	7	18	8
10	8	31	14

9. In a Chapter 8 exercise we found an educational psychology instructor trying to see if there was a relationship between scores on the midterm exam and course grades (A = 12, A− = 11, etc.) at the end of the term. She found a correlation of r = .74 between the midterm exam and course grades. The test score and course grade for each of her 40 students are shown below. Also shown are the column totals from Exercise 6 in Chapter 8.

a. Calculate $\overline{X}$, $\overline{Y}$, S_X, and S_Y for these data (use machine formulas).

b. Plot a scattergram of the data on a sheet of graph paper.

c. Calculate the regression equation and draw the regression line on the scattergram.

d. Jane scores 30 on the midterm exam. What would be our best prediction of her grade in the course?

e. Calculate the 68% CI for Jane's predicted score. What is the meaning of this interval?

Midterm Exam (X)	Course Grade (Y)	Midterm Exam (X)	Course Grade (Y)
36	7	41	12
35	8	40	10
43	9	42	9
45	9	44	10
35	9	47	12
37	8	46	12
44	9	37	9
44	12	43	11
34	6	38	8
41	8	31	4
25	5	39	9
43	12	40	7
40	6	38	8
37	6	36	8
41	9	42	9
42	12	40	5
41	9	44	12
32	9	44	11
30	6	43	10
33	6	36	6

$\Sigma X = 1{,}569$ $\quad \Sigma Y = 347$
$\Sigma X^2 = 62{,}451$ $\quad \Sigma Y^2 = 3{,}205$
$r = .74$

10. In the development of Skylab, NASA scientists were interested in the relationship between hand steadiness and body balance.† Performance on the hand steadiness test was the length of time in seconds that the subject could hold a stylus in an aperture before touching the side of the box (X). Body balance was measured by timing how many seconds subjects could stand on a one-inch wide board before losing their balance (Y). Twenty subjects gave the results shown below.
    a. Calculate $\overline{X}$, $\overline{Y}$, S_X, and S_Y for these data (use machine formulas).
    b. Calculate the Pearson r for these data (use the machine formula).
    c. Plot a scattergram of the data on a sheet of graph paper.
    d. Calculate the regression equation and draw the regression line on the scattergram.

†Adapted from R. F. Haines, A. E. Bartz, and J. R. Zahn (1972), "Human Performance Capabilities in a Simulated Space Station-Like Environment," *NASA Technical Memorandum X-62,101*. Ames Research Center, Moffett Field, Calif.

e. Pete scores 15 on the hand steadiness test. What result would we predict for his body balance test?
f. Calculate the 68% CI for Pete's predicted body balance test score.

Steadiness X	**Balance** Y	**Steadiness** X	**Balance** Y
13	8	3	3
12	7	9	9
18	10	14	9
12	8	3	4
13	11	15	11
9	6	6	8
16	10	11	7
10	6	5	7
8	5	11	10
12	9	14	8

CHAPTER 10

The Significance of the Difference Between Means

Much of the activity of the researcher in education and the behavioral sciences is directed toward comparing the performance of two groups. A sampling of recent research studies might show investigations of differences between men and women on mathematical ability, differences between only children and children from large families on introversion, and differences between college GPAs of marijuana users and those of marijuana nonusers. Whenever such differences are established, that is, verified by a number of independent investigators, they become part of our body of scientific knowledge and finally find their place in books on individual differences, child psychology, or drug abuse.

Sampling Error or Real Difference?

Verification by the independent work of other researchers is an important part of the scientific method, but how does a single investigator determine whether or not he or she has found a real difference in the performance of two groups? For example, suppose that our researcher is trying to see if there is a difference in word comprehension between fifth-grade boys and girls. Random samples are used with an appropriate comprehension test, and the following mean scores (number of words correctly identified) occur.

Boys	Girls
$\overline{X} = 49.21$	$\overline{X} = 49.23$

We certainly would agree that there appears to be no difference between boys and girls on word comprehension. The slight difference we do observe (obviously 49.21 and 49.23 *are* different) is attributed to "chance" or "sampling error."

But what if the researcher had obtained the following?

Boys	Girls
$\overline{X} = 49.21$	$\overline{X} = 49.27$

We would probably still conclude that the .06 difference between the two means was due to sampling error and there basically was no difference between boys and girls on word comprehension.

But just how far apart must the two means be before we can say that there is a *real* difference between the two groups? 49.21 and 49.50? 49.21 and 52.00? 49.21 and 55.00? We are faced with the same dilemma that we noted in the ESP example of Note 7.2. At what point do we draw the line between a sampling error and a real difference?

To help us answer this question, we will have to resort to the sampling distribution, a concept first introduced in Chapter 7. Our ultimate goal is to derive a technique that will enable us to make a probability statement regarding sampling error, just as we did in Chapter 7.

The Sampling Distribution of Differences Between Pairs of Means

As an example, let us say that we are interested in seeing if there is any difference between the mathematical ability of college chemistry majors and that of college biology majors. Using appropriate sampling techniques, we select a sample of 100 chemistry majors and 100 biology majors, administer a standardized mathematical aptitude test, and obtain the following results:

Chemistry	Biology
$\overline{X}_1 = 72.9$	$\overline{X}_2 = 68.4$

The obtained difference in the means of the two groups is $72.9 - 68.4 = 4.5$, and we would like to know if this indicates a real superiority in mathematical ability on the part of chemistry majors, or if it is simply the result of sampling error.

At this point we do a rather strange thing. We begin by assuming that there is *no difference* in the means of the populations from which our two samples were drawn. In other words, we assume that the mean mathematical ability of the population of chemistry majors is the same as the mean mathematical ability of the population of biology majors. If μ_1 is the mean ability of the population of chemistry majors and μ_2 is the mean ability of the population of biology majors, we are assuming that $\mu_1 = \mu_2$, or $\mu_1 - \mu_2 = 0$. *This statement is called a null hypothesis,* and it simply states that there is no difference between the means of the two populations from which we drew our two samples.

If this null hypothesis is true and $\mu_1 - \mu_2 = 0$, then our obtained difference of 4.5 is just sampling error. But to see if our assumption that $\mu_1 - \mu_2 = 0$ is true, let us indulge in another bit of fancy, just as we did in Chapter 7. We will conduct a hypothetical exercise by continuing to take pairs of samples and finding the differences (D) in their means. Keeping track of these pairs, we might get the results shown below (note that we always subtract the biology mean from the chemistry mean to obtain D):

	Chemistry	**Biology**	***D***
Sample 1	73.5	73.9	−0.4
Sample 2	74.6	72.1	2.5
Sample 3	72.8	71.3	1.5
Sample 4	73.4	74.6	−1.2
Sample 5	72.9	72.9	0

and so on.

As we ran this hypothetical exercise (all the while assuming a null hypothesis, $\mu_1 - \mu_2 = 0$), we would note that some of the differences were negative, some were positive, and most would be fairly close to 0.

If we continued to do this for an *infinite* number of pairs of samples, the following would occur:

1. A batch of differences, called a *sampling distribution of differences between means,* would result.
2. The mean of this sampling distribution $\overline{D}_{\overline{X}}$ (say "mean of the differences"), would be 0.
3. The batch of differences would be normally distributed around the mean of the distribution $\overline{D}_{\overline{X}}$, with a standard deviation of $\sqrt{\sigma^2_{\overline{X}_1} + \sigma^2_{\overline{X}_2}}$.

Let us take a closer look at these three results.

SAMPLING DISTRIBUTION OF DIFFERENCES BETWEEN MEANS. Even though we are assuming that the two populations have equal means, we do not expect that a sample drawn from one population will have a mean exactly identical to the mean of a sample drawn from the second population. We have grown accustomed to sampling error as a fact of life, and we would take for granted that in pairs of means, $\overline{X}_1$ will sometimes be greater than $\overline{X}_2$, and $\overline{X}_2$ will sometimes be greater than $\overline{X}_1$. This, of course, results in a distribution of differences, of which some will be negative and some positive.

THE MEAN OF THE DIFFERENCES. The mean of this distribution of differences, $\overline{D}_{\overline{X}}$, will be 0 under our assumption that $\mu_1 - \mu_2 = 0$. As indicated in the last paragraph, we expect that some differences will be negative and others positive, with the net effect being a mean that is exactly 0.

NORMAL SAMPLING DISTRIBUTION. Since the distribution of differences is normal (if our samples are large enough—say 50 or more), we can use the characteristics of the normal curve to show, for example, that

1. 68.26% of all the differences would fall between −1 and 1 standard deviation from the mean, $\overline{D}_{\overline{X}}$.
2. 95% of all the differences would fall between −1.96 and 1.96 standard deviations from the mean, $\overline{D}_{\overline{X}}$.
3. 99% of all the differences would fall between −2.58 and 2.58 standard deviations from the mean, $\overline{D}_{\overline{X}}$.

Similarly, the probability statements we made concerning the normal curve would apply to this sampling distribution as well; for example:

1. $p = .68$ that any difference would fall between $\overline{D}_{\overline{X}} \pm 1$ standard deviation.
2. $p = .95$ that any difference would fall between $\overline{D}_{\overline{X}} \pm 1.96$ standard deviations.
3. $p = .05$ that any difference would fall *outside* the same interval, $\overline{D}_{\overline{X}} \pm 1.96$ standard deviations.

With these probability statements in mind, let us get back to our original problem—that of deciding whether our obtained difference of 4.5 between the mean mathematical ability of chemistry and biology majors is simply due to sampling error. The first step in determining if it *is* in fact due to sampling error is to establish *where* in the sampling distribution of differences our obtained difference is located. Figure 10.1 shows this hypothetical sampling distribution, which, under the null hypothesis of $\mu_1 - \mu_2 = 0$, has a mean, $\overline{D}_{\overline{X}}$, of 0.

Using the same logic that was introduced in Chapter 7, we need only find *where* in the sampling distribution of differences our obtained difference falls. If it is located toward the center of the distribution (i.e., between −1.96 and 1.96 standard deviations from the mean), we say that our obtained difference was caused by sampling error. On the other hand, if our difference is so large that it falls way out in either tail of the distribution (labeled "Reject" in Figure 10.1, we know that it would

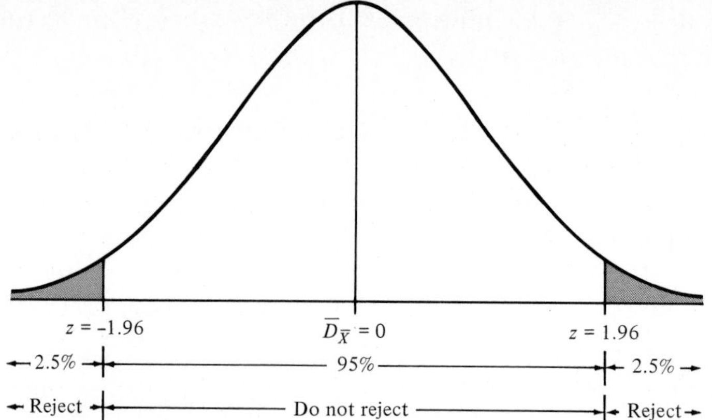

FIGURE 10.1. **Sampling distribution of differences between means assuming the null hypothesis ($\mu_1 - \mu_2 = 0$).**

be an extremely rare occurrence, and we would wonder whether our difference is really due to sampling error after all.

The procedure is simple indeed. We need only calculate the z score for our difference and find its location in the sampling distribution of differences, as in Figure 10.1. Under the null hypothesis that $\mu_1 - \mu_2 = 0$, this sampling distribution of differences has $\overline{D}_{\bar{X}} = 0$. Now if the z score of our difference were, for example, 1.23 or -0.67, we would say that our difference of 4.5 points between the mean mathematical ability of chemistry and biology majors could possibly be due to sampling error. However, if the z score of our difference were, for example, 1.98 or -2.37, we would say that it is a rare occurrence in a distribution that has $\overline{D}_{\bar{X}} = 0$ and would conclude that μ_1 is not equal to μ_2 (or $\mu_1 - \mu_2 \neq 0$), $\overline{D}_{\bar{X}}$ is not 0, and our obtained difference is a real difference and not sampling error.

Since that last sentence above is crucial to understanding this chapter, let us briefly examine the concepts involved by referring to Figure 10.1.

1. If there is no difference in the populations from which the samples came, the probability that any obtained difference would deviate by ± 1.96 standard deviations or more is .05. (Remember that only 5% of the differences are in the two tails of the distribution beyond ± 1.96 standard deviations.) These differences are in the shaded portions of Figure 10.1.
2. If, in fact, our obtained difference falls in the *unshaded* portion of the distribution, that is, between -1.96 and 1.96 in Figure 10.1, we say that our difference could be due to sampling error. Since differences with z scores between -1.96 and 1.96 standard deviations

happen 95% of the time just by sampling error, we have no reason to believe that our difference is anything but sampling error. We say that the difference is not significant.

3. However, if our obtained difference falls in the region of ± 1.96 standard deviations or more (the shaded portions of Figure 10.1 labeled "Reject"), we are faced with two possibilities.
   a. $\overline{D}_{\overline{X}}$ really *is* 0, and our difference is one of those rare occurrences that happen 5% of the time or less due to sampling error.
   b. $\overline{D}_{\overline{X}}$ is *not* 0, but some other value, and our difference tells us that there is a real difference between the two populations from which we drew our samples. In this example, we would say that the chemistry majors are higher in mathematical ability than biology majors.
4. In light of the statistical conventions mentioned in Chapter 7, we will conclude that any difference that happens 5% of the time or less by sampling error, that is, $p \leq .05$ (read "probability is equal to or less than .05"), is not due to chance at all but represents a real difference. We say that the difference is significant at the .05 level, and we reject the null hypothesis ($\mu_1 - \mu_2 = 0$) in favor of $\mu_1 - \mu_2 \neq 0$.

Standard Error of the Difference Between Means

In order to know where our obtained difference is located in the hypothetical sampling distribution of differences, we need to know the standard deviation of this distribution. We can then convert our difference to a z score and locate it precisely in the sampling distribution.

The standard deviation of this sampling distribution is called *the standard error of the difference between means*, and it is denoted by the symbol $\sigma_{D_{\overline{X}}}$. This statistic is to be interpreted as our ordinary, garden variety standard deviation. For example, we know that 95% of the differences fall between -1.96 and 1.96 standard errors of the difference, or between $\overline{D}_{\overline{X}} \pm 1.96\sigma_{D_{\overline{X}}}$.

When we calculate a z score for the sampling distribution of differences, we must modify the usual z score formula, $z = (X - \overline{X})/S$. The "score" of X is now the difference, or $\overline{X}_1 - \overline{X}_2$. The mean of the sampling distribution is now $\overline{D}_{\overline{X}}$, but, as you have seen, $\overline{D}_{\overline{X}} = 0$ under our assumption of the null hypothesis. The standard deviation of the sampling distribution is now the standard error of the difference, $\sigma_{D_{\overline{X}}}$. So our new formula for calculating a z score in a sampling distribution of differences is

$$z = \frac{(\overline{X}_1 - \overline{X}_2) - 0}{\sigma_{D_{\overline{X}}}} \qquad (10.1)$$

Let us leave the computation of $\sigma_{D_{\bar{X}}}$ for a later section and examine now the use of $\sigma_{D_{\bar{X}}}$ in determining the significance of an obtained difference in the following hypothetical example. Suppose that in our study of the mathematical ability of chemistry and biology majors, we had obtained the following results.

Chemistry	**Biology**
$\bar{X}_1 = 72.6$	$\bar{X}_2 = 67.6$
$\sigma_{D_{\bar{X}}} = 5.0$	

Is this difference between the means due to sampling error, or is it such a rare occurrence that we would suspect a real difference in the mathematical ability of the two groups? Our first step is to calculate a z score for the difference, which would be

$$z = \frac{(\bar{X}_1 - \bar{X}_2) - 0}{\sigma_{D_{\bar{X}}}} = \frac{72.6 - 67.6}{5} = \frac{5}{5} = 1.0$$

Since by sampling error it is just as easy to get $\bar{X}_2$ greater than $\bar{X}_1$, we will consider both tails of the curve and ask the question "What is the probability of obtaining a difference of 5 points just by sampling error from a sampling distribution whose mean is 0?" As we just noted, a difference of 5 points results in a z score of 1 (or -1 if $\bar{X}_2$ is greater than $\bar{X}_1$), and we can locate a difference of 5 points in the sampling distribution of differences as shown in Figure 10.2a. By using Table B we would see that 68.26% of the differences would fall between $-1z$ and 1z, and 31.74% of the differences would result in z scores larger than 1 *if the null hypothesis is true* ($\mu_1 - \mu_2 = 0$, or $\bar{D}_{\bar{X}} = 0$). Converting the percentages to probabilities, we see that $p = .3174$ that *any* obtained difference would deviate by 5 points or more just by sampling error. Since anything that happens 32% of the time is hardly a "rare occurrence," we conclude that the difference of 5 points in mathematical ability between chemistry and biology majors is possibly just due to sampling error.

But suppose, instead, that we had obtained the following results, shown graphically in Figure 10.2b.

Chemistry	**Biology**
$\bar{X}_1 = 72.6$	$\bar{X}_2 = 65.1$
$\sigma_{D_{\bar{X}}} = 5.0$	

$$z = \frac{(\bar{X}_1 - \bar{X}_2) - 0}{\sigma_{D_{\bar{X}}}} = \frac{72.6 - 65.1}{5} = \frac{7.5}{5} = 1.5$$

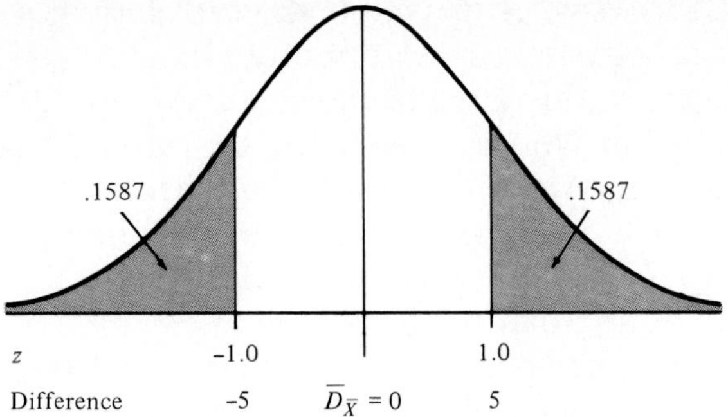

(a) Probability of difference larger than 5 is $p = .3174$.

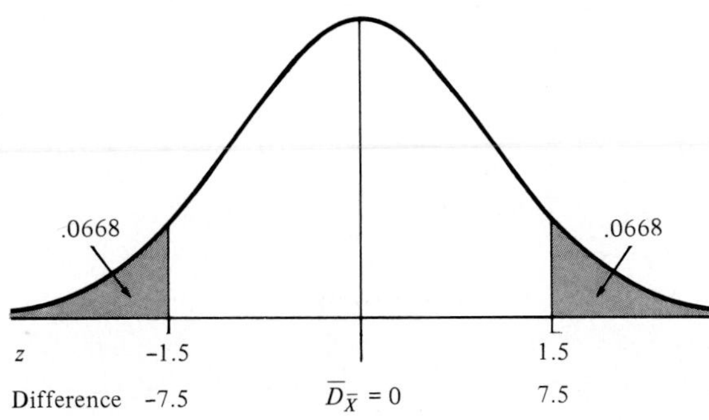

(b) Probability of difference larger than 7.5 is $p = .1336$.

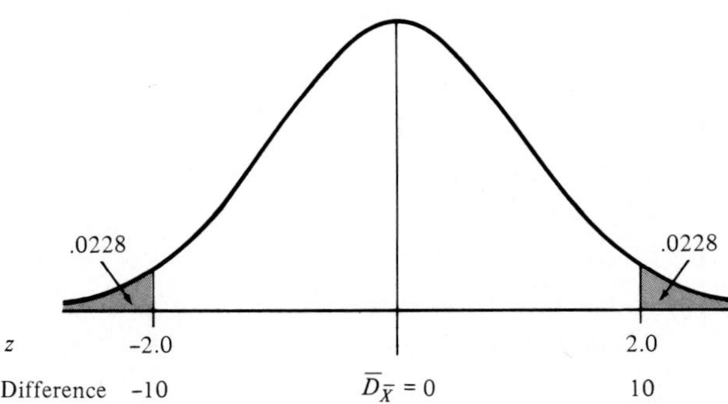

(c) Probability of difference larger than 10 is $p = .0456$.

FIGURE 10.2. **Probabilities of obtaining differences as large as or larger than a result by sampling error.**

Our calculations show that our obtained difference of 7.5 between the two means results in a z score of 1.5. What is the probability of obtaining a difference this large or larger just by sampling error from a sampling distribution whose mean is 0? By referring to Figure 10.2b and using Table B in Appendix 4, we see that differences as large as 7.5 and larger will happen only 13.36% of the time just by sampling error, or $p = .1336$ that *any* obtained difference will be that large or larger. Since something that happens 13% of the time just by chance is not really a "rare occurrence," we would state that a difference of 7.5 points between the means of the two groups may simply be due to sampling error. But note that if the null hypothesis is *really* true ($\mu_1 - \mu_2 = 0$, or $\overline{D}_{\overline{X}} = 0$) a difference as large as 7.5 is less likely to occur just by chance than is a difference of 5.0 (13% vs. 32%).

As a final example, let us suppose that we had obtained the following results on the mathematical ability of chemistry and biology majors.

Chemistry	**Biology**
$\overline{X}_1 = 72.6$	$\overline{X}_2 = 62.6$

$$\sigma_{D_{\overline{X}}} = 5.0$$

$$z = \frac{(\overline{X}_1 - \overline{X}_2) - 0}{\sigma_{D_{\overline{X}}}} = \frac{72.6 - 62.6}{5} = \frac{10}{5} = 2$$

What is the probability of obtaining a difference of 10 points or more just by sampling error? Figure 10.2c and Table B show that we expect only 4.56% of the differences to be 10 points or more, if the null hypothesis that $\mu_1 - \mu_2 = 0$ is true. In other words, less than 5% of the differences would be this large or larger, so $p < .05$ that *any* obtained difference would be this large or larger. Since this is a relatively rare occurrence, we reject the null hypothesis that $\overline{D}_{\overline{X}}$ is really 0 and state that there is a significant difference between the average mathematical ability of the population of chemistry majors and that of the population of biology majors.

In summary, the steps involved in determining the significance of a difference are

1. Assume the null hypothesis—that the means of the two populations from which the samples are drawn are equal ($\mu_1 = \mu_2$, or $\mu_1 - \mu_2 = 0$).
2. Calculate the z score for your obtained difference between the means of the two samples and locate it in the sampling distribution of differences whose mean, $\overline{D}_{\overline{X}}$, is 0.

3. If the resulting z score falls between $\overline{D}_{\overline{X}} \pm 1.96\sigma_{D_{\overline{X}}}$, assume that the difference in sample means is due to sampling error.
4. If the resulting z score falls *outside* the interval $\overline{D}_{\overline{X}} \pm 1.96\sigma_{D_{\overline{X}}}$, reject the null hypothesis, assume that $\mu_1 - \mu_2$ is *not* 0, and state that your obtained difference is significant at the .05 level.

Calculating the Standard Error of the Difference Between Means

The standard error of the difference between means, as we noted earlier, is the standard deviation of the hypothetical sampling distribution of differences, and it is given by the following formula:

$$\sigma_{D_{\overline{X}}} = \sqrt{\sigma_{\overline{X}_1}^2 + \sigma_{\overline{X}_2}^2} \tag{10.2}$$

where

$\sigma_{D_{\overline{X}}}$ is the standard error of the difference between means
$\sigma_{\overline{X}_1}$ is the standard error of the mean of one distribution
$\sigma_{\overline{X}_2}$ is the standard error of the mean of the other distribution

However, we noted in Chapter 7 that the standard error of the mean, $\sigma_{\overline{X}}$, is calculated from the *population* standard deviation, so formula 10.2 is of little value to us, since we rarely know the value of σ. As a result, we must have a formula for the standard error of the difference that permits us to use the standard deviations of our samples.† The formula that we shall use extensively is

$$s_{D_{\overline{X}}} = \sqrt{\frac{N_1S_1^2 + N_2S_2^2}{N_1 + N_2 - 2}\left(\frac{1}{N_1} + \frac{1}{N_2}\right)} \tag{10.3}$$

where

S_1 is the standard deviation of the first sample
S_2 is the standard deviation of the second sample
N_1 and N_2 are the sizes of the respective samples

Testing for Significant Differences. The *t* Test for Two Independent Samples

We must note that $s_{D_{\overline{X}}}$ is an *estimate* of $\sigma_{D_{\overline{X}}}$, and because $s_{D_{\overline{X}}}$ is based on the standard deviations of the samples, it is subject to sampling error.

†Note that the symbol for the standard error of the difference is $\sigma_{D_{\overline{X}}}$ when the standard error of the mean, $\sigma_{\overline{X}}$, is calculated from the population σ. When it is calculated from the sample standard deviation, S, we will use the symbol, $s_{D_{\overline{X}}}$.

When the standard deviations of the populations are not known (so $\sigma_{D_{\bar{X}}}$ cannot be calculated), the normal distribution can no longer be used, which, of course, means that the z score values of 1.96 and 2.58 cannot automatically be used to describe the region of rejection in the tails of the distribution.

So, instead of the normal distribution with the usual z values, we use what is called the *t* distribution, and the statistical procedure is known as the *t* test. We calculate a *t* value and locate its position in the *t* distribution, just as we did earlier for the z score. The procedure is the same as before, but we no longer use 1.96 and 2.58 as dividing lines for acceptance or rejection of the null hypothesis. The *t* statistic, in fact, has the same formula as the z score used earlier, except the denominator is now $s_{D_{\bar{X}}}$.

$$t = \frac{(\bar{X}_1 - \bar{X}_2) - 0}{s_{D_{\bar{X}}}} \tag{10.4}$$

The *t* Distribution

The *t* distribution (sometimes called "Student's *t*," after W. S. Gossett, who published under the pseudonym "Student") looks very much like the normal curve, except that the tails for the *t* distribution are higher and we must go farther out to find the *t* values that mark off the 5% and 1% regions in the distribution. This is illustrated graphically in Figure 10.3.

Note that the normal curve has 95% of its area between the usual z scores of −1.96 and 1.96 while the *t* distribution shown in Figure 10.3 for a combined sample size of 6 has 95% of its area between −2.78 and 2.78. As the sample size gets larger, the *t* distribution looks more and more like the normal curve, and, when the *N* of the combined samples

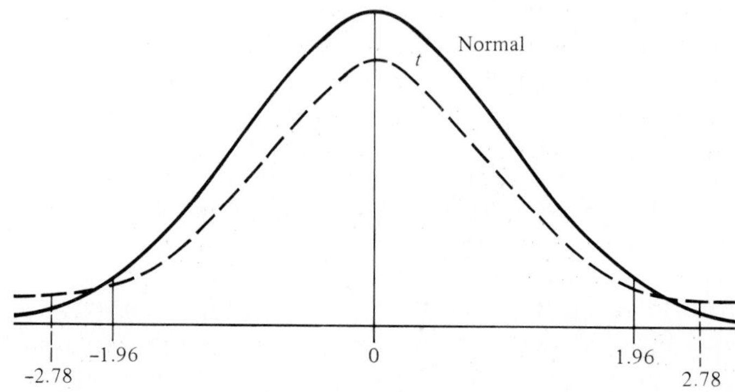

FIGURE 10.3. **Comparison of a normal distribution and a *t* distribution ($N = 6$) showing the 95% limits for each distribution.**

is about 30, the curves are approximately identical. What this means, of course, is that, as the size of your samples gets smaller, you need a larger t value to reject the null hypothesis.

Since the normal curve and the t distribution are exactly identical only with an *infinitely large* sample size, it has become conventional always to use the t distribution, *regardless of sample size*. This means that we can ignore the z values of 1.96 and 2.58 in favor of a specific t value which is determined by the size of our samples or, more properly, the *degrees of freedom* in our samples.

The Degrees of Freedom (df) Concept

Before we can look at the different t distributions, we must become acquainted, at least superficially, with a concept known as *degrees of freedom*. Although a knowledge of advanced statistical theory is necessary to completely understand this concept, we can say that the degrees of freedom in a sample is *the number of observations that are free to vary*.

For example, if we put a restriction on a set of five numbers, such that $\Sigma X = 12$, we may pick any *four* numbers we like, but the *fifth* one is not free to vary, since it, added to the others, must meet the requirement of $\Sigma X = 12$. If we choose 3, 2, 4, and 7, we *must* use -4 as the fifth number in order to meet this requirement, since $3 + 2 + 4 + 7 = 16$ and only a -4 would yield $\Sigma X = 12$. In this case we have $N - 1$ degrees of freedom, since only one observation is restricted in meeting the requirement. Stated more formally, $df = N - 1$.

In the calculation of a t test, each of the two samples has $N - 1$ degrees of freedom. This means that the total df for *both* samples would be $(N_1 - 1) + (N_2 - 1)$, which would sum to $N_1 + N_2 - 2$. Thus the total df for our samples is 2 less than the combined sample size. For example, if we had 20 students in one group and 10 students in the second group, our total df would be $(20 - 1) + (10 - 1) = 19 + 9 = 28$.

Using the t Table

As we noticed earlier, there is a different t distribution for every sample size. This would mean that we could have a table of t values similar to a normal curve table, for every sample size from 2 to infinity! In the interest of brevity, we do not usually consider all areas of the curve, but only those points that indicate the usual significance levels, such as .05, .01, and .001. Table E in Appendix 4 lists the t values that must be equaled or exceeded for these usual significance levels for various sample sizes (in terms of degrees of freedom). Note that the table gives both one-tailed and two-tailed values. Let us consider for now only the two-tailed values. We will interpret the one-tailed values later on in the chapter.

For the example above, with 28 degrees of freedom, how large must our t value be in order to be significant at the .05 level? From Table E, we see that for a df of 28, the tabled value of t is 2.048. We know that our calculated t must be at least 2.048 before we can reject the null hypothesis at the .05 level.

As another example, suppose that you had 7 students in one group and 12 in a second group. How large would your t value have to be in order for it to be significant at the .01 level? Again, from Table E we see that for 17 degrees of freedom ($df = 6 + 11$), we would need a t of at least 2.898 in order to reject the null hypothesis at the .01 level.

Calculating the t Test for Independent Samples[†]

Although the formula for $s_{D_{\overline{X}}}$ looks somewhat complicated, it is relatively easy to use in practice, since it requires only the standard deviations of the samples and the N of each sample. Table 10.1 shows the calculation of $s_{D_{\overline{X}}}$ and its use in determining the significance of a difference between the tested IQs of seventh-grade girls and boys.

TABLE 10.1. **Testing for Significant Differences in Mean IQ for Boys and Girls**

Boys	**Girls**
$\overline{X}_1 = 108.6$	$\overline{X}_2 = 110.2$
$S_1 = 14.8$	$S_2 = 13.9$
$N = 50$	$N = 40$

$$s_{D_{\overline{X}}} = \sqrt{\frac{N_1S_1^2 + N_2S_2^2}{N_1 + N_2 - 2}\left(\frac{1}{N_1} + \frac{1}{N_2}\right)}$$

$$= \sqrt{\frac{50(14.8)^2 + 40(13.9)^2}{50 + 40 - 2}\left(\frac{1}{50} + \frac{1}{40}\right)}$$

$$= \sqrt{\frac{10{,}952 + 7728.4}{88}(0.02 + 0.025)}$$

$$= \sqrt{212.28(0.045)}$$

$$= \sqrt{9.5526}$$

$$s_{D_{\overline{X}}} = 3.09$$

$$t = \frac{(\overline{X}_1 - \overline{X}_2) - 0}{s_{D_{\overline{X}}}} = \frac{108.6 - 110.2}{3.09} = \frac{-1.6}{3.09} = -0.52$$

t_{05}, $df = 80$, is 1.990
Not significant, $p > .05$

[†]By *independent samples* we mean that the selection of a member of one group does not influence the selection of any member of the second group.

Following the four steps listed earlier, we note in the example of Table 10.1 that a null hypothesis would state that the mean IQ for a population of seventh-grade boys is the same as the population mean for seventh-grade girls, or $\mu_1 - \mu_2 = 0$. A sample of 50 boys and 40 girls from these two populations shows a mean difference of $108.6 - 110.2 = -1.6$.

After calculating t to be -0.52, we consult Table E for 88 degrees of freedom. There are no values listed between $df = 80$ and $df = 90$, so we choose the value for 80—the more conservative value—and note that a t value of ± 1.990 marks off the 5% region of rejection. Since our calculated value of -0.52 is well within the central region (as shown in Figure 10.4), it is hardly a "rare occurrence" and we have no reason to believe that the null hypothesis should be rejected. We pronounce the obtained difference not significant, since the probability is greater than .05 ($p > .05$) that this could happen by sampling error alone.

Table 10.1 illustrated a typical research study, with the statistical logic behind the decision to conclude that an observed difference was not significant. Let us consider a recent study that led the investigators to conclude that the differences were significant. The mean GPA was calculated for a sample of 30 marijuana users and 30 nonusers on a college campus. The calculations are shown in Table 10.2.

In analyzing the data of Table 10.2, we would first state a null hypothesis, that the mean college GPA for the population of marijuana users would be the same as the population mean for nonusers, $\mu_1 - \mu_2 = 0$. A sample of 30 marijuana users and 30 nonusers from these two populations shows a mean difference in college GPA of $3.05 - 2.75 = 0.30$.

After calculating a t of 2.31, we note from Table E that ± 2.009 marks off the 5% region of rejection (we are again conservative and use the

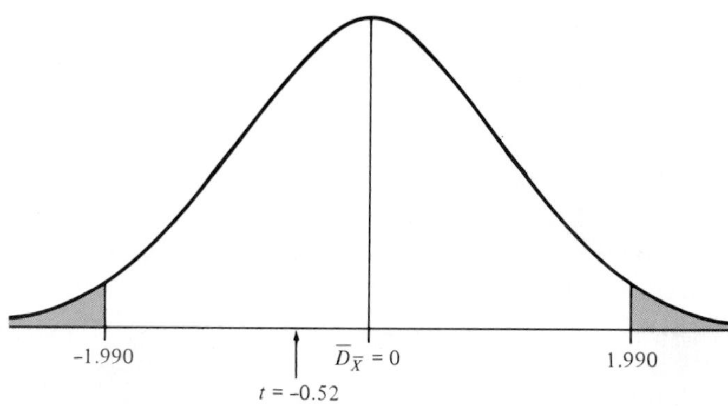

FIGURE 10.4. Nonsignificant difference in IQ scores of seventh-grade boys and girls, with a *t* of −.52.

TABLE 10.2. **Testing for Significant Differences in College GPA Among Marijuana Users and Nonusers**

Nonusers	**Users**
$\overline{X}_1 = 3.05$	$\overline{X}_2 = 2.75$
$S_1 = 0.60$	$S_2 = 0.40$
$N_1 = 30$	$N_2 = 30$

$$s_{D\overline{X}} = \sqrt{\frac{N_1S_1^2 + N_2S_2^2}{N_1 + N_2 - 2}\left(\frac{1}{N_1} + \frac{1}{N_2}\right)}$$

$$= \sqrt{\frac{30(0.60)^2 + 30(0.40)^2}{30 + 30 - 2}\left(\frac{1}{30} + \frac{1}{30}\right)}$$

$$= \sqrt{\frac{10.8 + 4.8}{58}(0.067)}$$

$$= \sqrt{0.2690\ (0.067)}$$

$$= \sqrt{0.018}$$

$$s_{D\overline{X}} = 0.13$$

$$t = \frac{(\overline{X}_1 - \overline{X}_2) - 0}{s_{D\overline{X}}} = \frac{3.05 - 2.75}{0.13} = \frac{0.30}{0.13} = 2.31$$

t_{05}, $df = 50$, is 2.009
Significant, $p < .05$

tabled value for $df = 50$, since the value for $df = 58$ is not listed). Note that our calculated t of 2.31 is outside the interval, as shown in Figure 10.5. Since this is a rare occurrence, happening less than 5% of the time by sampling error ($p < .05$), we conclude that it is not likely that $\mu_1 - \mu_2 = 0$. We reject this null hypothesis in favor of $\mu_1 - \mu_2 \neq 0$ and say that there is a significant difference in the grade-point averages of marijuana users and nonusers.

Significance Levels

In the preceding discussion and in the examples of Tables 10.1 and 10.2, we have seen that any difference occurring 5% of the time or less through sampling error is labeled a "significant" difference. Our concluding statement is that the difference is "significant at the .05 level," or "the null hypothesis is rejected at the .05 level." As you saw in Chapter 7, there is nothing magic about the 5% point: it is simply an arbitrary designation that has become a convention or rule of thumb. Presumably, the 6% or 4% levels could be similarly justified.

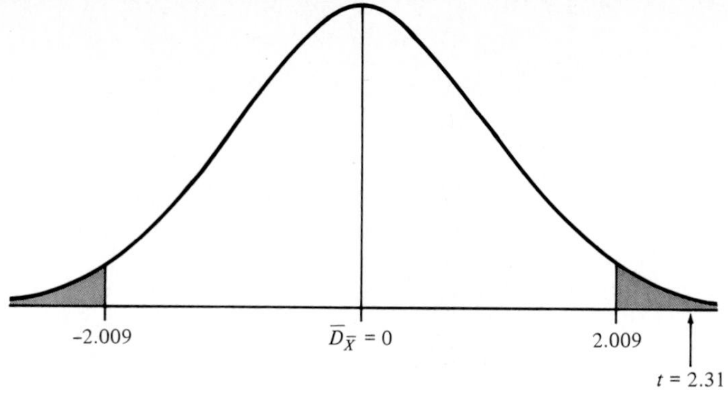

FIGURE 10.5. **Significant difference in college GPA of marijuana users and nonusers with a *t* of 2.31.**

However, many statisticians feel that there should be some way to indicate just how rare a rare occurrence is under the null hypothesis, so the .01 level and .001 level have been added to the familiar .05 level. If a difference is significant at the .01 level, we are saying that the obtained difference would happen by sampling error 1% of the time or less. Similarly, if a difference is stated as significant at the .001 level, the obtained difference would happen by sampling error 0.1% of the time or less. To summarize:

1. If a difference would happen 5% of the time or less by sampling error but is not large enough so that it reaches the 1% level, we state that the difference is significant at the .05 level, or $p \leq .05$ that the obtained difference would happen by sampling error, or $.05 > p > .01$ (probability of the obtained difference happening by sampling error is between .05 and .01).
2. If a difference would happen 1% of the time or less by sampling error but is not large enough so that it reaches the 0.1% level, we state that the difference is significant at the .01 level, or $p \leq .01$ that the difference would happen by sampling error, or $.01 > p > .001$.
3. If a difference would happen 0.1% of the time or less by sampling error, we state that the difference is significant at the .001 level, or $p \leq .001$.

We will have occasion in a later section to examine these various significance levels in greater detail and to show how they are involved in making decisions regarding sampling error.

NOTE 10.1

STATISTICAL SIGNIFICANCE VERSUS PRACTICAL SIGNIFICANCE

When we use the term *significant difference*, our definition of *significant* is just as we have explored in the last few pages—it is simply a statement of an arbitrarily chosen probability level.

However, a statistically significant difference may have no social or practical significance whatsoever. As an example, a colleague of mine studying peripheral vision noted that some of his subjects had faster reaction times to stimuli on the left side of their visual field on Mondays and Tuesdays, while other subjects had faster reactions in their right visual field during the latter part of the week. The difference was on the order of 1 millisecond (0.001 seconds) and was significant at the .05 level! Although such individual differences may be of academic interest, the practical importance of a daily variation of 1 millisecond applied to everyday situations seems somewhat trivial. We must distinguish, then, between a trivial difference and one that has practical significance.

The Null Hypothesis and Other Hypotheses

We have pursued the simplest approach in testing the null hypothesis. We eventually conclude that either $\mu_1 - \mu_2 = 0$ or that $\mu_1 - \mu_2 \neq 0$. These statements are sometimes designated as H_0 for the null hypothesis and H_A for the alternate hypothesis. In other words, we assume H_0, $\mu_1 - \mu_2 = 0$, but, if the observed difference between the sample means is large enough to be significant, we reject H_0 in favor of H_A.

It is common error to confuse the alternate hypothesis ($\mu_1 - \mu_2 \neq 0$) with a researcher's *experimental* hypothesis. An important maxim in statistical decision making is that the *rejection of the null hypothesis does not necessarily make an experimental hypothesis true.* For example, a researcher might hypothesize that elementary school pupils who had attended nursery school would get better grades than those who had not attended nursery school, due to the socializing influence of the nursery school setting.

The researcher first assumes the null hypothesis that there is no difference between the academic performance of both groups and then gathers data and runs a t test. If there is a significant difference between the two groups, the researcher can conclude only that $\mu_1 - \mu_2 \neq 0$. It cannot be said, for example, that the children with a nursery school background get better grades because they are more conforming, or more classroom-oriented, or more socially adjusted, on the basis of the statistical test alone. The statistical procedure allows us only to reject the null hypothesis, since such a decision is based on the probability of a difference occurring by sampling error. The researcher cannot go any farther than that and use this approach to prove why a difference exists.

While we are on the subject of null hypotheses, we must note that we do not always assume a null hypothesis H_0: $\mu_1 - \mu_2$ that is equal to zero. Occasionally, we may wish to test our obtained difference, $\overline{X}_1 - \overline{X}_2$, against some other null hypothesis. For example, the usual difference in yields between two varieties of corn might be 17 bushels per acre as reported from different experimental seed plots in Iowa and Nebraska. However, a test of the two varieties under different rainfall and soil conditions in several western states might have produced a sample difference of $\overline{X}_1 - \overline{X}_2 = 26$ bushels per acre. A *t* test could then be run on this difference to see whether the null hypothesis H_0: $\mu_1 - \mu_2 = 17$ bushels, should be rejected.

If other null hypotheses than $\mu_1 - \mu_2 = 0$ are tested, the formula for *t* needs to be altered to include your hypothesized value of $\mu_1 - \mu_2$ in place of the 0 in formula 10.4. But, it is probably true that most null hypotheses that we test in education and the behavioral sciences are of the $\mu_1 - \mu_2 = 0$ variety.

Assumptions Underlying the *t* Test

We noted in Chapter 9 that there were certain restrictions or assumptions about the data that were necessary before a Pearson *r* could be interpreted meaningfully. In a similar fashion, the *t* test has four restrictions that must be met before the *t* can be interpreted in the manner described in the preceding sections. These are

1. The scores must be interval or ratio in nature.
2. The scores must be measures on random samples from the respective populations.
3. The populations from which the samples were drawn must be normally distributed.
4. The populations from which the samples were drawn must have approximately the same variability (homogeneity of variance).

Since these assumptions will be popping up in a number of other contexts, let us look at each one in some detail.

INTERVAL OR RATIO DATA. Since means and standard deviations are calculated in the *t* test, the data must be at least interval in nature.

RANDOM SAMPLES. The statistical theory underlying the *t* test rests heavily on an assumption of random sampling, and this assumption must be met.

NORMALLY DISTRIBUTED POPULATIONS. In most cases, this is a reasonable assumption, since we are fairly certain that the majority of psychological and physiological characteristics are normally distributed. Note that this is *not* the same as saying that the *samples* are normally distributed.

HOMOGENEITY OF VARIANCE. There are statistical techniques available to test the samples for homogeneity of variance, but these techniques are beyond the scope of this book. If there are approximately the same number of observations in both groups, we can "eyeball" the standard deviations, and, if one is more than twice as large as the other, we should question whether the homogeneity of variance assumption has been met.

In summarizing the four assumptions, we are happy to note that the *t* test will give fairly accurate results, even if these assumptions have been violated to a certain degree. The *t* test is extremely resistant to departures from normality and, to a lesser degree, departures from homogeneity of variance. If you are in doubt about whether you can run a *t* test on some data because these assumptions may not have been met, it may be possible to do so simply by requiring a more stringent significance level (e.g., .01 instead of .05).

NOTE 10.2

THE ORIGIN OF "STUDENT'S *t* DISTRIBUTION"

In the 1900s an Englishman by the name of W. S. Gossett was using sampling techniques to ensure quality control at a brewery. The typical sampling procedure involved using a normal curve distribution to set up a confidence interval for estimating the population mean, μ. This procedure required that the researcher take large samples in order to have a sample variance that was a reliable estimate of the population variance, σ^2. Presumably, Gossett tired of the tedious effort involved in taking large samples and developed the *t* distribution, which enabled the researcher to estimate μ with much smaller samples. He published his theory in 1908 under the pseudonym "Student," and the distribution he developed is still occasionally called "Student's *t* distribution."

Testing for Significant Differences. The *t* Test for Two Correlated Samples

The *t* test that was described in previous sections is intended for samples that are *independent* samples or *uncorrelated* samples. But there are times when we would like to see if there is a significant difference between the means of *correlated* samples, and this section describes a *t* test for either *matched samples* or *repeated measurements of the same subject*.

Matched Samples

We noted in the *t* test for *independent* samples that subjects were assigned at *random* to one of two groups or *random* samples were taken from two populations. With matched samples, we do not depend on random assignment to get our samples but construct them carefully to make sure they are equal on any variables that affect what we are measuring. We can do this by matched pairs, split litters, and co-twin controls.

MATCHED PAIRS. When matched pairs are used, they are usually matched on the basis of some variable that correlates highly with what we are measuring. For example, if we wanted to see if college freshmen that come from metropolitan high schools get better grades in college than those that come from rural schools, we would probably decide to use college entrance exam scores as the matching variable. For example, a person with a high entrance exam score from an urban school would be matched with a person with a high score from a rural school. In this manner, we would feel quite certain that both groups of matched pairs are equal in ability at the start and that any differences in future college grades would be due to the environmental setting from which they came.

SPLIT LITTERS. Many animal experiments involve this method, which resembles the matched-pair method above. One member of a litter of kittens, for example, may be placed in an experimental group and another of the same litter placed in the control group. This method would ensure that both groups are alike with respect to certain hereditary and prenatal factors that might influence the variable that we are measuring.

COTWIN CONTROLS. The purpose of the cotwin control is very similar to that of the split-litter technique, except that human identical twins are used. In many studies of maturation and development, one child is assigned to a control group and its twin to the experimental group. Any differences between the groups as a result of some experimental procedure would then be attributed to the independent variable, since both groups are similar in hereditary and maturational factors.

One way to ensure that both groups are equal before some experimental procedure is attempted would be to have the same person serve in both groups. For example, if we were to investigate whether a person has greater hand steadiness with the preferred or the nonpreferred hand, we could test each subject under both conditions: once with the preferred hand and once with the nonpreferred hand. At least we do not have to worry about both groups being equal at the start, since both "groups" consist of the same people!

Calculating a Correlated *t* Test: Direct Difference Method

Whether we use matched pairs or repeated measurements, we wind up with one measurement in each condition. This is *correlated* data, since if one member of a pair scores high in one condition the matched partner would have a tendency to score high in the other condition. Obviously, the same would hold true for repeated measures, where a person scoring high in one condition would tend to score high in the other condition.

There are a number of computational methods for calculating a correlated t test, but one of the simplest is called the *direct difference* method, and it is shown in Table 10.3. A matched-group design was used to investigate reaction time under two different conditions. One group of subjects (two-choice condition) pressed a switch with their thumb whenever a red light came on and pushed another switch with their little finger whenever a green light came on. The other group (three-choice condition) had a similar task, but they also pressed a third switch with their middle finger whenever an amber light came on. The 20 subjects were grouped into 10 pairs on the basis of a pretest in which they pressed a button in response to a tone. Their reaction times for the two-choice and three-choice conditions are shown in hundredths of seconds in Table 10.3.

Note that in the direct difference method, we are working only with the differences between the pairs of observations in the difference column (D). The formula for the mean of the differences would be

$$\overline{D} = \frac{\Sigma D}{N} \tag{10.5}$$

where

ΣD is the sum of the differences (D) column
N is the number of pairs

In Table 10.3, the algebraic sum of the D column is 27, so the mean of

TABLE 10.3. **Direct Difference Method for Correlated *t* Test**

Subject Pair	Two-Choice	Three-Choice	D	D^2
A	37	39	2	4
B	39	45	6	36
C	35	34	−1	1
D	41	43	2	4
E	32	37	5	25
F	35	38	3	9
G	36	35	−1	1
H	39	45	6	36
I	40	42	2	4
J	38	41	3	9
	372	399	27	129
	$\overline{X}_1 = 37.2$	$\overline{X}_2 = 39.9$	$\overline{D} = 2.7$	

Mean of the Differences:

$$\overline{D} = \frac{\Sigma D}{N} = \frac{27}{10} = 2.7$$

Standard Deviation of the Differences:

$$S_D = \sqrt{\frac{\Sigma D^2}{N} - \overline{D}^2} = \sqrt{\frac{129}{10} - (2.7)^2}$$

$$= \sqrt{12.9 - 7.29} = \sqrt{5.61}$$

$$S_D = 2.37$$

Standard Error of the Mean for the Differences:

$$s_{\overline{X}_D} = \frac{S_D}{\sqrt{N-1}} = \frac{2.37}{\sqrt{10-1}} = \frac{2.37}{3} = 0.79$$

Correlated t Test:

$$t = \frac{\overline{D}}{s_{\overline{X}_D}} = \frac{2.7}{0.79} = 3.42$$

t_{01}, $df = 9$, is 3.250
Significant, $p < .01$

the differences, $\overline{D}$, is 2.7. As you can see, this is the same as the difference between the means of the scores, 39.9 and 37.2.

You calculate the standard deviation of the differences in the usual way, by squaring each difference to obtain the D^2 column and substituting ΣD^2 into the slightly modified version of our ordinary standard deviation formula,†

†The machine formula for S (formula 5.4) could be used as well, with similar modifications for ΣD and ΣD^2.

$$S_D = \sqrt{\frac{\Sigma D^2}{N} - \overline{D}^2} \tag{10.6}$$

where
ΣD^2 is the sum of the D^2 column
$\overline{D}$ is the mean of the differences
N is the number of pairs

The standard deviation of the differences, S_D, in Table 10.3 is 2.37.

We next calculate the standard error of the mean for the differences, $s_{\overline{X}_D}$, using formula 10.7.

$$s_{\overline{X}_D} = \frac{S_D}{\sqrt{N - 1}} \tag{10.7}$$

where
S_D is the standard deviation of differences
N is the number of pairs

Note that this formula is similar to an earlier one, for the standard error of the mean, where a standard deviation is divided by $\sqrt{N - 1}$. In Table 10.3, $s_{\overline{X}_D}$ is calculated to be 0.79.

Finally, we calculate the correlated t test by formula 10.8,

$$t = \frac{\overline{D}}{s_{\overline{X}_D}} \tag{10.8}$$

where
$\overline{D}$ is the mean of the differences
$s_{\overline{X}_D}$ is the standard error of the mean of the differences

For the data of Table 10.3, $t = 3.42$. For 9 degrees of freedom and a two-tailed test, a t of 3.250 is significant at the .01 level, and so we declare our t significant, $p < .01$.

Errors in Making Decisions: The Type I and Type II Errors

On several occasions, both in this chapter and in Chapter 7, we have examined in detail the problem of deciding whether a particular experimental result was due to sampling error or whether the observed difference between the sample means represented a *real* difference in the variables under study. We have noted that if a result could occur by *chance* 50% of the time, or 13%, or 6%, we will still attribute the results to sampling error. And, as was pointed out earlier, statisticians have conventionally used the 5% level as the cutting point; that is, if a certain result happens 5% of the time or less by chance, we will say that it is *not* sampling error but the result is a real one.

These statements should be very familiar to you by now, but let us look closely at the problems that arise when we are confronted with decisions between sampling error and real results. Suppose that we run a study to see if there is any difference in reading proficiency between left-handers and right-handers. As usual, we assume the null hypothesis ($\mu_1 - \mu_2 = 0$), gather a sample of left- and right-handed school children, administer a reading proficiency test, and perform the necessary calculations. We now have to decide whether our findings are "significant."

Let us pause a moment to consider the choices confronting us. As usual, we are faced with the two familiar possibilities:

1. There is no difference between left- and right-handers (H_0: $\mu_1 - \mu_2 = 0$ is true), and the obtained difference is one of those rare occurrences when a difference that large is due to sampling error.
2. There is indeed a difference between left- and right-handers (H_0: $\mu_1 - \mu_2 = 0$ is *not* true).

If we find that our t value will occur 5% of the time or less just by sampling error, we will very likely choose (2) above and decide that there is a real difference between left- and right-handers, rejecting the null hypothesis. *But we could be wrong!* We just might have been unlucky enough to get one of those rare occurrences that happen 5% of the time or less even when the null hypothesis is true. If there is no difference between the means of the populations from which the samples were drawn, then $\mu_1 - \mu_2 = 0$ and we have made a mistake by calling our difference a significant one instead of sampling error. This mistake is called a *Type I error.* Formally defined, a Type I error is committed *if the null hypothesis is rejected when it actually is true.*

So, what are we to do in order to avoid making a Type I error? In an earlier section, we noted that there are different significance levels, and as cautious, conservative researchers, we might very well demand that our t value be significant at the .01 level, or even the .001 level, in order to be more certain that we do not label as significant a difference that really is only sampling error.

But look what happens when we do this. Suppose we choose (1) above and decide that our finding is due to sampling error—*but it may be a real difference!* That is to say, there may be a difference between the means of the two populations that we have decided to attribute to sampling error. This error is called a *Type II error.* A Type II error is committed *if the null hypothesis is accepted when actually it is false.*

We obviously have a real dilemma here, because the more we try to avoid making a Type I error by demanding greater significance, the greater are the chances of our making a Type II error. And, if we scrupulously avoid the taint of a Type II error by not being quite so strict, the possibility of our making a Type I error is back again. Before looking at ways in which this problem is handled, let us review this complex decision process in the form of a *decision matrix* (shown in Table 10.4) and consider each of the four decision possibilities.

TABLE 10.4. **Decision Matrix for Rejecting or Not Rejecting H_0**

	Decision on the Basis of Sampling	
	Reject H_0	Accept H_0
H_0 is True in population	Type I error $p = \alpha$	correct
H_0 is False in population	correct	Type II error $p = \beta$

H_0 IS REJECTED WHEN IT IS, IN FACT, TRUE. The upper left square of the matrix illustrates the Type I error, where the researcher decides that the t value is significant and rejects the null hypothesis when it really is true. The probability of making a Type I error is the same as the value of the significance level that the researcher chooses. (The Greek letter alpha, α, is used to indicate significance levels, so α can be .05, .01, .001, etc.) Thus, if $\alpha = .05$, for example, we say that the probability of a Type I error occurring is .05. A moment's reflection should show why this is true. Since 5% of the sampling distribution (with $\overline{D}_{\overline{X}} = 0$) is in the region of rejection when the researcher chooses $\alpha = .05$, if all differences were labeled as significant, he or she would be wrong 5% of the time (i.e., whenever the difference fell in this region). One thing must be noted: The probability of a Type I error is under our direct control, since we are responsible for setting the significance level.

H_0 IS ACCEPTED WHEN IT IS TRUE. The upper right square indicates a decision to accept H_0, which was a correct decision, since H_0 is, in fact, true. We must be careful to note that a finding of nonsignificance does not *prove* that the null hypothesis is true. It is evidence, but it is not conclusive proof.

H_0 IS ACCEPTED WHEN IT IS FALSE. The lower right square of the decision matrix illustrates the Type II error, where the researcher decides to accept the null hypothesis, when, in fact, the null hypothesis is false. The probability of making a Type II error is noted by the Greek letter beta, β. In an introductory textbook it is just not possible to go into the characteristics of β, but we should note that the relationship between α and β is not a simple one. However, it is true that as we decrease the probability of a Type I error we increase the probability of a Type II error.

H_0 IS REJECTED WHEN IT IS FALSE. The lower left square indicates the correct decision to reject the null hypothesis when it is false. We gave only a passing comment to the other correct decision (accepting H_0 when it was true), but we need to devote more time to the correct rejection of the null hypothesis. The typical researcher is very much concerned with correct rejection of the null hypothesis, since we basically are seeking to establish significant relationships. In other words, if there is a difference between two variables out in the population, we want to find it. Therefore, we want to use a statistical test that is sensitive to these differences. *The ability of a test to reject the null hypothesis when it is false is called the "power" of a test.* Power (defined as $1 - \beta$) is a characteristic of any statistical test; the more power a statistical test has, the more likely it is to detect significant differences if they exist. So, all things being equal, a researcher will choose a statistical test that has the greatest amount of power. We will have occasion to examine other tests besides the *t* test in later chapters, and we will note the power of these tests.

Resolving the Sampling Error–Real Difference Dilemma

You have probably gathered by now that there is no easy way of determining whether an observed difference is sampling error or a real difference. But we can ask ourselves if we prefer a Type I error or a Type II error! Our choice of which error to risk is determined by the nature of the subject matter being studied.

On important theoretical issues, such as demonstrating the existence of ESP, we might wish to set our significance levels very low ($\alpha = .01$ or .001). We do not want to risk our laboratory or scientific career by stating that we found some weird or bizarre result to be significant when it was really due to sampling error. In this instance we definitely prefer to avoid a Type I error (by reducing our α to .01 or .001) and to risk a Type II error.

If, on the other hand, a researcher is examining two poultry disinfectants for germ-killing ability and chicken farmers everywhere are in dire need of some kind of disinfectant, we do not really care very much if we pronounce brand A better than brand B. If we say that one is better than the other, when actually there is no difference, we are making a Type I error, but since some kind of disinfectant is sorely needed, no great problem arises from our error. Either brand A or brand B will do the job.

One very practical approach to the problem involves setting up a middle ground or buffer zone between sampling error and real differences. Some researchers suggest that any difference that will occur by chance 1% of the time or less ($p \leq .01$) be called "significant" and any difference that happens by chance more than 10% of the time ($p > .10$) be labeled "not significant." Any result that is in between the .01 level and the .10 level should lead to a state of "suspended judgment," which demands repetition of the experiment and verification by other researchers. Although such an approach has not been universally accepted, it has a lot to recommend it, and it may eventually become one of our statistical conventions.

Correlated Versus Uncorrelated *t* Tests and Type II Errors

If instead of using the direct difference method on the data of Table 10.3 we had mistakenly used the *t* test for random samples, our *t* value would have been 1.79, which is not significant ($p > .05$), rather than 3.42, which is significant at the .01 level. The reason for the larger *t* when we are using the correlated *t* test is that the method has a built-in correction for the amount of correlation between the matching variable and the actual measures. If it turned out that the correlation between the two were actually 0, both the direct difference method and the *t* test for random samples would yield the same value of *t*.

The larger *t* with the correlated *t* test means that you are more likely to reject the null hypothesis and *less likely to make a Type II error*. You are more likely to pick up a difference, if it exists, with a correlated *t* test.

Then why not use a correlated design with its matched pairs all the time? The answer to this question lies in the number of degrees of freedom for the two different approaches. As we saw earlier in the chapter, the degrees of freedom for a *t* test for independent samples was $df = (N_1 - 1) + (N_2 - 1)$. Note that this is twice as many degrees of freedom as we have in the correlated *t* test, where $df = N - 1$ and N is the number of pairs. As a result, a somewhat higher *t* value is needed for significance at a given level with the correlated *t* test.

Consequently, there should be a substantial correlation between the variable that is used for matching the pairs (the independent variable) and the variable that is used for the measurements to be made (the dependent variable). For example, a researcher studying the reading comprehension of adults in the Midwest versus adults in the Pacific Northwest would not match pairs on the basis of height or weight, since there is no relationship between reading comprehension and these physical measures. As explained in the previous paragraph, using these as a matching variable would only reduce the number of degrees of freedom in the *t* test.

One-Tailed Versus Two-Tailed Tests

The significance tests that we have discussed so far have all been *two-tailed* tests. This means that the investigator proposes a null hypothesis that $\mu_1 = \mu_2$, so, if he decides to reject the hypothesis, it may be that either $\mu_1 > \mu_2$ or $\mu_2 > \mu_1$. In other words, the null hypothesis will be rejected if the *t* value (or *z*, or other statistics to be discussed in future chapters) is either to the extreme left of the sampling distribution or to the extreme right. In Figure 10.1 you saw that the region of rejection at the .05 level was to the left of a *z* of -1.96 or to the right of a *z* of 1.96. Since both of the "tails" of the sampling distribution are involved in this decision, such an approach is called a *two-tailed* test. The null hypothesis states that $\mu_1 = \mu_2$, and the rejection of it leads us to accept the alternate hypothesis that $\mu_1 \neq \mu_2$. We are *not* stating the *direction* of the difference when we say that $\mu_1 \neq \mu_2$, only that there *is* a difference. If we are using the .05 level of significance, our *t* value is significant in a two-tailed test if it falls in either the top 2.5% or the bottom 2.5% of the sampling distribution of differences.

The One-Tailed *t* Test

If an investigator predicts before collecting her data that μ_1 is greater than μ_2, the alternate hypothesis is not $\mu_1 \neq \mu_2$ but $\mu_1 > \mu_2$. Note that this not only states that there will be a difference but predicts the direction of the difference as well. Hence it is sometimes called a *directional* hypothesis. Since our interest is in only those $\overline{X}_1 - \overline{X}_2$ differences that are positive, the region of rejection is now confined to the right end of the sampling distribution. This region of rejection is at one end of the sampling distribution only, and we have what is called a *one-tailed test* (as shown in Figure 10.6).

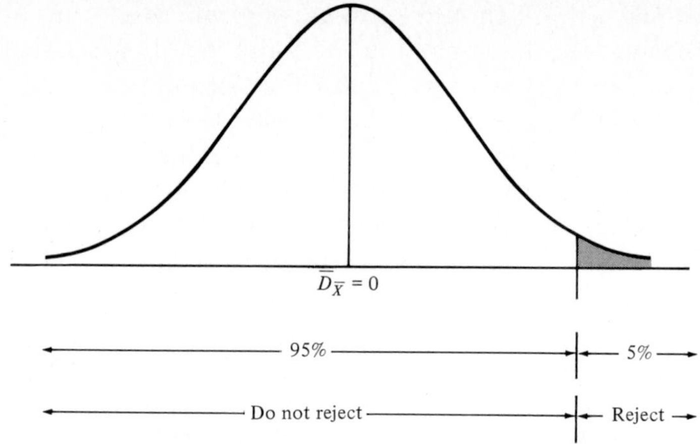

FIGURE 10.6. **Region of rejection for a one-tailed test.**

When is an investigator able to predict that the difference, if it exists, will be only in one direction? The answer is not always that obvious, but in some cases a difference in the opposite direction would be virtually impossible because of prior information regarding physiology, developmental stages, or maturation. For example, if we were investigating the effect of vitamin C on the incidence of common colds, we would certainly have a directional hypothesis and would specify a one-tailed test. The alternate hypothesis would be $\mu_{NoC} > \mu_C$, that is, that those not taking vitamin C would have more colds than those taking the vitamin. It would seem unreasonable to hypothesize that $\mu_C > \mu_{NoC}$, that those taking the vitamin would have more colds! As a result, we are more efficient by specifying $\mu_1 > \mu_2$ as an alternate hypothesis instead of $\mu_1 \neq \mu_2$, since this approach places the region of rejection all in one tail of the sampling distribution of differences.

If a one-tailed approach is used, the t test is computed by *always subtracting the sample mean that was predicted to be the smaller from the sample mean that was predicted to be the larger*. In other words, the one-tailed test makes use of positive $\overline{X}_1 - \overline{X}_2$ differences only. In the two-tailed test it was immaterial whether we subtracted $\overline{X}_1$ from $\overline{X}_2$ or $\overline{X}_2$ from $\overline{X}_1$, but in the one-tailed test we *always* subtract in the direction of our prediction. Of course, if the difference turns out to be negative, this means that our results are opposite to what we predicted, and we can forget about calculating the t test entirely.

Let us examine some of these concepts through the use of an example. An educational psychologist is interested in the effects of knowledge of results in a classroom setting. She assigns 10 children at random to the knowledge-of-results (KR) group and 10 to the no-knowledge-of-results (NKR) group. In the KR group each child's arithmetic paper is scored and returned immediately after the daily test is completed, while in the NKR group children do not get their tests back until the next day. After a 2-week period, a unit exam is given over the same material. The errors for each student are shown in Table 10.5. Is there significant difference in errors between the two groups?

TABLE 10.5. **Calculating the One-Tailed *t* Test for Error Scores as a Function of Knowledge of Results**

KR Group	NKR Group
1	4
2	3
2	1
2	2
4	4
2	7
2	3
1	2
2	1
2	4
$\Sigma X = 20$	$\Sigma X = 31$
$\Sigma X^2 = 46$	$\Sigma X^2 = 125$
$\overline{X}_1 = 2.0$	$\overline{X}_2 = 3.1$

$$S_1 = \frac{1}{N}\sqrt{N\Sigma X^2 - (\Sigma X)^2} \qquad S_2 = \frac{1}{N}\sqrt{N\Sigma X^2 - (\Sigma X)^2}$$

$$= \frac{1}{10}\sqrt{10(46) - (20)^2} \qquad = \frac{1}{10}\sqrt{10(125) - (31)^2}$$

$$= \frac{\sqrt{60}}{10} = \frac{7.74}{10} \qquad = \frac{\sqrt{289}}{10} = \frac{17}{10}$$

$$S_1 = 0.77 \qquad S_2 = 1.7$$

$$s_{D\overline{X}} = \sqrt{\frac{N_1S_1^2 + N_2S_2^2}{N_1 + N_2 - 2}\left(\frac{1}{N_1} + \frac{1}{N_2}\right)} = \sqrt{\frac{10(0.77)^2 + 10(1.7)^2}{10 + 10 - 2}\left(\frac{1}{10} + \frac{1}{10}\right)}$$

$$s_{D\overline{X}} = \sqrt{\frac{5.9 + 28.9}{18}(0.2)} = \sqrt{.3867} = 0.62$$

$$t = \frac{\overline{X}_2 - \overline{X}_1}{s_{D\overline{X}}} = \frac{3.1 - 2.0}{.62} = 1.77$$

t_{05}, $df = 18$, is 1.73
Significant, $p < .05$

The researcher will likely use a one-tailed test in this situation, since KR has been shown in previous research to aid the learning process. Certainly, we would not expect KR to retard learning! So, the usual null hypothesis would be $\mu_{NKR} = \mu_{KR}$, and the alternate hypothesis will be $\mu_{NKR} > \mu_{KR}$ (i.e., the mean number of errors for the population receiving NKR will be greater than that for the population receiving KR).

The calculations for the one-tailed t test are shown in Table 10.5. The procedure is identical to that for the two-tailed tests of Tables 10.1 and 10.2. The only difference is that $\overline{X}_1$ must be subtracted from $\overline{X}_2$ in computing t, since that was the direction hypothesized by the researcher.

After t is calculated, we enter Table E of Appendix 4 with $df = 18$ and obtain the one-tailed values. The tabled value at the .05 level is 1.73, and our t of 1.77 is larger, so we conclude that there is a significant difference in the number of errors between the NKR and KR groups.

It should be noted that a one-tailed t has more *power* than a two-tailed t; that is, with the one-tailed t test, the null hypothesis is more likely to be rejected if it should be rejected. This is true, however, *only if the direction of the difference is predicted in advance of collecting the data*. If the researcher on knowledge of results originally wanted to see if there was *any* kind of difference but, after seeing the first three exam papers, proposed a directional hypothesis, this would be an unacceptable foundation for a one-tailed test. A two-tailed test should have been run under those conditions. It is probably safe to conclude that a two-tailed test should be run unless we are willing to retain the null hypothesis even when the results are extreme in the *unexpected* direction. If we are willing to adhere to this vow, we may use a one-tailed test, but, in actual practice, it is a rare research situation where we are willing to relinquish the right to report on a significant result opposite to our expectations.

Testing Hypotheses About a Single Sample Mean

Back in Chapter 7 we had occasion to test whether there was a significant difference between the mean of a sample, $\overline{X}$, and some hypothesized population mean, μ. For example, in Sample Problem 1 in Chapter 7, a group of researchers assessed the effect of early childhood intervention on the IQ tested later on in the upper elementary grades. A well-known IQ test with a mean of 100 and a standard deviation of 16 was used to assess intelligence (i.e., $\mu = 100$ and $\sigma = 16$). A group of 40 ten-year-olds from an economically disadvantaged area of a midwestern city was given this IQ test and a mean of 82 was calculated.

We first calculated the standard error of the mean:

$$\sigma_{\overline{X}} = \frac{\sigma}{\sqrt{N}} = \frac{16}{\sqrt{40}} = \frac{16}{6.3246} = 2.53$$

We then set up a 95% CI about the sample mean:

$$\overline{X} \pm 1.96\sigma_{\overline{X}}$$
$$82 \pm 1.96(2.53)$$
$$82 \pm 4.96$$
$$77.04 \text{ to } 86.96$$

We were then able to test the hypothesis that the population mean IQ, μ, was 90. However, we noted early in Chapter 7 that this technique was limited to situations *where we knew the population* standard deviation, σ. We are now ready to consider situations where we do *not* know σ, and we must make two changes in the method described above. First, we will use a slightly different formula for the standard error of the mean. We resort to an alternate formula that permits us to use our *sample* standard deviation, S, and we can then estimate the standard deviation of the sampling distribution of means. The formula for this estimate is

$$s_{\overline{X}} = \frac{S}{\sqrt{N-1}} \tag{10.9}$$

where
$s_{\overline{X}}$ is the standard error of the mean[†]
S is the standard deviation of our *sample*
N is the size of the sample

Note that the above formula calls for $N - 1$ in the denominator, which "corrects" for the fact that the sample standard deviation, S, tends to be smaller than the population standard deviation, σ.

Second, instead of relying on the familiar normal curve with z values of 1.96 and 2.58 for our confidence intervals, we must use the *t* distribution and *t* values from Table E in Appendix 4. The procedure can best be illustrated by an example.

[†]Note that the symbol for the standard error of the mean is $\sigma_{\overline{X}}$ when calculated from the population σ and is $s_{\overline{X}}$ when calculated from the sample standard deviation, S.

Table 10.6 shows the results of a random sample of grade-point averages (GPA) taken from 41 second-semester freshmen at a large university. The sample mean was 2.92 with a standard deviation of 0.60. Notice that we follow the same steps we did in Chapter 7—first calculating the standard error of the mean and then forming the confidence interval.

TABLE 10.6. **Estimating the Grade-Point Average of University Freshmen: Calculating the Standard Error of the Mean When σ Is Unknown**

Sample Data:

$$\overline{X} = 2.92 \qquad S = 0.60 \qquad N = 41$$

Standard Error of the Mean:

$$s_{\overline{X}} = \frac{S}{\sqrt{N-1}} = \frac{0.60}{\sqrt{41-1}} = \frac{0.60}{6.3246} = .09$$

95% Confidence Interval:

$$\overline{X} + t_{05}s_{\overline{X}} = 2.92 \pm 2.021(.09)$$

$$2.92 \pm .18$$

$$2.74 \text{ to } 3.10$$

The only change in procedure is in the confidence interval itself. Since we can no longer assume a normal curve and the z values of 1.96 and 2.58, we must use the *t* distribution and the *t* values associated with the size of the sample. The 95% confidence interval would be

95% CI: $$\overline{X} \pm t_{05}\, s_{\overline{X}} \qquad (10.10)$$

where

$\overline{X}$ is the sample mean

t_{05} is the value of *t* at the .05 level for the associated degrees of freedom

$s_{\overline{X}}$ is the standard error of the mean calculated from the sample standard deviation

The 99% confidence interval would be

99% CI: $$\overline{X} \pm t_{01}\, s_{\overline{X}} \qquad (10.11)$$

where all the terms are as previously defined, except t_{01} would be the value of *t* at the .01 level for the associated degrees of freedom.

So, the only change as far as computation goes, is that we must look up the t value for the degrees of freedom in our sample. For the GPA data of 41 freshmen in Table 10.6, we note that in Table E the t value at the .05 level for $df = 40$ ($df = N - 1 = 41 - 1 = 40$) is 2.021. As you can see, the 95%CI becomes 2.74 to 3.10. If the Director of Institutional Research at this university had hypothesized that the average GPA for the entire freshman class should be 3.00 (i.e., $\mu = 3.00$), we would have no reason to reject the hypothesis, since a mean GPA of 3.00 is in the 2.74 to 3.10 interval.

We constructed confidence intervals for μ on a number of occasions in Chapter 7 but were always limited because we had to know the value of the population standard deviation, σ, in order to calculate the standard error of the mean. But with the t distribution and the formula for the standard error of the mean using the *sample* standard deviation, we are now able to use the confidence interval approach in many practical applications.

Significant Differences Between Other Statistics

This entire chapter has been devoted to testing the significance of the difference between means. A similar approach is used in testing the significance of a difference between medians, proportions, standard deviations, and other statistics. For example, we might want to know if our local political candidate has any chance to win the election if a sample of voters shows that 48% will vote for her. Since she needs 50% of the vote to win, we ask if 48% is only sampling error (we hope!) or if it is the real population value and our candidate stands to be defeated. A test for the significance of a proportion would help us answer our question.

A number of these significance tests are beyond the intended scope of this introductory textbook, and you are referred to one of the more advanced textbooks listed in the References.

SAMPLE PROBLEM 1

A social psychologist is investigating the development of "generosity" in preschool children and would like to see if girls are more generous than boys at age 4. Each child at a day care center is given 16 small pieces of candy and is asked to "put some in a sack for your very best friend." The numbers of candies set aside for "friends" by 12 girls and 10 boys follow. Calculate a t test for random samples to see if there is a significant sex difference in generosity.

Girls X	Girls X^2	Boys X	Boys X^2
7	49	2	4
3	9	3	9
6	36	5	25
9	81	3	9
3	9	4	16
8	64	2	4
6	36	2	4
7	49	1	1
5	25	2	4
9	81	3	9
8	64	27	85
7	49		
78	552		

Means:

$$\overline{X}_1 = \frac{78}{12} = 6.5 \qquad \overline{X}_2 = \frac{27}{10} = 2.7$$

Standard Deviations:

$$S_1 = \frac{1}{N}\sqrt{N\Sigma X^2 - (\Sigma X)^2} = \frac{1}{12}\sqrt{12(552) - (78)^2} = \frac{\sqrt{540}}{12} = \frac{23.24}{12}$$

$$S_1 = 1.94$$

$$S_2 = \frac{1}{N}\sqrt{N\Sigma X^2 - (\Sigma X)^2} = \frac{1}{10}\sqrt{10(85) - (27)^2} = \frac{\sqrt{121}}{10} = \frac{11}{10}$$

$$S_2 = 1.10$$

Standard Error of the Difference:

$$s_{D_{\overline{X}}} = \sqrt{\frac{N_1S_1^2 + N_2S_2^2}{N_1 + N_2 - 2}\left(\frac{1}{N_1} + \frac{1}{N_2}\right)}$$

$$= \sqrt{\frac{12(3.75) + 10(1.21)}{12 + 10 - 2}\left(\frac{1}{12} + \frac{1}{10}\right)}$$

$$= \sqrt{\frac{45 + 12.1}{20}(0.08 + 0.10)} = \sqrt{\frac{57.1(0.18)}{20}} = \sqrt{\frac{10.278}{20}}$$

$$s_{D_{\overline{X}}} = \sqrt{0.5139} = 0.72$$

t Test:

$$t = \frac{(\overline{X}_1 - \overline{X}_2) - 0}{s_{D_{\overline{X}}}} = \frac{6.5 - 2.7}{0.72} = \frac{3.8}{0.72} = 5.28$$

SAMPLE PROBLEM 1 (cont.)

$t_{.001}$, $df = 20$, is 3.850
Significant, $p < .001$

Since our calculated t of 5.28 is greater than the value for the .001 level for 20 degrees of freedom in Table E, we would conclude that the girls set aside significantly more candies than the boys and that the difference in means was significant at the .001 level.

SAMPLE PROBLEM 2

Eight novice bowlers are dissatisfied with their present bowling averages and decide to take a lesson from a professional bowler. For a month after the lesson they calculate their bowling averages again and compare them with the averages before the lesson. Use a t test for correlated samples to see if there was a significant difference between the bowling averages before the lesson and after the lesson.

Bowler	Before	After	D	D^2
A	144	151	7	49
B	126	120	−6	36
C	132	137	5	25
D	143	154	11	121
E	133	132	−1	1
F	128	131	3	9
G	152	149	−3	9
H	126	130	4	16
	1,084	1,104	20	266
	$\overline{X}_1 = 135.5$	$\overline{X}_2 = 138$	$\overline{D} = 2.5$	

Mean of the Differences:

$$\overline{D} = \frac{\Sigma D}{N} = \frac{20}{8} = 2.5$$

Standard Deviation of the Differences:

$$S_D = \sqrt{\frac{\Sigma D^2}{N} - \overline{D}^2} = \sqrt{\frac{266}{8} - (2.5)^2}$$

$$= \sqrt{33.25 - 6.25} = \sqrt{27}$$

$$S_D = 5.20$$

Standard Error of the Mean for the Differences:

$$s_{\overline{X}_D} = \frac{S_D}{\sqrt{N-1}} = \frac{5.20}{\sqrt{8-1}} = \frac{5.20}{2.65} = 1.96$$

Correlated t Test:

$$t = \frac{\overline{D}}{s_{\overline{X}_D}} = \frac{2.5}{1.96} = 1.28$$

t_{05}, $df = 7$, is 2.365
Not significant, $p > .05$

Since our calculated t of 1.28 is less than that required for significance at the .05 level for a two-tailed test, we conclude that there was no significant difference in the average bowling scores after the lesson.

STUDY QUESTIONS

1. How different must two means be before you know whether or not you have a "real" difference?
2. Under the null hypothesis, the sampling distribution of differences is normally distributed about $\overline{D}_{\overline{X}} = 0$ with a standard deviation of $\sqrt{\sigma_{\overline{X}_1}^2 + \sigma_{\overline{X}_2}^2}$. What does this mean?
3. A null hypothesis is often stated as $\mu_1 - \mu_2 = 0$. How is this statement related to $\overline{D}_{\overline{X}} = 0$ in Study Question 2?
4. Sketch a sampling distribution of differences with $\overline{D}_{\overline{X}} = 0$ and show the regions of rejection and nonrejection.
5. What do we mean when we say a difference is "significant at the .05 level."
6. A difference is labeled "not significant." What does this statement mean?
7. How does the t distribution differ from the normal curve?
8. What is the difference between $\sigma_{D\overline{X}}$ and $s_{D\overline{X}}$?
9. What is the distinction between statistical significance and practical significance?
10. List the assumptions for the t test.
11. Distinguish between independent samples and correlated samples.
12. Distinguish between a Type I error and a Type II error.
13. Under what conditions would you be willing to settle for one type of error rather than the other?

14. What is the difference between a one-tailed test and a two-tailed test? Which would we be most likely to use?

EXERCISES

1. Weight gains are calculated for 12 rats in one experimental group and 17 in another. A researcher finds a *t* of 2.52 for these data. Is there a significant difference between the means at the .05 level? At the .01 level?

2. A child psychologist has 7 children in an experimental group and 9 children in a control group. She obtains a *t* value of 2.08. Is there a significant difference between the means at the .05 level? At the .01 level?

3. We noted in Chapter 5 that in the National Institutes of Health study on the effects of diet and exercise on weight loss, a group of 17 women both dieted and exercised while another group of 34 women dieted only. The means and standard deviations of the amount of weight lost in pounds over a 15-week period are shown below. Was there a significant difference between the mean weight loss for the two groups? (Use a two-tailed test.)

Diet and Exercise	**Diet Only**
$N_1 = 17$	$N_2 = 34$
$\overline{X}_1 = 18.0$	$\overline{X}_2 = 10.5$
$S_1 = 5.62$	$S_2 = 3.48$

4. The College Life Questionnaire mentioned earlier also surveyed the prevalence of Type A behavior and grade-point averages. Shown below are the means and standard deviations of the GPAs for a sample of 15 Type A students and 20 who were judged to be Type B. Was there a significant difference in the GPA's of these two groups? (Use a two-tailed test.)

Type A GPA	**Type B GPA**
$N_1 = 15$	$N_2 = 20$
$\overline{X}_1 = 3.2553$	$\overline{X}_2 = 2.8975$
$S_1 = 0.2792$	$S_2 = 0.6589$

5. A relationship between Type A behavior and physical illness in college students was discovered recently (Woods and Burns, 1984) by having Type A and Type B students fill out the Physical Symptoms Checklist (PSC). The PSC is a list of physical illness symptoms, and the score is determined by the number of symptoms and the frequency of each symptom noted during the last 12 months. In a follow-up study, six Type A and five Type

B college men were given the PSC. Use a two-tailed test to see if there was a significant difference between the PSC scores of the two groups.

Type A	Type B
21	15
29	6
42	6
30	17
26	29
37	

6. A wheat farmer wanted to see if there was any difference between using expensive certified seed and using his own wheat. He planted 7 fields with the certified seed and 6 fields with his own wheat. At harvest he determined the yield in bushels per acre for the 13 fields. Was there any difference between the certified seed and the farmer's own wheat? (Use a two-tailed test.)

Certified	Own
33	30
34	26
34	29
29	30
35	27
31	32
35	

7. The resting heart rates were measured in a sample of women smokers and nonsmokers at Pennsylvania State University. There were 8 smokers and 17 nonsmokers in the sample. Was there a significant difference in the resting heart rates of the two groups? (Use a two-tailed test.)

Smokers	Nonsmokers	
78	72	62
100	82	66
88	62	68
62	84	96
94	61	58
88	68	87
76	72	80
90	64	78
	76	

8. A self-disclosure inventory was used by two researchers to see if male college students would report intimate and personal details differently to a male interviewer than a female interviewer (Reynolds and McDuffy, 1985). Twelve college men filled out the inventory while talking to a female interviewer and another 12 completed the inventory while talking to a male interviewer. Was there a difference in the number of details reported by the two groups? (Use a two-tailed test.)

Female Interviewer		Male Interviewer	
20	20	18	16
20	18	16	14
20	16	19	16
16	20	20	14
16	20	17	20
17	16	16	20

9. On the first day of class, 10 students in a wellness-fitness course measured their heart rate before exercise and again after stepping up and down on their desk chair 3 times. The results are shown below. Did the short amount of exercise cause a significant increase in heart rate? (Use a two-tailed test.)

Student	Before	After	Student	Before	After
A	73	80	F	96	108
B	70	99	G	61	69
C	69	76	H	62	60
D	76	100	I	62	59
E	76	84	J	84	87

10. The depth perception test mentioned earlier where subjects attempted to align two vertical rods from a distance of 10 feet was given to 12 members of a high school driver education class. In one condition they held their heads steady (Fixed) and in the other, they could move their heads from side to side (Moving). The amount of error in millimeters is shown below for the two conditions. Was there a significant difference in the amount of error in the two conditions? (Use a two-tailed test.)

Student	Fixed	Moving	Student	Fixed	Moving
A	16	8	G	7	10
B	14	10	H	15	9
C	16	15	I	7	9
D	17	16	J	31	17
E	14	10	K	28	15
F	21	13	L	9	8

CHAPTER 11

One-Way Analysis of Variance

The t tests described in the last chapter have one very serious limitation—they are restricted to tests of the significance of the difference between only *two* groups. Certainly, there are many times when we would like to see if there are significant differences among three, four, or even more groups. For example, we may want to investigate which of three teaching methods (lecture, discussion, or programmed textbook) is best for a first-year algebra class or which of four brands of pain killers works best for dental trauma. In these cases we cannot use the ordinary t test, because more than two groups are involved.

We *cannot* solve the problem by running a t test on two groups at a time. If we have three means, $\overline{X}_1$, $\overline{X}_2$, and $\overline{X}_3$, we cannot use the ordinary t test first on $\overline{X}_1$ and $\overline{X}_2$, then on $\overline{X}_1$ and $\overline{X}_3$, and finally on $\overline{X}_2$ and $\overline{X}_3$. The reason this is an invalid procedure is that the probabilities associated with obtaining various t values given in Table E in Appendix 4 are for pairs of means from *random* samples. If we have a number of pairs of means to be compared, we definitely are *not* choosing two of them at random for the t test since, if we have chosen to compare, say, $\overline{X}_1$ with $\overline{X}_2$ first, then our next two choices must necessarily be $\overline{X}_1$ with $\overline{X}_3$ and $\overline{X}_2$ with $\overline{X}_3$. Thus the probability values given in Table E in Appendix 4 are not applicable.

It is for this reason that we now consider one of the most useful techniques in statistics—the analysis of variance (abbreviated AOV or ANOVA). This technique allows us to compare two or more means to see if there are significant differences between or among them. The analysis of variance is used in a wide variety of applications and in varying degrees of complexity by researchers in such diverse fields as psychology, agriculture, education, and industrial engineering. It is such an important part of the professional's repertoire that at least superficial acquaintance with ANOVA is essential for anyone in education and the behavioral sciences. Since there are entire textbooks and two-semester courses devoted to ANOVA, we will just barely scratch the surface in applying this statistical tool. But we will become acquainted with the introductory concepts and build a foundation for further course work. In this chapter we will consider the most elementary form of ANOVA—the *simple analysis of variance*, sometimes called the *one-way classification analysis of variance.*

The Concept of Variance Revisited

Before beginning the discussion, it might be a good idea to review briefly the concept of variance. We noted in Chapter 5 that the variance is a measure of variability based on the squared deviations from the mean. The numerator of the formula is the familiar $\Sigma(X - \overline{X})^2$, which is called the *sum of squares*. You will recall from Chapter 5 that the sum of squares is the result of subtracting the mean from each score to obtain the deviation $(X - \overline{X})$, squaring each deviation $(X - \overline{X})^2$, and finally summing the squared deviations to obtain $\Sigma(X - \overline{X})^2$. In the example of Table 11.1, $\Sigma(X - \overline{X})^2 = 118$.

TABLE 11.1. **Calculating the Unbiased Estimate of the Population Variance**

X	$(X - \overline{X})$	$(X - \overline{X})^2$
12	6	36
7	1	1
9	3	9
2	−4	16
10	4	16
7	1	1
1	−5	25
3	−3	9
4	−2	4
5	−1	1
60	0	$\Sigma(X - \overline{X})^2 = 118$

$$\overline{X} = 6$$

$$s^2 = \frac{\Sigma(X - \overline{X})^2}{N - 1} = \frac{118}{9} = 13.11$$

The denominator of the variance formula, for the unbiased estimate, is the degrees of freedom, $N - 1$. As you can see from Table 11.1, the variance, s^2, is obtained by division of the sum of squares by the degrees of freedom, or $\frac{118}{9} = 13.11$. This value is a measure of variability for these 10 scores, and s^2 would be smaller for a group of scores that deviated less from the mean and larger for a group that deviated more.

Note that the sum of squares, $\Sigma(X - \overline{X})^2$, is not itself a measure of variability, since the size of $\Sigma(X - \overline{X})^2$ depends not only on the extent of the deviations from the mean but also on the size of the sample. Thus it is necessary to divide by the degrees of freedom to obtain a sort of average.

Sources of Variation

Let us use a hypothetical example to show graphically the basic structure of the one-way classification ANOVA. A researcher wants to know which of four methods of teaching introductory psychology produces the best results: (1) lecture, (2) films and videotapes, (3) discussion groups, or (4) self-study with a programmed text. A total of 200 college students is available for the research project, and 50 are assigned at random to each of the four groups. After the semester's course work is completed under each of the four different techniques, all students are given a final exam covering basic psychological principles.

To make our graphic analysis easier, let us assume that the 50 scores in each group are normally distributed and that there is no overlapping of any of the groups. This nonoverlapping, of course, would never happen in practice, but it is easier to see the *sources of variation* in this way.

Figure 11.1 shows the four frequency polygons for the final examination scores. Note that the self-study group had the highest scores and

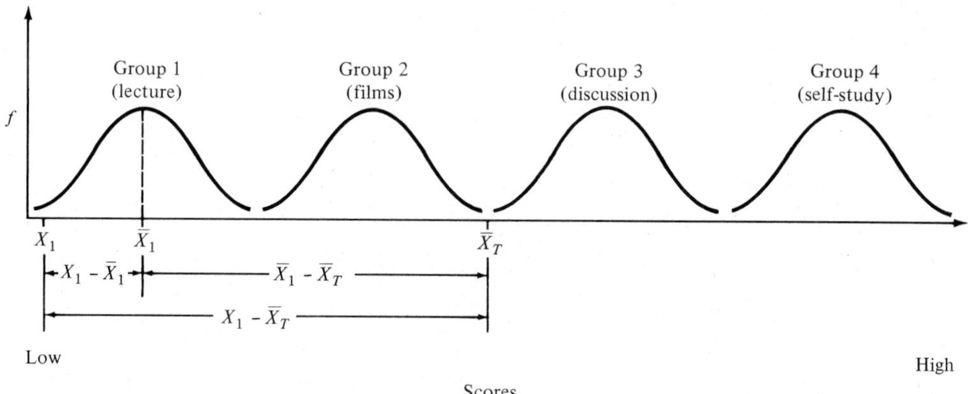

FIGURE 11.1. **Within-groups variability and between-groups variability equals total variability.**

the lecture group (this is a hypothetical example!) had the lowest exam scores. Also shown on the polygon is a score (X_1) made by a student in group 1, the mean of group 1 ($\overline{X}_1$), and the mean of all 200 students, called the total mean ($\overline{X}_T$). Examining Figure 11.1 carefully, we see that there are really three kinds of variability shown.

Total Variability

We can see that there is variability in the distribution of all 200 scores, which is exemplified by the deviation of a score in group 1 from the total mean, or $X_1 - \overline{X}_T$. This $X_1 - \overline{X}_T$ shown in Figure 11.1 would be the contribution to *total* variability of the single score X_1. There would, of course, be 199 others ($X_2 - \overline{X}_T$, $X_3 - \overline{X}_T$, etc.), and each deviation of a score from the total mean contributes to the *total variability* in the combined distribution.

But note in Figure 11.1 that the deviation of a single score from the total mean, $X_1 - \overline{X}_T$, can be broken down into two separate components. The first component is the deviation of a given score from its group mean, $X_1 - \overline{X}_1$, and the second component is the deviation of the group mean from the total mean, $\overline{X}_1 - \overline{X}_T$. Note that these are additive, both graphically in Figure 11.1 and algebraically, since $(X_1 - \overline{X}_1) + (\overline{X}_1 - \overline{X}_T) = X_1 - \overline{X}_T$.

Variability Within Groups

The deviation of a score from its group mean is part of the variability *within groups;* that is, the amount each score in any group deviates from its own mean. The deviation of $X_1 - \overline{X}_1$ in Figure 11.1 would be one part of the variability within group 1, and each of the other 49 scores would contribute its deviation to the variability within group 1. There would be a similar variability in the other three groups.

Variability Between Groups

The deviation of a group mean from the total mean contributes to the *variability between groups.* Again looking at Figure 11.1, we see that the deviation of the mean of group 1 from the total mean, $\overline{X}_1 - \overline{X}_T$, is part of the variability between groups. Note that the term *between groups* is used despite the fact that the actual deviation is not between the individual means themselves but between the group mean and the total mean ($\overline{X}_1 - \overline{X}_T$, $\overline{X}_2 - \overline{X}_T$, etc.).

The aim of our analysis, of course, is to determine whether there are significant differences between the group means. *We will do this eventually by comparing the variability between groups with the variability within groups.* We know that in any single distribution of scores there will be variability, so we will always expect to find *variability within groups.*

However, if there are no significant differences between the means of the groups, we expect that there will be very little variability between groups—only the small amount we expect to find by sampling error.

This point is illustrated in Figure 11.2 for three distributions. Note that in Figure 11.2a the three groups differ very little from each other and their means differ only slightly due to sampling error. There is still variability *within each group,* of course, but there is much less *variability between groups* than in Figure 11.2b, where the means of the groups differ markedly. Eventually, we will be able to use the comparison of the *variability between groups* to the *variability within groups* to show whether there are significant differences between the means.

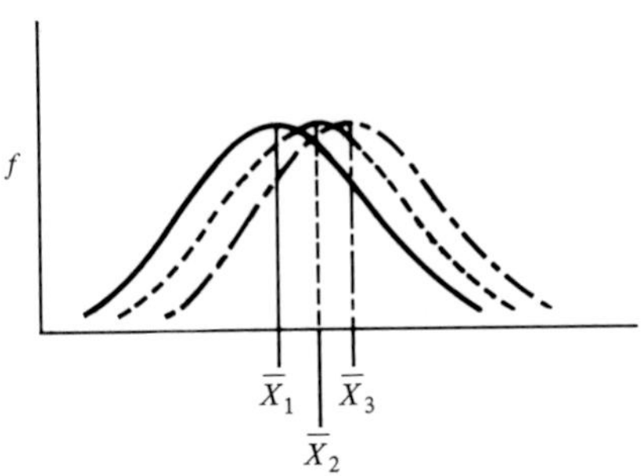

(a) No significant differences between the means

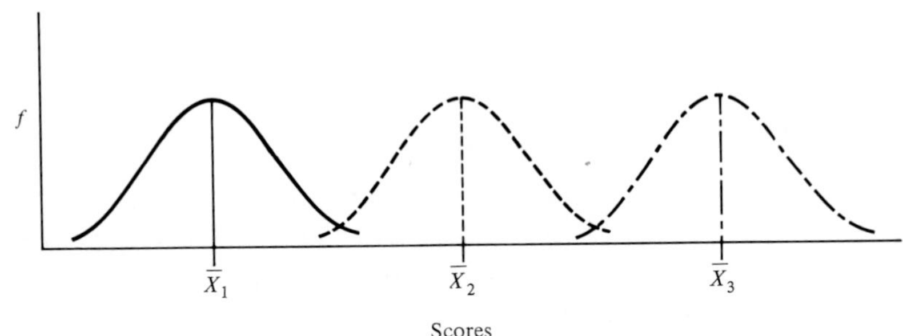

(b) Significant differences between the means

FIGURE 11.2. **Possible relationship between three group means for significant and nonsignificant differences.**

Before we can calculate some variances to analyze, we need to develop formulas to calculate the sums of squares, which will be the numerators of the various variance formulas. Remembering that the *total variability* is composed of variability *within groups* and variability *between groups*, we can express this as we did before, in Figure 11.1, as

$$(X - \overline{X}_T) = (X - \overline{X}_1) + (\overline{X}_1 - \overline{X}_T)$$

As we noted before, this expresses the contribution to the total variability of *one* score in group 1, and the deviation of this score from the total mean is composed of the deviation of the score from its group mean plus the deviation of the group mean from the total mean.

However, the deviation from the mean must be squared before we can arrive at the sums of squares, so we must square both sides of the above expression to obtain

$$(X - \overline{X}_T)^2 = (X - \overline{X}_1)^2 + 2(X - \overline{X}_1)(\overline{X}_1 - \overline{X}_T) + (\overline{X}_1 - \overline{X}_T)^2$$

This represents the squared deviation of *one* score in group 1. We now have to sum for *all* of the scores in group 1. Summing both sides of the equation gives

$$\sum^{N_1}(X - \overline{X}_T)^2 = \sum^{N_1}(X - \overline{X}_1)^2 + 2(\overline{X}_1 - \overline{X}_T)\sum^{N_1}(X - \overline{X}_1) + N_1(\overline{X}_1 - \overline{X}_T)^2$$

Note that $\sum^{N_1}$ means that we are summing the various deviations for all the scores in group 1. Note also that in the second term on the right-hand side of the equation, both 2 and $(\overline{X}_1 - \overline{X}_T)$ are constants, so they appear in front of the summation sign, since the sum of a constant times a variable is equal to the constant times the sum of the variable. Note also that in the third term on the right-hand side, we finish up with $N(\overline{X}_1 - \overline{X}_T)^2$, since the sum of a constant is equal to N times the constant.

Examining the second term again, we note that it contains $\sum^{N_1}(X - \overline{X}_1)$. This is nothing more than our $\Sigma(X - \overline{X})$, which we know is equal to 0 for a given group of scores. Thus the entire second term drops out, and we are left with

$$\sum^{N_1}(X - \overline{X}_T)^2 = \sum^{N_1}(X - \overline{X}_1)^2 + N_1(\overline{X}_1 - \overline{X}_T)^2$$

In a verbal description of this formula, along with the graphical representation of Figure 11.1, we would say that the squared deviations of the scores in group 1 from the total mean are made up of the squared deviations of the scores from the mean of group 1 plus the size of group 1 times the squared deviation of the mean of group 1 from the total mean.

There remains one additional step. The formula given is for scores in one group only. We now need to sum over all the groups (k = number of groups), to obtain

Total Sum of Squares	Within-Groups Sum of Squares	Between-Groups Sum of Squares

$$\sum^{k}\sum^{N_G}(X - \overline{X}_T)^2 = \sum^{k}\sum^{N_G}(X - \overline{X}_G)^2 + \sum^{k}N_G(\overline{X}_G - \overline{X}_T)^2 \tag{11.1}$$

Note that the subscript G refers to the number of the group. For example, N_3 refers to the number of scores in group 3, while $\overline{X}_2$ would be the mean of group 2.

Formula 11.1 is one of the most complex formulas we have encountered in this book, but it is probably one of the most important, so let us examine each component in detail. Remember that each is a sum of squares (abbreviated SS), that is, the sum of squared deviations from a mean, $\Sigma(X - \overline{X})^2$.

Total Sum of Squares (SS_T)

$$SS_T = \sum^{k}\sum^{N_G}(X - \overline{X}_T)^2 \tag{11.2}$$

This term represents the squared deviations of *all* scores from the total mean. The expression $(X - \overline{X}_T)^2$ is the squared deviation of a score from the total mean. $\sum^{N_G}$ indicates that these squared deviations are to be summed for all the scores in each group, and $\sum^{k}$ indicates that the sums of the squared deviations from each group are to be added together finally to give the total sum of the squared deviations.

Within-Groups Sum of Squares (SS_{WG})

$$SS_{WG} = \sum^{k}\sum^{N_G}(X - \overline{X}_G)^2 \tag{11.3}$$

This term represents the squared deviations of scores from their respective group means. $(X - \overline{X}_G)^2$ is the squared deviation of a score from its group mean. $\sum^{N_G}$ indicates that the squared deviations are to be summed for each group; and $\sum^{k}$ indicates that the sums for each group are to be added together to give the within-groups sum of squares for all groups.

Between-Groups Sum of Squares (SS_{BG})

$$SS_{BG} = \sum^{k} N_G(\overline{X}_G - \overline{X}_T)^2 \qquad (11.4)$$

This term represents the squared deviation of each group mean from the total mean. $(\overline{X}_G - \overline{X}_T)^2$ indicates this squared deviation. N_G indicates that each squared deviation is to be multiplied by the size of its group (this is called "weighted by N"); and $\sum^{k}$ indicates that the sums for each group are to be added together to give the weighted squared deviations for all groups.

So we conclude that the sum of squares from the total mean is composed of the sum of squares within the groups plus the sum of squares between groups, or

$$SS_T = SS_{WG} + SS_{BG}$$

Let us see how the formulas that we have developed work in a hypothetical example. Table 11.2 shows three groups of scores and the calculation of SS_T, SS_{BG}, and SS_{WG}.

Note that the calculations in Table 11.2 are simply the arithmetic counterpart of the logic of ANOVA developed in the previous sections. The calculation of SS_T is the sum of the squared deviations of each score from the total mean of 5. The SS_{BG} value shows the deviation of each group mean from the total mean, each deviation being squared and multiplied by its sample size N. Finally, the SS_{WG} value shows that within each group the deviation of each score from its mean is squared and the squares are summed for the entire group and these sums are themselves summed over all three groups. Of course, the last statement in Table 11.2 shows that $SS_{BG} + SS_{WG} = SS_T$.

Calculating the Variances

The previous sections dealt with the sums of squares, which we noted earlier are the numerators of the variances we wish to calculate. Before

TABLE 11.2. **Calculation of Sums of Squares: Deviation Formulas**

I		II		III	
3		11		2	
1	$\overline{X}_1 = 2$	6	$\overline{X}_2 = 8$	6	$\overline{X}_3 = 4$
2		5		4	
6		10		12	
		32			

$$\overline{X}_T = \frac{\Sigma X_T}{N_T} = \frac{6 + 32 + 12}{10} = 5$$

SS_T:

$$\sum^{k}\sum^{N_G}(X - \overline{X}_T)^2 = (3-5)^2 + (1-5)^2 + (2-5)^2 = 29$$
$$(11-5)^2 + (6-5)^2 + (5-5)^2 + (10-5)^2 = 62$$
$$(2-5)^2 + (6-5)^2 + (4-5)^2 = 11$$
$$SS_T = 29 + 62 + 11 = 102$$

SS_{BG}:

$$\sum^{k} N_G(\overline{X}_G - \overline{X}_T)^2 = N_1(\overline{X}_1 - \overline{X}_T)^2 + N_2(\overline{X}_2 - \overline{X}_T)^2 + N_3(\overline{X}_3 - \overline{X}_T)^2$$
$$SS_{BG} = 27 + 36 + 3 = 66$$

SS_{WG}:

$$\sum^{k}\sum^{N_G}(X - \overline{X}_G)^2 = (3-2)^2 + (1-2)^2 + (2-2)^2 = 2$$
$$(11-8)^2 + (6-8)^2 + (5-8)^2 + (10-8)^2 = 26$$
$$(2-4)^2 + (6-4)^2 + (4-4)^2 = 8$$
$$SS_{WG} = 2 + 26 + 8 = 36$$

Summary:

$$SS_T = SS_{BG} + SS_{WG}$$
$$102 = 66 + 36$$

we can calculate these variances, we need to deal now with the denominators of the variances, or the *degrees of freedom*, in order to have a variance of the form $\frac{\Sigma(X - \overline{X})^2}{N - 1}$. Just as we found that the total sum of squares could be partitioned into a between-groups and a within-groups sum of squares, we now observe that the total degrees of freedom can be partitioned in the same way.

TOTAL *df*. The degrees of freedom associated with the entire group of observations is equal to $N_T - 1$, where N_T is the total number of observations. The data of Table 11.2 consisted of 10 scores, so the total $df = 9$.

BETWEEN-GROUPS *df*. The degrees of freedom associated with the between-groups component is equal to $k - 1$, where k is the number of groups. For the data of Table 11.2, the *df* for between groups would be 2.

WITHIN-GROUPS *df*. The degrees of freedom within *one* group would be one less than the N of that group. Combining them for all groups would give $(N_1 - 1) + (N_2 - 1) + (N_3 - 1)$ and so on for as many groups as required, which would equal $N_T - k$. For the example in Table 11.2, the within-groups $df = 10 - 3 = 7$.

Note that the degrees of freedom are additive in the same way that the sums of squares were; that is,

$$df_T = df_{BG} + df_{WG}$$
$$N_T - 1 = (k - 1) + (N_T - k)$$

For the data of Table 11.2, $2 + 7 = 9$.

The Mean Squares

We are now ready to calculate the variance estimates associated with the between-groups component and the within-groups component. These variance estimates are called *mean squares*. Mean squares are variance estimates, and they consist of a sum of squares divided by the appropriate degrees of freedom. Since we will eventually want to compare the variance estimate based on the between-groups component with the variance estimate based on the within-groups component, we will focus our attention on the mean square between groups, MS_{BG}, and the mean square within groups, MS_{WG}. The formulas for these variance estimates are

$$MS_{BG} = \frac{SS_{BG}}{k - 1} \tag{11.5}$$

$$MS_{WG} = \frac{SS_{WG}}{N_T - k} \tag{11.6}$$

For the data of Table 11.2, the mean squares would be

$$MS_{BG} = \frac{66}{3 - 1} = 33$$

$$MS_{WG} = \frac{36}{10 - 3} = 5.14$$

Variance Estimates and the Null Hypothesis

Now that we have spent considerable time and energy on the calculation of MS_{WG} and MS_{BG}, just what are they? What do they represent? Since they have a sum of squares in the numerator and a *df* term in the denominator, they certainly must be variance estimates of some sort.

MS_{WG} and MS_{BG} as Variance Estimates

We have already stated that the sums of squares within groups, SS_{WG}, was composed of squared deviations from the various group means. For a *single* group, then, SS_{WG_1} would be the sum of squared deviations from the mean of group 1. And if we calculated

$$MS_{WG_1} = \frac{SS_{WG_1}}{N_1 - 1}$$

we would have the mean square within group 1, *which is an unbiased estimate of the population variance from which group 1 came.* To refer back to Figure 11.1, the lecture group, group 1, is a random sample from a population of available college students, and MS_{WG_1} is an estimate of the variance of that population.

We could also calculate MS_{WG_2} by dividing SS_{WG_2} by $N_2 - 1$. This would be an unbiased estimate of the variance of the population from which group 2 came.

We could do this in a similar manner for MS_{WG_3} and MS_{WG_4} or for as many groups as we had. Each one is an unbiased estimate of the variance of the different populations from which the samples came. *However, under the null hypothesis, these various estimates are all estimating the same thing!* Since the null hypothesis states that $\mu_1 = \mu_2 = \mu_3 = \mu_4$, we assume that the populations from which the various samples have been drawn are identical and that MS_{WG_1}, MS_{WG_2}, MS_{WG_3}, and MS_{WG_4} are just estimates of the same population variance. Since we are rather knowledgeable on the topic of sampling by now, we know that a *mean* of these estimates would give us a very accurate overall estimate of the population variance *if the null hypothesis is true.* So, if we take an average of MS_{WG_1}, MS_{WG_2}, and so on, we wind up with our familiar MS_{WG}, the average unbiased estimate of the variance of the population from which the samples came.

Let us get back to the mean square between groups, MS_{BG}. It can be shown that, if the null hypothesis is true ($\mu_1 = \mu_2 = \mu_3$, etc.), then MS_{BG} is *also an unbiased estimate of the variance of the population from which the samples came.* The mathematical proof for this statement is beyond the scope of this book, but we will accept it on faith.

The Ratio MS_{BG}/MS_{WG}

Since MS_{BG} and MS_{WG} are estimates of the same population variance, we would expect that if the null hypothesis is true the ratio MS_{BG}/MS_{WG} will be equal to 1.0. Of course, we do not expect it to be *exactly* 1.0, since we know that there will be some fluctuations due to sampling error.

However, if the ratio is quite a bit larger than 1.0, we reject the null hypothesis and conclude that there is a real difference between the means of the populations from which the samples were drawn ($\mu_1 \neq \mu_2 \neq \mu_3$, etc.). If the ratio is such that it would happen less than 5% or 1% or 0.1% of the time by sampling error alone, we conclude that there is a significant difference between the means. The statistical procedure for determining whether the ratio MS_{BG}/MS_{WG} is significant is called the *F* test.

The *F* Test

The *F* test (named after a British statistican, Sir Ronald Fisher) consists of examining the ratio of two variances to see if the departure from 1.0 is sufficiently large so that it is not likely to be due to sampling error. The *F* test (or *F* ratio), as indicated above, is the ratio

$$F = \frac{MS_{BG}}{MS_{WG}} \tag{11.7}$$

But how large must *F* be before we reject the null hypothesis? We noted earlier with the *t* test that the value to be equaled or exceeded in Table E depended upon the degrees of freedom on which the *t* test had been computed. The *F* value to be equaled or exceeded depends upon the degrees of freedom also, but *both* the numerator and denominator of the *F* test determine the degrees of freedom. You will notice in Table

F in Appendix 4 a list of various values of F that need to be equaled or exceeded for the given significance levels. As with the *t* distributions, these values are points that mark off the 5% and 1% points in the sampling distribution of *F*. These values are determined by the degrees of freedom associated with both the numerator and the denominator of the *F* ratio. For example, if you have 4 degrees of freedom in the numerator and 20 degrees of freedom in the denominator, you would need an *F* value of at least 2.87 for the ratio to be significant at the .05 level and 4.43 for it to be significant at the .01 level. For the data shown in Table 11.2 the *F* value would be

$$F = \frac{MS_{BG}}{MS_{WG}} = \frac{33}{5.14} = 6.42$$

Remembering that the *df* for MS_{BG} was $k - 1 = 2$ and the *df* for MS_{WG} was $N_T - k = 7$, we would proceed to Table F and note that for 2 and 7 degrees of freedom F_{05} is 4.74 and F_{01} is 9.55. We would then conclude that $p < .05$ that our *F* of 6.42 happened by sampling error and would state that there is a significant difference between the means of the three groups.

Computational Formulas

The deviation formulas used in Table 11.2 demonstrate the logic of ANOVA very clearly, but they are rather cumbersome to work with. Subtracting a mean from each score and squaring the deviation is very time-consuming, and for this reason computational formulas have been developed. (You may remember we did the same thing for the standard deviation in Chapter 5.) These formulas for the sums of squares deal with the raw scores rather than the deviations. The *raw score* formula for the total sum of squares is as follows:

$$SS_T = \sum^{N_T} X^2 - N_T \overline{X}_T^2 \qquad (11.8)$$

In this formula, $\sum^{N_T} X^2$ indicates that *all* scores are squared and then summed, and $N_T \overline{X}_T^2$ indicates that the total mean is squared and multiplied by the total number of observations. This result is then subtracted from $\sum^{N_T} X^2$.

$$SS_{BG} = N_1 \overline{X}_1^2 + N_2 \overline{X}_2^2 + N_3 \overline{X}_3^2 + \cdots - N_T \overline{X}_T^2 \qquad (11.9)$$

This formula for the between-groups sum of squares indicates that each group mean is squared and then multiplied by the size of that group and that these products are added together for as many groups as there are. The quantity $N_T\overline{X}_T^2$, which is the same value calculated previously in the formula for SS_T, is then subtracted.

$$SS_{WG} = \left(\sum^{N_1} X^2 - N_1\overline{X}_1^2\right) + \left(\sum^{N_2} X^2 - N_2\overline{X}_2^2\right) + \left(\sum^{N_3} X^2 - N_3\overline{X}_3^2\right) + \cdots \quad (11.10)$$

The formula for the within-groups sum of squares shows that the mean of group 1 is squared, multiplied by the size of group 1, and subtracted from the sum of the squared scores in group 1. This procedure is repeated for as many groups as there are, and the sums of squares for each group are then added together to yield SS_{WG}.

To illustrate the use of the computational formulas, let us consider the data of Table 11.3. An investigator was studying the learning abilities of four species of laboratory animals. She taught 5 squirrels, 6 cats, 3 guinea pigs, and 6 rats a simple maze response and kept track of the errors each animal made. Were there significant differences in the learning ability of these four species?

In applying the raw score formulas for the sums of squares, we should note the starred items in Table 11.3:

1. *Remember that you cannot simply take an average of the four means to obtain $\overline{X}_T$. You must divide the total ΣX by the total N.
2. **The total ΣX^2 is the sum of the individual ΣX^2 columns.
3. ***Note that the quantities $N_1\overline{X}_1^2$, $N_2\overline{X}_2^2$, and so on, have already been calculated in the previous SS_{BG} formula. So, each component is simply the ΣX^2 of each group minus the $N\overline{X}^2$ for that group.

The main interest, of course, is in the F value, and we see that the calculated F is 5.24, which is greater than the tabled value at the 5% level for 3 and 16 degrees of freedom. We would conclude that there are significant differences between the means of the maze errors committed by the different laboratory animals.

The ANOVA Summary Table

In the professional literature, you will find that, when ANOVA is used for determining the significance of the differences among means, all the

TABLE 11.3. **Using the Computational Formulas for ANOVA**

Squirrels		Cats		Guinea Pigs		Rats	
X	X^2	X	X^2	X	X^2	X	X^2
9	81	17	289	12	144	10	100
9	81	16	256	10	100	8	64
3	9	14	196	14	196	7	49
5	25	15	225	36	440	9	81
6	36	10	100			12	144
32	232	6	36			8	64
		78	1,102			54	502
$\overline{X}_1 = 6.4$		$\overline{X}_2 = 13$		$\overline{X}_3 = 12$		$\overline{X}_4 = 9$	

$$^{*}\overline{X}_T = \frac{\Sigma X_T}{N_T} = \frac{32 + 78 + 36 + 54}{5 + 6 + 3 + 6} = \frac{200}{20} = 10.0$$

$$^{**}\overset{N_T}{\Sigma} X^2 = 232 + 1102 + 440 + 502 = 2276$$

Sums of Squares:

$$SS_T = \overset{N_T}{\Sigma} X^2 - N_T\overline{X}_T^2 = 2276 - 20(10.0)^2$$
$$= 2276 - 2000$$
$$SS_T = 276$$

$$SS_{BG} = N_1\overline{X}_1^2 + N_2\overline{X}_2^2 + N_3\overline{X}_3^2 + N_4\overline{X}_4^2 - N_T\overline{X}_T^2$$
$$= 5(6.4)^2 + 6(13)^2 + 3(12)^2 + 6(9)^2 - 2000$$
$$= 204.8 + 1014 + 432 + 486 - 2000$$
$$SS_{BG} = 136.8$$

$$^{***}SS_{WG} = \left(\overset{N_1}{\Sigma} X^2 - N_1\overline{X}_1^2\right) + \left(\overset{N_2}{\Sigma} X^2 - N_2\overline{X}_2^2\right)$$
$$+ \left(\overset{N_3}{\Sigma} X^2 - N_3\overline{X}_3^2\right) + \left(\overset{N_4}{\Sigma} X^2 - N_4\overline{X}_4^2\right)$$
$$= (232 - 204.8) + (1102 - 1014) + (440 - 432) + (502 - 486)$$
$$= 27.2 + 88 + 8 + 16$$
$$SS_{WG} = 139.2$$

Check:

$$SS_{BG} + SS_{WG} = SS_T$$
$$136.8 + 139.2 = 276$$

Mean Squares and the F Test:

$$MS_{BG} = \frac{SS_{BG}}{k - 1} = \frac{136.8}{3} = 45.6$$

$$MS_{WG} = \frac{SS_{WG}}{N_T - k} = \frac{139.2}{16} = 8.7$$

$$F = \frac{MS_{BG}}{MS_{WG}} = \frac{45.6}{8.7} = 5.24$$

F_{05}, $df = 3/16$, is 3.24
Significant, $p < .05$

computational steps of Table 11.3 are not shown. Instead, a *summary table* is used that lists the *source of variance, degrees of freedom, sums of squares,* and *mean squares,* as well as the value of *F*. Such a summary table has been prepared for the data of Table 11.3 and is shown in Table 11.4. Note that this table contains all the necessary information for the reader of a book or a journal article, including a probability statement on the value of *F*.

TABLE 11.4. **ANOVA Summary Table for Animal Learning Experiment**

Source of Variance	df	SS	MS	F
Between groups	3	136.8	45.6	5.24*
Within groups	16	139.2	8.7	
Total	19	276.0		

*$p < .05$

Testing for Differences Among Pairs of Means

After a significant *F* has been obtained, we are faced with the question of *which* differences between means are significant. If there are three means, $\overline{X}_1$, $\overline{X}_2$, and $\overline{X}_3$, it is possible that they are all significantly different from each other. On the other hand, maybe $\overline{X}_1$ and $\overline{X}_2$ are about the same, but $\overline{X}_3$ is significantly different from those two. With a little imagination, you can see how these comparisons could become complex with as many as five or six or even more groups to be compared.

Clearly, what is needed is a technique that will enable us to determine which differences between means are significant and which are not. A number of techniques have been developed, and you may run across a reference to Duncan's multiple-range test, the Newman–Keuls procedure, the Scheffé method, Tukey's procedure, and others. Each of these methods has been developed for a particular purpose, and your instructor may have a personal bias for or against a particular technique. We will confine our discussion to Tukey's procedure, at the same time remembering that your instructor or an advanced textbook might prefer another approach. The Tukey procedure may be used in all cases where a significant F was obtained in the ANOVA calculation. And even when F was not significant, the procedure may be used on those mean differences *predicted to be significant prior to collection of the data.*

The Tukey Method. Unequal *N*'s

The statistic that will enable us to evaluate differences among pairs of means is called the *studentized range statistic, q,* whose general formula is

$$q = \frac{\overline{X}_L - \overline{X}_S}{\sqrt{\frac{MS_{WG}}{2}\left(\frac{1}{N_L} + \frac{1}{N_S}\right)}} \qquad (11.11)$$

where

$\overline{X}_L$ is the larger of the two means
$\overline{X}_S$ is the smaller of the two means
MS_{WG} is the mean square within groups from the ANOVA calculations
N_L is the size of the group with the larger mean
N_S is the size of the group with the smaller mean

If q is large enough, we can reject the hypothesis that the difference between two means is due only to sampling error, and we pronounce the difference as significant. How large does q have to be? Table M in Appendix 4 lists the values for the .05 and .01 significance levels (i.e., for α). If these tabled values are equaled or exceeded by our calculated q, they are significant at the stated level. Table M is entered by using (1) the appropriate value of k, the number of means in the ANOVA, and (2) the df for the MS_{WG}, the number of degrees of freedom in the calculation of MS_{WG}.

Let us use the animal learning data shown in Table 11.5 to help clarify the procedure. Note that each mean is paired with every other mean, and the calculated q is compared with the tabled values from Table M. Since we have four means and $df = 16$ for the MS_{WG}, we enter Table M to find that $q_{05} = 4.05$ and $q_{01} = 5.19$. In Table 11.5 we see

TABLE 11.5. The Tukey Method for Animal Learning Data (Errors)

Squirrels	**Cats**	**Guinea Pigs**	**Rats**
$\overline{X} = 6.4$	$\overline{X} = 13$	$\overline{X} = 12$	$\overline{X} = 9$
$N = 5$	$N = 6$	$N = 3$	$N = 6$

$$MS_{WG} = 8.7$$

$$df = N_T - k = 16$$

Squirrels–Cats:

$$q = \frac{\overline{X}_L - \overline{X}_S}{\sqrt{\frac{MS_{WG}}{2}\left(\frac{1}{N_L} + \frac{1}{N_S}\right)}} = \frac{13 - 6.4}{\sqrt{\frac{8.7}{2}\left(\frac{1}{6} + \frac{1}{5}\right)}} = \frac{6.6}{\sqrt{4.35\ (.367)}}$$

$$= \frac{6.6}{\sqrt{1.596}} = \frac{6.6}{1.26} = 5.24$$

Significant, $p < .01$

Squirrels–Guinea Pigs:

$$q = \frac{12 - 6.4}{\sqrt{\frac{8.7}{2}\left(\frac{1}{3} + \frac{1}{5}\right)}} = \frac{5.6}{\sqrt{4.35\ (.533)}} = \frac{5.6}{\sqrt{2.319}} = \frac{5.6}{1.52} = 3.68$$

Not significant, $p > .05$

Squirrels–Rats:

$$q = \frac{9 - 6.4}{\sqrt{\frac{8.7}{2}\left(\frac{1}{6} + \frac{1}{5}\right)}} = \frac{2.6}{\sqrt{4.35\ (.367)}} = \frac{2.6}{\sqrt{1.596}} = \frac{2.6}{1.26} = 2.06$$

Not significant, $p > .05$

Cats–Guinea Pigs:

$$q = \frac{13 - 12}{\sqrt{\frac{8.7}{2}\left(\frac{1}{6} + \frac{1}{3}\right)}} = \frac{1}{\sqrt{4.35\ (.5)}} = \frac{1}{\sqrt{2.175}} = \frac{1}{1.47} = .68$$

Not significant, $p > .05$

Cats–Rats:

$$q = \frac{13 - 9}{\sqrt{\frac{8.7}{2}\left(\frac{1}{6} + \frac{1}{6}\right)}} = \frac{4}{\sqrt{4.35\ (.333)}} = \frac{4}{\sqrt{1.449}} = \frac{1}{1.20} = 3.33$$

Not significant, $p > .05$

Guinea Pigs–Rats:

$$q = \frac{12 - 9}{\sqrt{\frac{8.7}{2}\left(\frac{1}{3} + \frac{1}{6}\right)}} = \frac{3}{\sqrt{4.35\ (.5)}} = \frac{3}{\sqrt{2.175}} = \frac{3}{1.47} = 2.04$$

Not significant, $p > .05$

that the mean error score for squirrels ($\overline{X} = 6.4$) is significantly less than the mean error score for cats ($\overline{X} = 13$), with a q value of 5.22, $p < .01$. However, we would have to conclude that the rest of the means are not significantly different, $p > .05$.

The Tukey Method. Equal *N*'s

The procedure described above is greatly simplified if all the groups are of equal size. If $N_1 = N_2 = N_3$, and so on, then formula 11.11 reduces to

$$q = \frac{\overline{X}_L - \overline{X}_S}{\sqrt{\frac{MS_{WG}}{N_G}}} \tag{11.12}$$

where N_G is the size of any group and the rest of the terms are as described earlier. When the groups are of equal size, we begin the Tukey procedure by selecting the *largest* difference between means, applying formula 11.12, and evaluating our value of q in Table M. We then repeat the procedure with the next largest difference and continue until our q is no longer significant. Obviously, since the denominator remains the same, once we have a difference between means that is not large enough to yield a significant q, it is pointless to test smaller differences. This procedure is illustrated in the Sample Problem at the end of this chapter.[†]

It must be emphasized that the Tukey method is only one of many methods for making multiple comparisons, and your instructor may wish to pursue one or more of the other methods mentioned earlier. An excellent resource is a summary article by Hopkins and Anderson (1973).

Assumptions for the Analysis of Variance

In order for the F test to be a valid procedure for determining the significance of the differences between means, the following assumptions or restrictions must be met. These assumptions are identical to those listed in Chapter 10 for the t test.

1. The scores must be interval or ratio in nature.

[†]It is possible to find a significant F value in your ANOVA computation and not find any q values to be significant. Such an occurrence is quite rare, however.

2. The scores must be measures on random samples from the respective populations.
3. The populations from which the samples were drawn must be normally distributed.
4. The populations from which the samples were drawn must have approximately the same variability (homogeneity of variance).

Concluding Remarks

This elementary introduction to the analysis of variance barely touches on the use of a highly popular and versatile statistical tool. Almost any professional journal in education or the behavioral sciences will contain one or more studies where ANOVA has been used for the data analysis. The reasons for its popularity have not been obvious in our examination of the single classification method, which may appear to be nothing more than an extension of the *t* test for more than two means. One of the unique features of the more complex ANOVA designs is its measurement of an *interaction* effect, the relationship that one variable has to another variable in producing a significant difference. This form of ANOVA will be treated in detail in the next chapter.

SAMPLE PROBLEM

An industrial psychologist was investigating three different training methods for speeding up assembly line production. Thirty workers were assigned at random to three groups, and each group was trained in a different method for completing the assemblies. Each worker's performance was then measured in number of units assembled per hour. These measurements are shown below for each of the three different methods. Was there a significant difference in the workers' performance in the three training methods? If so, which differences were significant?

Method A		Method B		Method C	
X	X^2	X	X^2	X	X^2
52	2,704	61	3,721	76	5,776
54	2,916	62	3,844	65	4,225
49	2,401	68	4,624	66	4,356
62	3,844	58	3,364	76	5,776
45	2,025	43	1,849	84	7,056
47	2,209	39	1,521	83	6,889
31	961	41	1,681	78	6,084
35	1,225	50	2,500	66	4,356

Method A		Method B		Method C	
X	X^2	X	X^2	X	X^2
41	1,681	50	2,500	73	5,329
40	1,600	53	2,809	62	3,844
456	21,566	525	28,413	729	53,691

$$\overline{X}_1 = \frac{456}{10} = 45.6 \qquad \overline{X}_2 = \frac{525}{10} = 52.5 \qquad \overline{X}_3 = \frac{729}{10} = 72.9$$

$$\overline{X}_T = \frac{\Sigma X_T}{N_T} = \frac{456 + 525 + 729}{30} = \frac{1{,}710}{30} = 57.0$$

$$\sum^{N_T} X^2 = 21{,}566 + 28{,}413 + 53{,}691 = 103{,}670$$

Sums of Squares:

$$SS_T = \sum^{N_T} X^2 - N_T\overline{X}_T^2 = 103{,}670 - 30(57)^2$$

$$= 103{,}670 - 97{,}470$$

$$SS_T = 6{,}200$$

$$SS_{BG} = N_1\overline{X}_1^2 + N_2\overline{X}_2^2 + N_3\overline{X}_3^2 - N_T\overline{X}_T^2$$

$$= 10(45.6)^2 + 10(52.5)^2 + 10(72.9)^2 - 97{,}470$$

$$= 20{,}793.6 + 27{,}562.5 + 53{,}144.1 - 97{,}470$$

$$SS_{BG} = 4{,}030.2$$

$$SS_{WG} = \left(\sum^{N_1} X^2 - N_1\overline{X}_1^2\right) + \left(\sum^{N_2} X^2 - N_2\overline{X}_2^2\right) + \left(\sum^{N_3} X^2 - N_3\overline{X}_3^2\right)$$

$$= (21{,}566 - 20{,}793.6) + (28{,}413 - 27{,}562.5)$$

$$+ (53{,}691 - 53{,}144.1)$$

$$= 772.4 + 850.5 + 546.9$$

$$SS_{WG} = 2{,}169.8$$

Check:
$SS_{BG} + SS_{WG} = SS_T$
$4{,}030.2 + 2{,}169.8 = 6{,}200$

Mean Squares and the F Test:

$$MS_{BG} = \frac{SS_{BG}}{k - 1} = \frac{4{,}030.2}{2} = 2{,}015.1$$

$$MS_{WG} = \frac{SS_{WG}}{N_T - k} = \frac{2{,}169.8}{27} = 80.36$$

$$F = \frac{MS_{BG}}{MS_{WG}} = \frac{2{,}015.1}{80.36} = 25.08$$

Source of Variance	df	SS	MS	F
Between groups	2	4,030.2	2,015.1	25.08*
Within groups	27	2,169.8	80.36	
Total	29	6,200.0		

*$p < .01$.

Since an F value of 25.08 would happen less than 1% of the time by sampling error, we conclude that there is a significant difference between the means of the three training groups.

Tukey's Procedure

Method A	Method B	Method C
$\overline{X}_1 = 45.6$	$\overline{X}_2 = 52.5$	$\overline{X}_3 = 72.9$

$$MS_{WG} = 80.36;\ df = 27;\ k = 3$$

Methods A and C:

$$q = \frac{\overline{X}_L - \overline{X}_S}{\sqrt{\frac{MS_{WG}}{N}}} = \frac{72.9 - 45.6}{\sqrt{\frac{80.36}{10}}} = \frac{27.3}{\sqrt{8.036}}$$

$$= \frac{27.3}{2.83} = 9.65$$

Significant, $p < .01$

Methods B and C:

$$q = \frac{72.9 - 52.5}{2.83} = \frac{20.4}{2.83} = 7.21$$

Significant, $p < .01$

Methods A and B:

$$q = \frac{52.5 - 45.6}{2.83} = \frac{6.9}{2.83} = 2.44$$

Not significant, $p > .05$

We conclude that method C is superior to methods A and B; however, there appears to be no significant difference between methods A and B. Note that we used formula 11.12 for Tukey's procedure, since there were an equal number of subjects in the three groups.

STUDY QUESTIONS

1. If you are testing to see if there are significant differences between three means, why is it incorrect to use three separate t tests with $\overline{X}_1$ and $\overline{X}_2$, $\overline{X}_1$ and $\overline{X}_3$, and $\overline{X}_2$ and $\overline{X}_3$?

2. What is meant by the term *sum of squares*?
3. Why is the sum of squares not used as a measure of variability?
4. In your own words describe what is meant by the expression $(X_1 - \overline{X}_1) + (\overline{X}_1 - \overline{X}_T) = (X_1 - \overline{X}_T)$.
5. "We will always expect to find variability *within* groups, but there may not be variability *between* groups." What does this mean?
6. What does MS_{WG} represent? Under the null hypothesis what does MS_{BG} represent?
7. Why should the ratio MS_{BG}/MS_{WG} equal 1.0 under the null hypothesis?
8. What information does an ANOVA summary table contain?
9. Why is Tukey's procedure needed after a significant value of F has been found?
10. What are the assumptions of the one-way analysis of variance?

EXERCISES

1. Shown below are error scores on a children's problem solving test for three different age groups. Use the deviation formulas (as in Table 11.2) to calculate SS_T, SS_{BG}, and SS_{WG} for these data.

Age 6	Age 8	Age 10
9	1	2
7	6	1
6	5	3
10		

2. Calculate MS_{BG}, MS_{WG}, and the F test for Exercise 1. Is there a significant difference in the number of errors made by these three age groups?
3. Use the computational formulas to calculate SS_T, SS_{BG}, and SS_{WG} on the data in Exercise 1, checking your results against the deviation method.
4. Arithmetic quiz scores for three groups of students are shown below. Use the deviation formulas (as in Table 11.2) to calculate SS_T, SS_{BG}, and SS_{WG}.

Group 1	Group 2	Group 3
7	8	12
4	10	12
9	8	10
3	9	7
2	10	9

5. Calculate MS_{BG}, MS_{WG}, and the F ratio for Exercise 4. Is there a significant difference between the quiz scores for these three groups?

6. Calculate SS_T, SS_{BG}, and SS_{WG} for Exercise 4 using the computational formulas, checking your results against the deviation formulas.

7. A physiological psychologist has 8 rats in group 1, 6 in group 2, and 7 in group 3. How large must her F value be to be significant at the .05 level?

8. A developmental psychologist is testing subjects under four different conditions. He has 6 children in one group, 4 in a second group, 9 in a third group and 12 in a fourth. How large must his F value be to be significant at the .01 level?

9. The researcher in Exercise 7 wishes to apply Tukey's procedure to see which means are significantly different. How large must her q value be to be significant at the .05 level?

10. You are running an analysis of variance on data from four groups of subjects with seven subjects in each group. If you use Tukey's procedure to test for significant differences among the means, how large a q value will you need for significance at the .05 level?

11. In a study to see how the difficulty level of a driving task was related to drivers' ability to react to events in their peripheral field of view, 21 college students were tested in a driving simulator (Bartz, 1976). Seven subjects each were tested in three different levels of driving difficulty—an easy task, one of moderate difficulty, and one of extensive difficulty. While they were engaged in their driving task, a light would come on in their peripheral field of view, and the time (in hundredths of a second) it took them to press a hand-held switch was recorded. The reaction time score (larger score = longer reaction time) for each subject is shown below for the three conditions. Was there a significant difference in reaction times to the peripheral lights between the three levels of driving difficulty? If so, use Tukey's procedure to find which differences were significant.

Difficult Task	**Moderate Task**	**Easy Task**
76	66	52
70	67	46
68	70	48
68	72	48
53	58	33
55	48	32
58	46	49

12. A sample of students at Pennsylvania State University was asked to fill out a questionnaire on fitness, health and exercise. They also were asked to have their blood pressure, heart rate and other physiological measures taken. Shown below are the resting heart rates of students who indicated

(1) very little, if any, exercise, (2) a moderate amount of exercise, and (3) a regular exercise routine. Was there a significant difference in the mean resting heart rate sampled from these three degrees of exercise? If so, use Tukey's procedure to find which means were significantly different.

Very Little	Moderate	Regular
66	76	84
74	74	60
90	82	70
62	62	74
90	68	72
84	62	64
82	60	48
90	61	78
78	66	64
80	68	76

13. An instructor in business statistics had 8 sophomores, 15 juniors and 10 seniors in her class and was curious about the relative performance of the three classes. At the end of the term, she compared the course grades (A = 12, A− = 11, B+ = 10, etc.) of the three classes, and the results are shown below. Was there a significant difference between the classes? If so, use Tukey's procedure to find which differences were significant.

Sophomores	Juniors	Seniors
7	9	12
6	12	12
9	10	6
10	12	11
6	7	6
6	12	9
9	12	12
9	11	6
	9	12
	7	10
	12	
	12	
	9	
	12	
	11	

14. A group of elderly subjects who had shown some evidence of memory loss and other mental impairments participated in an exercise program sponsored by the National Institutes of Health, the Veterans Administration and the University of Utah.[†] Volunteers were assigned at random to one of three groups—aerobic exercise (fast walking), nonaerobic exercise (calisthenics), and a non-exercise control group. A test of mental dexterity was one of several psychological and physiological characteristics assessed before the exercise program began and again four months later. The scores after four months for a sample of subjects are shown below. Was there a significant difference between the three groups? If so, which differences were significant?

Aerobic Exercise	Nonaerobic Exercise	No Exercise
69	51	59
74	41	56
72	34	54
78	56	39
61	61	44
64	61	49
49	56	34
54	46	64
59	51	34
59	36	69

15. The State-Trait Anxiety Inventory (STAI) referred to earlier in exercise 13 in Chapter 4 was given to 461 neuropsychiatric patients, 161 general medical–surgical patients and 212 prison inmates (Spielberger, 1983). Summary statistics for these three groups on the State-Anxiety scale are shown below. Was there a significant difference between these groups in state anxiety? If so, which differences were significant?

Neuro-psychiatric	Medical–Surgical	Prison Inmates
$N_1 = 461$	$N_2 = 161$	$N_3 = 212$
$\Sigma X = 22{,}008$	$\Sigma X = 6{,}823$	$\Sigma X = 9{,}744$
$\Sigma X^2 = 1{,}131{,}481$	$\Sigma X^2 = 319{,}783$	$\Sigma X^2 = 473{,}651$
$\overline{X}_1 = 47.74$	$\overline{X}_2 = 42.38$	$\overline{X}_3 = 45.96$

[†]Adapted from Dustman, R. E., et al. (1984). Aerobic exercise training and improved neuropsychological function of older individuals. *Neurobiology of Aging*, 5, 35–42.

CHAPTER 12

Two-Way Analysis of Variance

The preceding chapter was devoted entirely to the *one-way* ANOVA. If, for example, we want to see which of three teaching methods (lecture, discussion, or programmed workbook) is best for teaching freshman English to college students, we can assign freshmen at random to one of three groups. Each group can then learn under its designated instructional method, and we can use final test scores as our criterion measure. We can diagram the results schematically as below, where the blanks are final exam scores, and $\overline{X}_1$, $\overline{X}_2$, and $\overline{X}_3$ are the group means for the three instructional methods.

Lecture	**Discussion**	**Workbook**
______	______	______
______	______	______
______	______	______
______	______	______
$\overline{X}_1 =$	$\overline{X}_2 =$	$\overline{X}_3 =$

The null hypothesis would state that $\mu_1 = \mu_2 = \mu_3$ and any difference between $\overline{X}_1$, $\overline{X}_2$, and $\overline{X}_3$ is due to sampling error. This design is called a one-way or single-classification ANOVA because only *one* variable (teaching method) is being tested. In this example, we are comparing three levels (lecture, discussion, and workbook) of a single variable. We might have 3 or 7 or 15 different levels, but we still are dealing with only one variable, that of teaching method.

However, we may sometimes want to investigate the effects of *two* variables simultaneously. We may wish to see if the teaching methods above have different effects with superior students than with average students. We could do such an analysis by conducting two separate experiments—one with superior students and one with average students. This, however, would be inefficient, since it would require twice as much effort as a single experiment. Also, unfortunately, we would not be able to get a direct measure of the *interaction effect:* One teaching method might be better for superior students, while another method might work better for average students! In the material to follow we will develop a method for looking at the effects of two variables simultaneously, the two-way ANOVA.

Factorial Designs

The two-way ANOVA that we will be examining is called a factorial design, indicating that we are looking at the effects of two factors (variables) simultaneously. For example, let us suppose that we have 60 superior students (determined by their college entrance exam scores) and 60 average students, and we assign 20 of each at random to each of the three English teaching methods mentioned earlier. Since we have two ability levels (superior and average) and three teaching methods (lecture, discussion, and workbook), we have a 2 × 3 (say "two by three") factorial design. Factorial designs come in all shapes and sizes (3 × 7, 2 × 4, and so on), depending on how many levels we have for each of the two variables.

The 2 × 3 factorial design for the ability level and teaching method experiment is presented in Table 12.1. Each compartment is called a *cell* and will contain the final exam scores for the 20 students in that cell. For example, the upper left cell would have the final exam scores for the 20 superior students taught under the lecture method. After the data have been tabulated into rows and columns as in Table 12.1, a factorial design is always analyzed in terms of two components: main effects and interaction effects.

TABLE 12.1. **An Illustration of a Two-Factor Experiment**

		Teaching Method for Freshman English			
		Lecture	Discussion	Workbook	
Ability Level	Superior				$\overline{X}_{Sup}$ =
	Average				$\overline{X}_{Ave}$ =
		$\overline{X}_L$ =	$\overline{X}_D$ =	$\overline{X}_W$ =	

Main Effects

Obviously, we are interested in the effects of our variables or factors, and these are called *main* effects. In this example, we are looking at the possible effects of teaching method and ability level.

Let us first consider the variable of *ability level,* indicated by the row means. If we disregard which teaching method is used, we have a mean for 60 superior students ($\overline{X}_{Sup}$) and a mean for 60 average students ($\overline{X}_{Ave}$). Even if there were no *real* difference between the performance of superior and average students, we still would expect a slight difference in the sample means, $\overline{X}_{Sup}$ and $\overline{X}_{Ave}$, due to sampling error, so we will use the null hypothesis that these two sample means are random samples from populations with identical means. Later on, we will develop an *F* test to see if we can reject the null hypothesis that there is no difference in ability level (i.e., that there is a main effect of ability level).

Similarly, we now consider the variable of teaching method, indicated by the column means. We disregard the ability level of the students and note that we have three means—one for the 40 students in the lecture method ($\overline{X}_L$), one for the 40 in the discussion method ($\overline{X}_D$), and one for the 40 using the workbook ($\overline{X}_W$). Under the null hypothesis there is no real difference in the populations from which the sample means were drawn, and differences between $\overline{X}_L$, $\overline{X}_D$, and $\overline{X}_W$ are simply due to sampling error. Again, an *F* test is used to see if these differences are large enough to be significant, that is, if there is a main effect of teaching method.

Interaction Effects

An important purpose of the factorial design is to explore possible *interaction effects*. For example, variable A may have a different effect at one level of variable B than it does at another level of variable B. In the teaching method example, it might be that average students do better than superior students when using the programmed workbook (the superior students might find it boring and not do their best work) while superior students might do better than average students in the discussion group method.

This state of affairs is an interaction effect: One variable is behaving differently at one level of the other variable. When an interaction effect occurs, our interest in any main effects is diminished, since the effect of one main variable is dependent upon the level of the other main variable. In our example, if there were a significant interaction effect and someone asked, "Which teaching method is best for freshman English?" we would answer, "It depends on whether you are working with superior or average students." Conversely, if the question were which students do better in freshman English, we would have to say that it would depend on which teaching method is used.

Graphical Methods to Illustrate Interaction

A graph of the result of a factorial experiment is often helpful in understanding the concept of interaction. Let us first consider an experiment where there is no significant interaction effect.

AN EXAMPLE OF A NONSIGNIFICANT INTERACTION. A developmental psychologist is studying differences in reading ability between boys and girls and is also interested in whether these differences change with age. She administers a reading test to students in the fourth, fifth, and sixth grades. This is a 2×3 factorial design—two levels of sex (boys and girls) and three levels of grade placement (fourth, fifth, and sixth)—and the reading test scores are tabulated for each cell. The means for each cell, as well as row and column means, are shown in Table 12.2.

TABLE 12.2. **Table of Means for Reading Scores of Boys and Girls in Grades 4–6**

		Sex		
		Girls	Boys	
	4	$\bar{X} = 29$	$\bar{X} = 19$	$\bar{X}_{R_1} = 24$
Grade	5	$\bar{X} = 40$	$\bar{X} = 30$	$\bar{X}_{R_2} = 35$
	6	$\bar{X} = 42$	$\bar{X} = 32$	$\bar{X}_{R_3} = 37$
		$\bar{X}_{C_1} = 37$	$\bar{X}_{C_2} = 27$	

After completing our table of means, we construct a graph with the dependent variable (reading scores) on the y-axis and one of the main variables (let us use grade level) on the x-axis. The cell means are plotted for both boys and girls and the resultant graph is shown in Figure 12.1. Any main effects can easily be seen, so if sex differences are significant we conclude that girls had higher reading scores at all grade levels. Similarly, if there is a significant difference between grade levels, we conclude that reading scores are lowest at the fourth grade and highest at the sixth.

However, our main interest in drawing the graph is to examine any

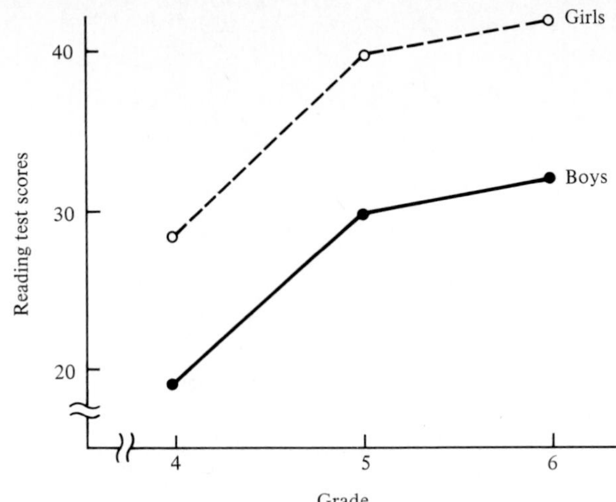

FIGURE 12.1. **Reading test scores. An example of a nonsignificant interaction.**

possible interaction effect. We note that there is a separation between the curves for girls and boys, but the separation is the same at all three grades; that is, the curves are parallel. If the curves are parallel, *there is no interaction effect,* since the reading scores are not affected by the sex of the student more at one grade level than another. An examination of the table of means in Table 12.2 is also helpful. We note that the differences between the column means (girls and boys) is 37 − 27 = 10 points, and this difference is the same at all grade levels. And this, by definition, means there is no interaction effect.

EXAMPLE OF A SIGNIFICANT INTERACTION. An educational psychologist was investigating the characteristics of creative children. One of her tests for creativity presented the subject with a number of small circles, and the subject was asked to draw as many things using these circles as possible. The psychologist wanted to know what effect there would be if the students were given examples to start them out. She then developed two sets of instructions for the creativity test—one in which the student was shown two examples of drawings from circles and one in which the student was shown no examples at all.

She also wanted to see what effect the two sets of instructions had on "creative" and "noncreative" children. She asked several elementary school teachers to identify which of their children were creative and which were not creative, based strictly on classroom observation.

This design is, of course, a 2 × 2 factorial design. One factor is *instructions* (the two levels are "examples" and "no examples") and

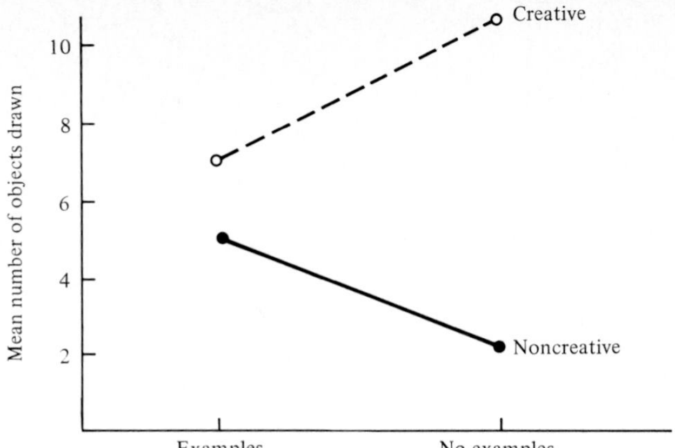

FIGURE 12.2. **Number of objects drawn. An example of a significant interaction.**

the other is *type of student* (the two levels are "creative" and "noncreative"). The elementary school teachers identified 20 creative and 20 noncreative students. Ten of each were assigned at random to the group that was given examples, and 10 of each were also assigned to the group that was not given examples. The creativity test was administered and scored for the number of objects drawn from the circles in 5 minutes. The mean number of objects drawn by each group is shown in Table 12.3 and Figure 12.2.

It is obvious from the column means of Table 12.3 that creative students produced more drawings ($\overline{X}_{C_1} = 9$) than noncreative students ($\overline{X}_{C_2} = 4$). But of greater interest is a possible interaction effect shown

TABLE 12.3. **Number of Original Drawings by "Creative" and "Noncreative" Students with Different Instructional Sets**

		Type of Student		
		Creative	Noncreative	
Instructions	Examples	$\overline{X} = 7$	$\overline{X} = 5$	$\overline{X}_{R_1} = 6$
	No examples	$\overline{X} = 11$	$\overline{X} = 3$	$\overline{X}_{R_2} = 7$
		$\overline{X}_{C_1} = 9$	$\overline{X}_{C_2} = 4$	

in Figure 12.2. When examples were used in the instructions, there was not much difference between creative and noncreative students (who had means of 7 and 5, respectively). However, when no examples were given, creative students produced more drawings while noncreative students were much less productive (with means of 11 and 3, respectively).

Here we have a clear interaction effect. The curves in Figure 12.2 are obviously not parallel. The instructions given had different effects, depending on the type of student. It is possible that the creative students were blocked or inhibited by the examples while the noncreative students were helped by the examples. However, when no examples were given, the noncreative students were definitely handicapped, while the creative students were free to use their inventive abilities to create new forms.

A Definition of Interaction

Interaction can be defined as a significant departure from a parallel relationship of two or more curves. Even if there is no real interaction between the two variables, we, of course, do not expect to get perfectly parallel curves. The cell means will fluctuate just by sampling error, causing some degree of divergence or convergence from a parallel relationship. However, at some point the deviation from parallelism may be so great that we reject the notion of no interaction and say that we have a significant interaction effect. The sections that follow describe the approach we will use to determine significance.

Calculating the Sums of Squares

As with the one-way ANOVA, we eventually need to calculate variance estimates of the form $\Sigma(X - \overline{X})^2/(N - 1)$, so we need to find various values of $\Sigma(X - \overline{X})^2$ or sums of squares. We will consider only the computational formulas in this chapter and not develop the deviation formulas as in the last chapter.

The formulas are easier to understand in terms of an example, so let us suppose that the topic of perceptual-motor skills with preferred or nonpreferred hand was investigated by an experimental psychologist. The motor skills performance was measured by pursuit rotor. The apparatus is something like a phonograph turntable that has a small disk about the size of a dime on it. The subject attempted to follow this disk with a stylus as it went around. An electronic timer kept track of the time that the stylus was in contact with the disk. This measurement, called "time on target," was recorded to the nearest second. The greater the time on target, the better the performance.

The experimenter was interested in seeing how performance was affected at different pursuit rotor speeds and chose 20, 40, and 60 revolutions per minutes (RPM) as the three levels of task difficulty. Since there were two levels of hand used (preferred vs. nonpreferred) and three levels of task difficulty (20, 40, and 60 RPM), this was a 2×3 factorial design.

Thirty subjects volunteered for the experiment and were assigned at random to *one* of the six combinations of pursuit rotor speed and hand used. Thus there were 5 subjects in each combination. Each subject's time on target during a 20-second trial is shown in Table 12.4. As we have done so many times before, we calculate ΣX, ΣX^2, and $\overline{X}$ for each column. For the first step, we will be using the computational formulas for the one-way ANOVA from the last chapter, so we also need the total mean ($\overline{X}_T$) and the sum of all the squared scores ($\overset{N_T}{\Sigma} X^2$).

TABLE 12.4. Time on Target (Seconds) for Preferred or Nonpreferred Hand at Three Pursuit Rotor Speeds

	Preferred Hand			Nonpreferred Hand		
	20	40	60	20	40	60
	8	9	6	5	1	2
	7	5	5	5	2	0
	8	7	4	7	4	1
	6	4	5	6	2	1
	6	5	5	7	6	1
ΣX:	35	30	25	30	15	5
ΣX^2:	249	196	127	184	61	7
$\overline{X}$:	7	6	5	6	3	1

$$\overline{X}_T = \frac{35 + 30 + 25 + 30 + 15 + 5}{30} = 4.67$$

$$\overset{N_T}{\Sigma} X^2 = 249 + 196 + 127 + 184 + 61 + 7 = 824$$

To begin our two-way ANOVA, we first calculate SS_T, SS_{BG}, and SS_{WG} as we did in the last chapter.

$$\begin{aligned}
SS_T &= \overset{N_T}{\Sigma} X^2 - N_T\overline{X}_T^2 = 824 - (30)(4.67)^2 \\
&= 824 - 654.27 = 169.73 \\
SS_{BG} &= N_1\overline{X}_1^2 + N_2\overline{X}_2^2 \ldots + N_6\overline{X}_6^2 - N_T\overline{X}_T^2 \\
&= 5(7)^2 + 5(6)^2 + 5(5)^2 + 5(6)^2 + 5(3)^2 + 5(1)^2 - 654.27 \\
&= 780 - 654.27 = 125.73 \\
SS_{WG} &= SS_T - SS_{BG} = 169.73 - 125.73 = 44.0
\end{aligned}$$

Note that we calculated SS_{WG} by simply subtracting SS_{BG} from SS_T. We could, of course, have calculated SS_{WG} directly using formula 11.10.

Now that we have our sums of squares separated into the familiar SS_{BG} and SS_{WG}, we need to consider their meaning in a two-way ANOVA. As in the one-way ANOVA, SS_{WG} will be used to calculate a variance estimate, an estimate of the variance of the population from which the samples came. We will have more to say about this estimate later. And what about SS_{BG}? In the one-way ANOVA of the last chapter, SS_{BG} was used to calculate a variance estimate which, under the null hypothesis, was estimating the same variance as that based on SS_{WG}.

But, in the two-way ANOVA, SS_{BG} reflects *both* the effects of the row variable (speed) and the column variable (hand used), *as well as* the interaction effect if any. This is easiest to understand if we present the results of Table 12.4 in the form of a table of means. This 2 × 3 table in Table 12.5 shows 5 subjects in each of the 6 combinations of hand used and pursuit rotor speed. Each of the 6 cells shows the N of 5 in the corner and the mean time on target for the 5 subjects. Note also the row and column means, which indicate performance by a particular row or column. For example, $\overline{X}_{R_2} = 4.5$ is the mean time on target for the 10 subjects at 40 RPM, while $\overline{X}_{C_1} = 6.0$ is the mean for the 15 subjects using preferred hands.

We are now ready to begin calculating the sums of squares that will be used directly in our two-way ANOVA. The formula for the sum of squares for the row variable, SS_R, is

TABLE 12.5. **Mean Time on Target as a Function of Hand Used and Pursuit Rotor Speed**

		Hand		
		Preferred	Nonpreferred	
Speed (revolutions per minute)	20	$\overline{X} = 7$ (N = 5)	$\overline{X} = 6$ (N = 5)	$\overline{X}_{R_1} = 6.5$
	40	$\overline{X} = 6$ (N = 5)	$\overline{X} = 3$ (N = 5)	$\overline{X}_{R_2} = 4.5$
	60	$\overline{X} = 5$ (N = 5)	$\overline{X} = 1$ (N = 5)	$\overline{X}_{R_3} = 3.0$
		$\overline{X}_{C_1} = 6.0$	$\overline{X}_{C_2} = 3.33$	$\overline{X}_T = 4.67$

$$SS_R = N_{R_1}\overline{X}_{R_1}^2 + N_{R_2}\overline{X}_{R_2}^2 + N_{R_3}\overline{X}_{R_3}^2 - N_T\overline{X}_T^2 \quad (12.1)$$

Plugging in the values from Table 12.5, we get

$$\begin{aligned} SS_R &= 10(6.5)^2 + 10(4.5)^2 + 10(3.0)^2 - 30(4.67)^2 \\ &= 422.5 + 202.5 + 90 - 654.27 \\ &= 60.73 \end{aligned}$$

The formula for the sum of squares for the column variable, SS_C, is

$$SS_C = N_{C_1}\overline{X}_{C_1}^2 + N_{C_2}\overline{X}_{C_2}^2 - N_T\overline{X}_T^2 \quad (12.2)$$

Plugging in the values from Table 12.5, we get

$$\begin{aligned} SS_C &= 15(6.0)^2 + 15(3.33)^2 - 30(4.67)^2 \\ &= 540 + 166.33 - 654.27 \\ &= 52.06 \end{aligned}$$

The formula for the sum of squares for interaction, $SS_{R \times C}$ (say "rows by columns"), is

$$SS_{R \times C} = SS_{BG} - SS_R - SS_C \quad (12.3)$$

Plugging in the values already obtained, we get

$$\begin{aligned} SS_{R \times C} &= 125.73 - 60.73 - 52.06 \\ &= 12.94 \end{aligned}$$

Calculating the Variances

The previous section dealt with the calculation of the sums of squares, which are the numerators of the variances we wish to calculate. We noted in the last chapter that variances are sums of squares divided by their respective degrees of freedom (*df*). We are now interested in the total *df* ($N_T - 1$) and also in the *df* associated with the two main variables (rows and columns) and the interaction between the two (rows by columns).

The *df* for the row variable is one less than the number of rows, or $r - 1$.

$$df_R = r - 1 \quad (12.4)$$

In the example, the row variable is pursuit rotor speed at three levels (20, 40, and 60 RPM), so the *df* for speed is

$$df_R = 3 - 1 = 2$$

The *df* for the column variable is one less than the number of columns, or $c - 1$.

$$df_C = c - 1 \qquad (12.5)$$

In the example, the column variable is the hand used with two levels (preferred or nonpreferred), so the *df* for hand used is

$$df_C = 2 - 1 = 1$$

The *df* for the interaction of the row variable and column variable is one less than the number of levels of the row variable multiplied by one less than the number of levels of the column variable, or $(r - 1)(c - 1)$. The formula for the *df* for interaction is

$$df_{R \times C} = (r - 1)(c - 1) \qquad (12.6)$$

In the example there are three speeds and two hands, so

$$df_{R \times C} = (3 - 1)(2 - 1) = 2$$

In the last chapter the formula for the within-groups *df* was given as $N_T - k$, where k was the number of groups. We use the same approach in the two-way ANOVA except that the number of groups is now the number of cells. Since we have r rows and c columns, the number of groups would be $r \times c$, or rc. Thus the within-groups *df* would be $N_T - rc$.

$$df_{WG} = N_T - rc \qquad (12.7)$$

In the example there were 30 subjects divided among three rows and two columns, so the within-groups *df* is

$$df_{WG} = 30 - (3)(2) = 24$$

We again expect that the addition of the various degrees of freedom should equal the total *df*, $N_T - 1$. Using the symbolic notation of the previous paragraphs, we would have $df_T = df_R + df_C + df_{R \times C} + df_{WG}$.

Gathering the numbers we just calculated for the pursuit rotor experiment, we have 2 + 1 + 2 + 24 = 29, which is the same as $N_T - 1 = 30 - 1 = 29$.

The Mean Squares

We can now calculate the variance estimates, or mean squares, by dividing each sum of squares by its appropriate *df*. We will have four mean squares in a two-way ANOVA, one each for rows, columns, rows by columns, and within groups. The formulas are

$$MS_R = \frac{SS_R}{r - 1} \tag{12.8}$$

$$MS_C = \frac{SS_C}{c - 1} \tag{12.9}$$

$$MS_{R \times C} = \frac{SS_{R \times C}}{(r - 1)(c - 1)} \tag{12.10}$$

$$MS_{WG} = \frac{SS_{WG}}{N_T - rc} \tag{12.11}$$

From the pursuit rotor data, we would calculate the following mean squares.

$$MS_R = \frac{60.73}{2} = 30.365$$

$$MS_C = \frac{52.06}{1} = 52.06$$

$$MS_{R \times C} = \frac{12.94}{(2)(1)} = 6.47$$

$$MS_{WG} = \frac{44}{30 - (3)(2)} = 1.83$$

The Meaning of the Variance Estimates

As in the last chapter, we stop momentarily after calculating the mean squares in order to reflect on their meaning. They obviously are variance estimates of some sort, since they have a sum of squares in the numerator and degrees of freedom in the denominator. But just what are they estimating?

MS_{WG}. In the one-way ANOVA of the last chapter, MS_{WG} was a combined estimate of the variance of the populations from which each sample came. Under the null hypothesis, the population means are identical, so MS_{WG_1}, MS_{WG_2}, and so on, were estimates of the same population variance, and MS_{WG} was the average of these individual estimates from each group. However, in a two-way ANOVA each cell corresponds to a single "group," and we again could consider MS_{WG} for cell 1, MS_{WG} for cell 2, and so on, as individual estimates of the same population variance. Thus MS_{WG} in a two-way ANOVA is a combined estimate of the variance of the population from which each sample came.

MS_R. The variance estimate based on the row means is *also* an estimate of the population variance *if the null hypothesis is true*. In other words, if $\mu_{R_1} = \mu_{R_2} = \mu_{R_3}$, and so on, the variance estimate (MS_R) derived from the sample row means ($\overline{X}_{R_1}$, $\overline{X}_{R_2}$, etc.) is estimating the same population variance as MS_{WG}. We expect, of course, that the sample row means will vary just by sampling error and MS_R will reflect this sampling variation among $\overline{X}_{R_1}$, $\overline{X}_{R_2}$, and so on. However, if the row variable is exerting a significant effect, the row means will vary more than what would be expected by chance, and MS_R will be larger than the variance estimate, MS_{WG}. With the pursuit rotor data of Table 12.5, we are saying under the null hypothesis that the sample row means of 6.5, 4.5, and 3.0 seconds on target for the various pursuit rotor speeds are different simply because of chance variation.

MS_C. Using the same logic we applied to MS_R, we note that under the null hypothesis, MS_C is also an estimate of the population variance calculated from the column means. If $\mu_{C_1} = \mu_{C_2} = \mu_{C_3}$, and so on, we expect that the column means ($\overline{X}_{C_1}$, $\overline{X}_{C_2}$, etc.) will vary only by sampling error, and the variance estimate, MS_C, will be an estimate of the same population variance as MS_{WG}. However, if the column variable is exerting a significant effect, the column means will vary more than what would be expected by chance, and MS_C will be larger than the variance estimate, MS_{WG}. In the example, we are saying under the null hypothesis that the sample column means of 6.0 and 3.33 for the preferred and nonpreferred hands are different only because of chance variation.

$MS_{R \times C}$. Before investigating what the remaining variance estimate, $MS_{R \times C}$, is estimating, let us review the concept of interaction. In Table 12.2 we saw that there was no interaction effect of reading test scores of boys and girls in fourth, fifth, and sixth grades. We stressed that the graph in Figure 12.1 showed *parallel* lines, indicating that the difference between boys and girls was consistent at the three grade levels. Note that, when there is no interaction, we can predict the value of the cell means from the column and row means alone. For example, the

girls' mean and boys' mean are 37 and 27, respectively. The total mean would be 32, and these two column means are 5 points on each side of the total mean. Now, looking at the row mean of 24 for fourth-graders, we note that if there is no interaction present at this level the cell means for fourth-grade girls and fourth-grade boys should be 5 points on each side of the row mean for fourth-graders. This would yield means of 29 and 19, respectively, and is exactly what is shown in the fourth-grade cells of Table 12.2. The same calculations could be done, of course, to obtain the means in the remaining cells. In general, we note that for any table of results we have now established what the cell means should be when there is no interaction present. We need only remember that the sample cell means could be expected to deviate from these expected values because of sampling error. $MS_{R \times C}$ is a variance estimate based on the deviations from these expected values, and under the null hypothesis (no significant interaction) $MS_{R \times C}$ is also an estimate of the variance of the population from which the samples came. And using our usual logic, if the deviations from these expected values are significant (i.e., if the interaction is exerting a significant effect), we note that $MS_{R \times C}$ will be larger than it would if the deviation of the cell means about their expected values were due to chance variation. As a result, the variance estimate of $MS_{R \times C}$ will be larger than the estimate given by MS_{WG}. In the pursuit rotor example of Table 12.5, we are saying under the null hypothesis that the sample cell means of 7, 6, 6, 5, 3, and 1 are deviating from their expected values by an amount due only to chance variation.

The *F* Tests

Finally, we are ready to attempt what we set out to do at the beginning of this chapter: to see if one or both of the variables or their interaction in a two-factor experiment has had a significant effect. As with the one-way ANOVA, we will set up a ratio of two variance estimates and see if the resulting F value is significant. Note that the denominator in each case is MS_{WG}, the estimate of the population variance that is *not* affected by the action of the row variable or column variable or by their interaction. The numerator in each case will be a mean square of the row variable or column variable or their interaction. As we noted in the preceding section, these variance estimates are also estimates of the population variance if the null hypothesis is true. However, if the variable is exerting a significant effect, the respective variance estimate will be larger than the estimate given by MS_{WG}, and the F value will be significantly greater than 1. The formula for the row effect is

$$F_R = \frac{MS_R}{MS_{WG}} \tag{12.12}$$

The formula for the column effect is

$$F_C = \frac{MS_C}{MS_{WG}} \tag{12.13}$$

The formula for the interaction effect is

$$F_{R \times C} = \frac{MS_{R \times C}}{MS_{WG}} \tag{12.14}$$

The calculations for the pursuit rotor data are, for the row effect:

$$F_R = \frac{30.365}{1.83} = 16.59$$

for the column effect:

$$F_C = \frac{52.06}{1.83} = 28.45$$

and for the interaction effect:

$$F_{R \times C} = \frac{6.47}{1.83} = 3.54$$

We now summarize the results of all our computational efforts in the typical ANOVA summary table shown in Table 12.6.

TABLE 12.6. ANOVA Summary Table for Pursuit Rotor Data

Source of Variance	df	SS	MS	F
Speed (R)	2	60.73	30.36	16.59*
Hand (C)	1	52.06	52.06	28.45*
Speed by hand ($R \times C$)	2	12.94	6.47	3.54**
Within groups	24	44.00	1.83	
Totals	29	169.73		

*Significant, $p < .01$.
**Significant, $p < .05$.

Each F value is evaluated by entering Table F in Appendix 4 with the df associated with the numerator and the df associated with the denominator. For example, the main effect of speed is tested by the F ratio of MS_R/MS_{WG}, with 2 and 24 df, respectively. In Table F we note that for 2 and 24 df, our F needs to exceed 3.40 at the 5% level and 5.61 at the 1% level. Since our calculated value of 16.59 exceeds 5.61, we conclude that our F is significant beyond the .01 level. We evaluate the other F values and note their probability levels as shown in Table 12.6.

We are now ready to analyze the results of the pursuit rotor experiment. Our first task is to graph the mean shown earlier in Table 12.5, and this graph is shown in Figure 12.3. We see from the ANOVA summary table that the main effect of pursuit rotor speed is significant, and a glance at Figure 12.3 shows that performance with both preferred and nonpreferred hands decreased as the speed increased from 20 to 60 RPM. The summary table also shows a significant effect of handedness, and the separation of the curves shows better performance for subjects using their preferred hand than subjects using their nonpreferred hand.

Finally, the interaction effect is significant ($p < .05$), which is shown on the graph by the curves diverging as pursuit rotor speed increases. The interaction simply says that at the 20 RPM speed there was not much difference in the performance between preferred and nonpreferred hands. However, as the speed increased, performance of the nonpreferred hand deteriorated more rapidly than performance with the

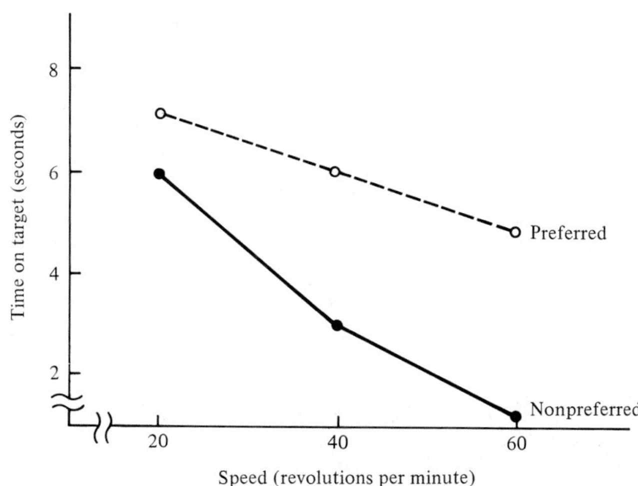

FIGURE 12.3. **Time on target as a function of hand used at three pursuit rotor speeds.**

preferred hand. As we noted earlier, a significant interaction effect always modifies our interpretation of the main effects. For example, in response to the question "Does an increase in pursuit rotor speed interfere with performance?" we would answer that the amount of interference depends on whether one is using the preferred or nonpreferred hand.

Assumptions for the Two-Way ANOVA

The assumptions for the two-way ANOVA are identical to those for the one-way ANOVA described in the last chapter, except that we are now talking about cells rather than groups.

1. Data must be in interval or ratio form.
2. Subjects must be assigned at random to cells.
3. The populations from which the samples are drawn must be normal.
4. The populations from which the samples are drawn must have equal variances.

In addition to the familiar assumptions listed, the ANOVA procedure described in this chapter assumes *an equal number of observations per cell*. In our pursuit rotor experiment shown in Table 12.5, there were 5 subjects in each of the 6 cells for a total of 30. It is essential for this ANOVA procedure that there be an equal number of subjects in each cell. There are methods for handling ANOVA with unequal numbers in the cells, but the procedures are beyond the scope of this book. If you are interested, consult an advanced statistics textbook for details of the method used.

Multiple Comparisons

After a significant F has been obtained with either of the two variables or with their interaction, a logical step is to investigate which differences between means are significant. Referring to the ANOVA summary table of our pursuit rotor results in Table 12.6, we note that there is a significant main effect of hand used. Looking at Table 12.5 and Figure 12.3, we note that the preferred hand was compared with the nonpreferred hand at three different speeds and that the difference between preferred and nonpreferred hand increases as pursuit rotor speed increases. But are the differences at all three speeds significant? There is not much of a difference at 20 RPM, and this difference might not be significant, while the differences at 40 and 60 RPM may indeed be significant. To test for significant differences among these and other pairs of differences, a test for *multiple comparisons* is needed. As was pointed out in the last chapter, tests developed by Scheffé, Duncan, and others can be used, and the article by Hopkins and Anderson (1973) will provide you with the necessary information about these tests.

Concluding Remarks

In two rather concentrated chapters we have covered only a small fraction of the possible ANOVA designs. Our interest in an introductory level statistics textbook is to become familiar with some of the basic concepts underlying this powerful statistical technique. Should you take another statistics course, you will undoubtedly spend much of the time on a variety of ANOVA designs, some of which are mentioned in Note 12.1. When you see how ANOVA can be used to analyze complex relationships among variables, it is easy to understand its wide popularity among researchers.

NOTE 12.1

SPLIT-PLOT DESIGNS?

A casual glance at journal articles, books, and other publications in education and the behavioral sciences shows a wide variety of applications of the analysis of variance—and it is easy to develop the misconception that the analysis of variance is the unique province of these fields. However, much of the theory and application of ANOVA grew out of agricultural and biological research. R. A. Fisher (the F test bears his initial) and his associates at the Rothamsted Experiment Station, in England, were initially responsible for developing much of the experimental design and analysis used in assessing the effects of soil conditions, types of fertilizer, time of planting, amount of moisture, and a host of other conditions related to agricultural productivity. *Randomized blocks, split-plots, factorial design,* and other such exotic terms were the result of early agricultural and biological research. Gradually, Fisher's applications and ideas were adopted and expanded by researchers in other fields, and present-day research efforts are greatly simplified by the early work of Fisher and the work of subsequent theorists.

SAMPLE PROBLEM

A researcher is interested in characteristics of obesity and is studying the amount of effort expended by obese subjects and normal-weight subjects to obtain food. A subject enters the laboratory and is told to fill out an attitude questionnaire. At the table where the subject sits is a bowl of peanuts. For half the subjects the bowl contains shelled peanuts, and for the other half the peanuts are in the shells. The experimenter chooses 20 obese and 20 normal-weight subjects from a pool of volunteers and records the number of peanuts each subject eats while filling out the questionnaire. This is a 2 × 2 factorial design (obese and normal subjects with shelled and unshelled peanuts), with 10 subjects of each weight assigned at random to the shelled or unshelled condition. The results are as follows.

	Obese		Normal	
	Shelled	**Unshelled**	**Shelled**	**Unshelled**
	16	4	10	7
	17	2	5	9
	12	2	9	6
	14	0	10	4
	13	1	9	3
	10	3	7	8
	9	2	6	7
	17	1	8	10
	16	1	5	4
	16	4	11	2
ΣX:	140	20	80	60
ΣX^2:	2,036	56	682	424
$\bar{X}$:	14	2	8	6

$$\bar{X}_T = \frac{140 + 20 + 80 + 60}{40} = 7.5$$

$$\sum^{N_T} X^2 = 2{,}036 + 56 + 682 + 424 = 3{,}198$$

	Shelled Nuts	Unshelled Nuts	
Obese	14 (n = 10)	2 (n = 10)	8
Normal	8 (n = 10)	6 (n = 10)	7
	11	4	7.5

Sums of Squares:

$$SS_T = \sum^{N_T} X^2 - N_T\overline{X}_T^2 = 3{,}198 - 40(7.5)^2 = 3{,}198 - 2{,}250 = 948$$

$$SS_{BG} = N_1\overline{X}_1^2 + N_2\overline{X}_2^2 + N_3\overline{X}_3^2 + N_4\overline{X}_4^2 - N_T\overline{X}_T^2$$

$$= 10(14)^2 + 10(2)^2 + 10(8)^2 + 10(6)^2 - 40(7.5)^2$$

$$= 3{,}000 - 2{,}250 = 750$$

$$SS_{WG} = SS_T - SS_{BG} = 948 - 750 = 198$$

$$SS_R = N_{R_1}\overline{X}_{R_1}^2 + N_{R_2}\overline{X}_{R_2}^2 - N_T\overline{X}_T^2$$

$$= 20(8)^2 + 20(7)^2 - 40(7.5)^2 = 10$$

$$SS_C = N_{C_1}\overline{X}_{C_1}^2 + N_{C_2}\overline{X}_{C_2}^2 - N_T\overline{X}_T^2$$

$$= 20(11)^2 + 20(4)^2 - 40(7.5)^2 = 490$$

$$SS_{R\times C} = SS_{BG} - SS_R - SS_C = 750 - 10 - 490 = 250$$

Mean Squares and F Tests:

$$MS_R = \frac{SS_R}{r - 1} = \frac{10}{1} = 10$$

$$MS_C = \frac{SS_C}{c - 1} = \frac{490}{1} = 490$$

$$MS_{R\times C} = \frac{SS_{R\times C}}{(r - 1)(c - 1)} = \frac{250}{1} = 250$$

$$MS_{WG} = \frac{SS_{WG}}{N_T - rc} = \frac{198}{40 - 4} = 5.5$$

$$F_R = \frac{MS_R}{MS_{WG}} = \frac{10}{5.5} = 1.82$$

$$F_C = \frac{MS_C}{MS_{WG}} = \frac{490}{5.5} = 89.10$$

$$F_{R\times C} = \frac{MS_{R\times C}}{MS_{WG}} = \frac{250}{5.5} = 45.45$$

ANOVA Summary Table

Sources of Variance	**df**	**SS**	**MS**	**F**
Weight (R)	1	10	10	1.82
Nut type (C)	1	490	490	89.10*
Weight by nut type ($R \times C$)	1	250	250	45.45*
Within groups	36	198	5.5	

*Significant, $p < .01$.

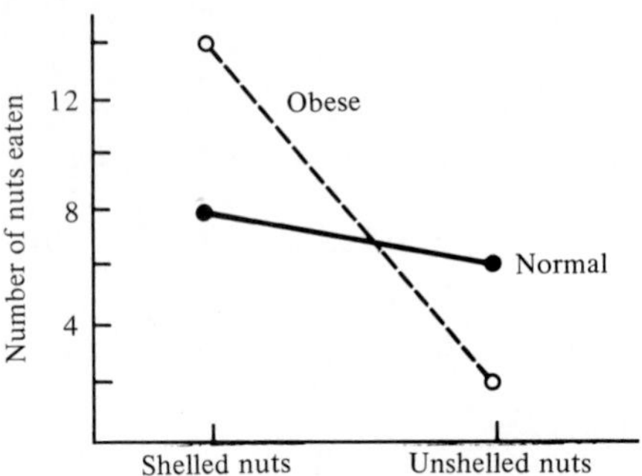

We see in the ANOVA summary table a significant F for nut type and note from the graph and the table of means that more shelled peanuts were eaten than unshelled peanuts. However, this is of minor importance because our main interest is in the interaction effect indicated by a significant F for weight by nut type. As you can see from the graph, when the peanuts were shelled the obese subjects ate more than the normal-weight subjects (means of 14 and 8, respectively). However, when the peanuts were not shelled the normal-weight subjects ate more than the obese subjects (means of 6 and 2, respectively).

STUDY QUESTIONS

1. Why is the two-way ANOVA considered more efficient than the one-way ANOVA?
2. What is meant by the term *main effect*? What is an *interaction effect*?
3. "When an interaction effect occurs, our interest in any main effects is diminished." Why is this so?
4. A friend of yours suggests that you use a 2 × 3 factorial design in your research paper. Describe what is meant by a "2 × 3 factorial design."
5. Describe the construction of a graph in a two-way ANOVA. How would that graph demonstrate that there was no interaction effect?
6. How would a graph demonstrate that there *was* an interaction effect?
7. How would the relative sizes of MS_R and MS_{WG} indicate whether there was a significant row effect? How about MS_C and MS_{WG}?
8. "If there is a significant interaction, the variance estimate of $MS_{R \times C}$ will be larger than the estimate given by MS_{WG}." What does this statement mean?
9. List the assumptions for the two-way ANOVA, including the size of cells assumption.

1. Children in a classroom for the emotionally disturbed (ED) and children in a regular classroom were rated for "creativity of play" when many toys were present or when there were just a few toys available. A rating of 1 indicated low creativity and 10 indicated high creativity. There were 10 children in each combination of diagnostic category and number of toys, with the mean ratings shown in the table below. Draw a graph similar to the one shown in the sample problem. On the basis of the table and graph, would you expect to find any significant main effects? Any interaction effect? Explain your answers.

		Category		
		ED Student	Regular	
Number of Toys	Few	$\bar{X} = 1.8$ (10)	$\bar{X} = 8.2$ (10)	$\bar{X} = 5.0$
	Many	$\bar{X} = 4.9$ (10)	$\bar{X} = 5.1$ (10)	$\bar{X} = 5.0$
		$\bar{X} = 3.35$	$\bar{X} = 6.65$	

2. Men and women subjects performed a manual dexterity test in a noisy room (stress condition) or a quiet room (no stress condition). There were 12 subjects in each combination of sex and noise level. The mean error scores for each group are shown below. Using a graph (similar to one in the sample problem) and the table of means, would you expect to find any significant main effects? Any interaction effect? Give reasons for your answers.

		Stress		
		Noise	Quiet	
Sex	Men	$\bar{X} = 24$ (12)	$\bar{X} = 15$ (12)	$\bar{X} = 19.5$
	Women	$\bar{X} = 23$ (12)	$\bar{X} = 14$ (12)	$\bar{X} = 18.5$
		$\bar{X} = 23.5$	$\bar{X} = 14.5$	

3. An experimental psychologist is studying the effects of stress on visual perception. In one experiment 80 decibels of white noise was used as the stressor while subjects proofread a prose passage projected on a screen. The projected image was sharp for some subjects and was slightly blurred for other subjects. Ten subjects were assigned to each of four combinations of stress and image quality—clear image with no stress, blurred image with no stress, clear image with stress, and blurred image with stress. The number of errors made in proofreading a 100-word paragraph by each subject is shown below. What effects of stress and image quality were demonstrated in this experiment? Calculate a two-way ANOVA, and show a table of means and a graph.

No Stress		Stress	
Clear	**Blurred**	**Clear**	**Blurred**
3	7	5	11
6	8	7	11
3	5	4	9
4	8	7	12
5	4	8	8
5	7	4	7
1	9	8	12
3	3	3	8
7	5	7	11
3	4	5	9

4. A researcher in social psychology studied sex differences in an individual's personal space. A male or female confederate stood in a large room while male or female subjects were asked to approach the person and start a conversation. The researcher (by means of a hidden optical system) measured the distance in inches between the subject and the stationary person to see how close together they were when conversation initiated. Twenty male and 20 female college students were assigned at random to one of four groups—male approaching male, male approaching female, female approaching male, and female approaching female. The distances between the approaching subjects and the stationary person are shown below in inches. What did the experiment show regarding the sex of the approaching person and the sex of the stationary person? Complete a two-way ANOVA, including a table of means and a graph.

Approaching Male		Approaching Female	
Stationary Male	Stationary Female	Stationary Male	Stationary Female
11	16	15	13
14	11	16	11
11	13	14	14
15	12	15	13
16	14	17	14
14	14	15	13
14	15	13	12
13	14	16	14
14	15	15	11
14	14	14	13

5. A study on the sex-typing of occupations by children (Blaske, 1984) was referred to earlier in the sample problem in Chapter 3. Sixty fourth-graders (30 boys and 30 girls) were presented flashcards containing pairs of proper names and occupations. Half of the children received 12 cards with traditional sex-typed pairs (e.g., Doctor-Bob) and the rest received nontraditional sex-typed pairs (e.g., Carpenter-Mary). The two factors in the experiment were sex (boys, girls) and degree of occupational sex-typing (sex-typed, nonsex-typed pairs). Fifteen boys received cards with sex-typed pairs, 15 boys received nonsex-typed pairs, 15 girls received sex-typed pairs and 15 girls received nonsex-typed pairs. Each flashcard was shown for 3 seconds with a 2-second interval in between. One minute after all 12 cards had been viewed, the children were asked to recall as many of the pairs as they could. The number of errors made by each child is shown below (perfect recall = 0 errors). After constructing a table of means and a graph from the data, calculate a two-way ANOVA. What did the experiment show regarding the sex of the child and memory for traditional and nontraditional sex-typed occupation-proper name pairs?

Traditional Sex-Typed		Nontraditional Sex-Typed	
Boys	Girls	Boys	Girls
0	2	4	8
1	0	5	8
4	2	2	6
4	2	6	6
2	1	4	8
2	0	4	8
2	0	3	6
0	2	4	2
8	4	0	0
2	0	4	8

Traditional Sex-Typed		Nontraditional Sex-Typed	
Boys	Girls	Boys	Girls
2	6	8	6
2	0	7	10
2	0	4	6
0	2	4	6
2	0	1	8

6. A developmental psychologist was studying how children form number and order concepts, and had third-graders and sixth-graders perform two different letter-digit substitution tasks. In these tasks children were given sheets of paper containing the letters A, B, C, D, and E (scrambled in different orders), and wrote a number beside each letter according to a "code" printed at the top of the sheet. For the "Natural" code, the numbers to be written were A = 1, B = 2, C = 3, D = 4, E = 5. For the "Random" code, A = 2, B = 5, C = 3, D = 1, and E = 4. Twenty third-graders and 20 sixth-graders were tested, with 10 of each grade assigned the natural code and the other 10 the random code. The scores below are the number of letters correctly coded in 45 seconds.

The researcher hypothesized that (1) sixth-graders would code more letters than third-graders, and (2) the *natural* code would not be any easier than the *random* code for the third-graders but would be easier for the sixth-graders. Construct a table of means and a graph, and calculate a two-way ANOVA to see if the researcher's hypotheses were supported.

Third Grade		Sixth Grade	
Natural Code	Random Code	Natural Code	Random Code
34	24	40	35
50	33	34	31
33	35	44	37
27	34	40	44
30	24	49	38
28	34	39	25
40	26	46	39
27	25	38	44
38	38	39	38
23	22	56	39

7. In the sample problem at the end of this chapter, obese and normal weight subjects were compared on the effort expended to obtain food. Another study on obesity used subjects of normal weight but who were classified as either *dieters* or *nondieters*.[†] Sixteen subjects who had previously dieted one or more times and 16 who had never dieted were asked to drink either no milkshake or two milkshakes during a laboratory test trial. They were then asked to sample several flavors of ice cream, and the amount of ice cream consumed (in grams) was recorded. The researchers hypothesized that those who had a history of dieting would lose control after two milkshakes and eat more ice cream than nondieters. Construct a table of means and a graph, and use a two-way ANOVA to see if the hypothesis is supported.

Dieters		**Nondieters**	
0 Milkshakes	**2 Milkshakes**	**0 Milkshakes**	**2 Milkshakes**
134	141	152	91
153	154	158	97
102	109	134	73
158	165	176	111
97	104	108	47
133	140	151	90
115	128	139	82
156	163	174	113

8. Medical personnel have long disagreed on how much information surgical patients should receive prior to their operation. Some researchers believe that the amount and type of detailed information a patient needs for a good adjustment depends on the patient's personality. In one study, 20 patients considering dental surgery were classified as "Internals" and 20 as "Externals" on the Internal–External Locus of Control Scale.[‡] Internals believe that they have control over the reinforcements they obtain in everyday life, while externals believe that factors outside their power (fate, luck, etc.) are responsible for what happens to them. Half of each group of patients watched a videotape with *specific* information on the exact details of the upcoming dental surgery, while the other half viewed a tape of *general* information about the clinic, its policies, and usual dental procedures. The criterion measure was a mean rating by the dentists of how well the patient adjusted to the procedure (low number = good adjustment).

The reserachers hypothesized that internals would make a better adjust-

[†]Adapted from J. A. Hibscher and C. P. Herman. (1977). Obesity, dieting, and the expression of "obese" characteristics. *Journal of Comparative and Physiological Psychology, 91*, 374–380.

[‡]Adapted from S. M. Auerbach, et al. (1976). Anxiety, locus of control, type of preparatory information, and adjustment to dental surgery. *Journal of Consulting and Clinical Psychology, 44*, 809–818.

ment to the surgery if they received specific information, while externals would adjust better if they received more general information. Construct a table of means and a graph, and use the two-way ANOVA to see if the hypothesis was supported.

Internal Control		**External Control**	
Specific	**General**	**Specific**	**General**
6	7	9	7
5	6	10	6
4	7	9	5
4	4	6	3
5	10	8	8
5	6	10	6
5	6	8	4
6	8	10	9
8	9	7	9
7	8	10	7

CHAPTER 13

Some Nonparametric Statistical Tests

The t test and the F test, described in the two previous chapters, are called *parametric* tests; that is, they assume certain conditions about the *parameters* of the populations from which the samples are drawn. We have spent considerable time discussing these conditions, noting, for example, that the populations must be normal and must have equal variances. These conditions are not ordinarily tested themselves but are assumed to hold, and the meaningfulness that we attach to a particular value of t or F depends on whether these assumptions are valid. In addition, we have also emphasized that the data must be at least interval in nature before the operations necessary to calculate means and standard deviations can be performed and a t or F value obtained.

But what is the researcher to do when he or she knows that a set of data cannot meet these requirements? Even if assured that the normality and homogeneity of variance assumptions are valid, the researcher may frankly admit that the data in question are only nominal or, at best, ordinal in nature. Some measurement purists will go so far as to say that the *majority* of data in education and the behavioral sciences does not reach the precision of an interval scale. We may be able to tell that individual A has more of something than individual B, but we are not sure just *how much* more of it individual A has! Data such as this would be ordinal in nature.

For these reasons, a number of *nonparametric* statistical methods have been developed that allow us to run significance tests on data that do not meet the assumptions of the parametric tests. There is a wide variety of nonparametric tests, but we will be able to include only a small sample of them—just enough to demonstrate their ease of computation and wide applicability. We will consider the appropriate techniques for two general classes of data: independent samples and correlated samples.

The nonparametric tests to be treated in this section require *independent* samples. This simply means that the placement of subjects in one category does not affect the occurrence of subjects in another category. If John Doe turns up in one category or if he is assigned at random to a particular group, the selection has no bearing on any other category. In short, our observation of John Doe is independent of any other observation.

The Chi Square: One-Way Classification

Suppose that we flip a coin 20 times and record the frequency of occurrence of heads and tails. We know from the laws of probability that we should expect 10 heads and 10 tails. We also know that because of sampling error, we could easily come up with 9 heads and 11 tails or 12 heads and 8 tails. As we asked ourselves in Note 7.1, at what point do we say that an observed deviation from a 50-50 split is *not* sampling error but is due to some other factor, such as a biased coin? At what point do we say that there is a *significant* deviation between the theoretical 50-50 split and our observed frequency distribution?

A technique that can be used to determine whether there is a significant difference between some *theoretical* or *expected* frequencies and the corresponding *observed* frequencies in two or more categories is the *chi square test* (chi is denoted by the Greek letter χ). The formula for the calculation of chi square is

$$\chi^2 = \sum \frac{(O - E)^2}{E} \tag{13.1}$$

where
O is the observed frequency in a given category
E is the expected frequency in a given category

Let us suppose our coin-flipping experiment yielded 12 heads and 8 tails. We would enter our expected frequencies (10-10) and our observed frequencies in a table resembling Table 13.1.

TABLE 13.1. **Calculating a χ^2: One-Way Classification**

	Observed	**Expected**	$(O - E)$	$(O - E)^2$	$(O - E)^2/E$
Heads	12	10	2	4	0.4
Tails	8	10	−2	4	0.4
	20	20			$\chi^2 = 0.8$

The calculation of χ^2 in a one-way classification is very straightforward. The expected frequency in a category (e.g., "heads") is subtracted from the observed frequency, the difference is squared, and the square is divided by its expected frequency. This is repeated for the remaining categories, and, as the formula for χ^2 indicates, these results are summed for all the categories.

THE CHI SQUARE DISTRIBUTION. How does a calculated χ^2 of 0.8 tell us if our observed results of 12 heads and 8 tails represent a significant deviation from an expected 10-10 split? To answer this question, we again resort to the concept of a sampling distribution.

In the same way that we had a sampling distribution of t or F, we are now concerned with the sampling distribution of chi square. The shape of the chi square sampling distribution depends upon the number of degrees of freedom, and the distribution for $df = 6$ shown in Figure 13.1 illustrates this. As with other sampling distributions, we can mark off the different values that would be equaled or exceeded by sampling error a given percentage of the time. Note in Figure 13.1 that for 6 degrees of freedom χ^2 values of 12.59 or greater happen less than 5% of the time by chance, so we conclude that a χ^2 of 12.59 or greater is significant at the .05 level.

Since the shape of the sampling distribution depends on the number of degrees of freedom, the χ^2 value to be equaled or exceeded a given percentage of the time also depends on the degrees of freedom. Table G in Appendix 4 shows these values for the usual .05, .01, and .001 significance levels. For example, for 1 degree of freedom, we see that our calculated χ^2 must equal or exceed 3.84 to be significant at the .05 level.

The degrees of freedom for a one-way classification χ^2 is $r - 1$, *where r is the number of categories.* Since the one-way classification χ^2 is often presented in the tabular form of Table 13.1, r can be thought of as the number of rows. In the coin-flipping experiment of Table 13.1, there are two categories ($r = 2$), so there would obviously be 1 degree of

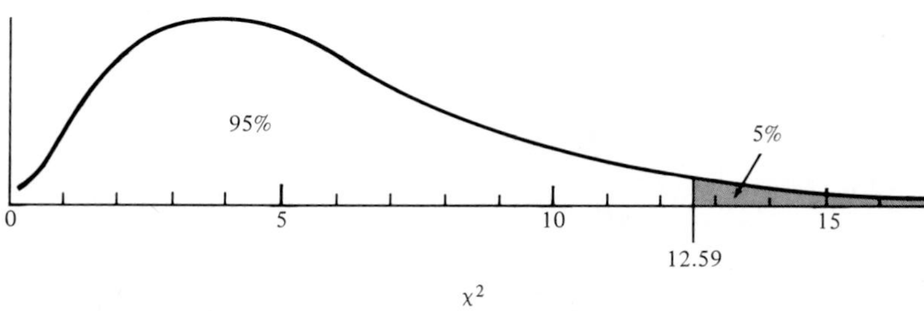

FIGURE 13.1. **Sampling distribution of chi square for 6 *df*.**

freedom. From Table G we see that a χ^2 of 3.84 or greater is needed for χ^2 to be significant at the .05 level, so we conclude that our χ^2 of 0.8 in the coin-flipping experiment could have happened by sampling error, and that the deviations between the observed frequencies and expected frequencies are not significant.

Let us consider another example that involves more than just two categories. A psychology instructor makes out his final grades for 200 students in an introductory psychology class. He is curious to see if his grade distribution resembles the "normal curve" and notes from the college catalog that in a normal distribution of grades 45% of them would be C's, 24% of them would be B's and 24% D's, and 3.5% of them would be A's and 3.5% F's. Table 13.2 shows the chi square table with the instructor's observed grade distribution and the distribution of letter grades that could be expected according to the normal curve model. Note that the professor obtained the expected distribution by multiplying the class size of 200 by the percentage for that letter grade (e.g., 3.5% × 200 = 7 A's).

TABLE 13.2. **Using χ^2 to Check "Normality" of a Grade Distribution**

Grades	Observed	Expected	$(O - E)$	$(O - E)^2$	$(O - E)^2/E$
A (3.5%)	15	7	8	64	$\frac{64}{7} = 9.14$
B (24%)	53	48	5	25	$\frac{25}{48} = 0.52$
C (45%)	87	90	−3	9	$\frac{9}{90} = 0.10$
D (24%)	33	48	−15	225	$\frac{225}{48} = 4.69$
F (3.5%)	12	7	5	25	$\frac{25}{7} = 3.57$
	200	200			$\chi^2 = 18.02$
					$\chi^2_{.01}$, $df = 4$, is 13.28
					Significant, $p < .01$

Note in the data of Table 13.2 that there are 4 degrees of freedom, since $df = r - 1$, where r is the number of categories. From Table G we see that for $df = 4$, a chi square value of 13.28 is needed for χ^2 to be significant at the .01 level, so we conclude that our χ^2 of 18.02 indicates that the professor's grade distribution deviates significantly from a normal distribution.

EFFECT OF SAMPLE SIZE ON ONE-WAY CLASSIFICATION χ^2. We must make certain in our calculation of χ^2 that when $df = 1$ (as in the coin-flipping data, where there are only two categories) the *expected frequencies are at least* 5. If values of E are less than 5, the χ^2 value is invalid.

When $df > 1$ (i.e., when you have more than two categories), the χ^2 technique should not be used when more than 20% of the expected frequencies are less than 5 or when any *single* expected frequency is less than 1. This problem can sometimes be eliminated by *combination* of adjacent categories. For example, if in the grade distribution data of Table 13.2 there had been only 100 students, there would have been two expected frequencies of less than 5 (3.5 in both the A and F categories). But the data could be combined into three categories instead of five (A with B, C, and D with F) and the problem of expected frequencies would be eliminated. There are a number of other restrictions governing the use of chi square, and we will consider these at length after the next section.

The Chi Square: Two-Way Classification

The two-way chi square is a convenient technique for determining the *significance of the difference* between the frequencies of occurrence in two or more categories with two or more groups. For example, we may ask if there is any difference in the number of freshmen, sophomores, juniors, or seniors as to their preference for spectator sports (football, basketball, or baseball). This is called a two-way classification, since we would need two bits of information from the students in our sample—their class and their sports preference.

Another use for a two-way classification chi square would be to see if there are sex differences for some variable. For example, Table 13.3 shows the educational level of 180 faculty members at a midwestern liberal arts college. The two elements of this two-way classification are, of course, sex and highest degree earned. This type of a two-way table is called a *contingency table,* and each entry is called a *cell.*

DETERMINING THE EXPECTED FREQUENCIES. In the one-way classification, the expected frequencies are determined by some *a priori* hypothesis: a 50-50 split in coin flipping or a normal distribution of grades. However, in the two-way classification, the expected values (sometimes called *independence values*) are calculated from the marginal totals of the contingency table. For example, the total number of faculty holding only a bachelor's degree is 15. Since there is a total of 180 faculty members, we know that $\frac{15}{180}$ or approximately 8% of the total group, both men and women, have only a bachelor's degree. Now, if the null hypothesis were true (i.e., if there were no difference in the frequency of men and women holding only a bachelor's degree), we would expect 8% of the men and 8% of the women to have only a bachelor's degree. Since there are 134 men, we simply multiply 8% by 134 to obtain 11. Thus, if there were no sex differences, we would expect 11 men to have only a bachelor's degree. Similarly, 8% times 46 yields an expected value of 4. We expect 4 women to have only a bachelor's degree. These

TABLE 13.3. **Sex Differences in Educational Level of Faculty Members**

		Highest Earned Degree			
		Bachelor's	Master's	Doctorate	Totals
Sex	Men	5 (11.17)	68 (75.19)	61 (47.64)	134
	Women	10 (3.83)	33 (25.81)	3 (16.36)	46
	Totals	15	101	64	N = 180

O	E	$(O - E)$	$\frac{(O - E)^2}{E}$
5	11.17	−6.17	3.41
10	3.83	6.17	9.94
68	75.19	−7.19	0.69
33	25.81	7.19	2.00
61	47.64	13.36	3.75
3	16.36	−13.36	10.91
		χ^2 =	30.70

$\chi^2_{.001}$, $df = 2$, is 13.82
Significant, $p < .001$

values (to two decimal places) are entered in parentheses for each cell of the contingency table.

A quick and handy way of calculating each of the expected values is to multiply the column total by the row total for each cell and divide by the total N. The expected values for each cell in Table 13.3 would be

$$\frac{15 \times 134}{180} = 11.17 \qquad \frac{101 \times 134}{180} = 75.19 \qquad \frac{64 \times 134}{180} = 47.64$$

$$\frac{15 \times 46}{180} = 3.83 \qquad \frac{101 \times 46}{180} = 25.81 \qquad \frac{64 \times 46}{180} = 16.36$$

After the expected values for each cell have been calculated, the same computational procedures for χ^2 are used as in the one-way classification. The differences between the observed and expected frequencies are found, the differences are squared, the squared difference is divided by its expected value, and the results of all the cells are summed. Notice that the *expected* values in each column or row add up to the same column or row total as the *observed* frequencies. This is an essential requirement for chi square, as we shall see in a later section.

DETERMINING THE DEGREES OF FREEDOM. It was easy to see that the *df* for the one-way classification was $r - 1$, because in order for the category frequencies to add up to the total number of cases all the frequencies could vary but one. For example, in the grade distribution data of Table 13.2, there were five categories, and $5 - 1 = 4$ degrees of freedom. That is, four of the category frequencies could take any value, while the fifth, in order for the total to be 200, would be dependent on the first four.

A similar techique is used in the two-way classification, but the number of degrees of freedom depends on both the number of rows *and* the number of columns. For the two-way classification chi square, $df = (r - 1)(c - 1)$, where r is the number of rows and c is the number of columns. For the example in Table 13.3, $df = (2 - 1)(3 - 1) = 2$. In other words, only two cell frequencies are free to vary; after two are given, the remaining frequencies are fixed so that the row and column totals will be correct. You may already have discovered this in examining the calculation of the expected values in Table 13.3. You would only have to calculate the expected values for Men-Bachelor's and Men-Master's, and the rest of the expected values could be obtained by subtraction. Although this is a tempting arithmetically sound shortcut, it might be a good idea to actually calculate several other expected values, in the event that you made a computational error on the first two!

WHAT DOES A SIGNIFICANT χ^2 MEAN? In Table G we find that with $df = 2$ a chi square of 13.82 is needed for the difference to be significant at the .001 level. Since chi squares of 13.82 or larger happen by chance less than 0.1% of the time, we conclude that our χ^2 of 30.70 is significant beyond the .001 level. Given the data in a contingency table, such as the sex and educational level tabulation of Table 13.3, what does a significant chi square mean? Besides indicating a significant deviation between observed and expected cell frequencies, it means that we can treat a significant χ^2 as either a *significant difference* between levels of one of the variables or as a *significant relationship* between the two variables.

In Table 13.3, our significant χ^2 tells us that there is a significant difference between men and women faculty members in their highest earned degrees. We can see that there are more women who have only a bachelor's degree than the number that would be expected by chance (10 vs. 4) and far fewer women who have a doctorate than the number that would be expected by chance (3 vs. 16).

Another way of describing our significant χ^2 would be to say that there is a significant relationship between the sex of a faculty member and the highest earned degree. An inspection of the contingency table shows that the relationship is due to the fact that more women than expected by chance have only a bachelor's degree, while more men and fewer women than expected by chance have a doctorate.

Although both of the above explanations are used, the *significant difference* approach is the more common.

THE 2 × 2 CONTINGENCY TABLE. When there are two levels of both variables in the two-way classification, the computational effort of calculating the χ^2 value is greatly reduced. The data are tabulated in a 2 × 2 (say "two by two") contingency table with the cell frequencies labeled *A* through *D*, along with the marginal totals and total *N*, as shown:

A	B	A + B
C	D	C + D
A + C	B + D	N

The computational formula is

$$\chi^2 = \frac{N(AD - BC)^2}{(A + B)(C + D)(A + C)(B + D)} \tag{13.2}$$

where

the letters *A* through *D* refer to the cell frequencies
N is the total number of observations
$(A + B)$ and $(C + D)$ are row totals
$(A + C)$ and $(B + D)$ are column totals

The degrees of freedom for the χ^2 calculated from this formula is always 1, since in a 2 × 2 table the quantity $(r - 1)(c - 1)$ is, of course, 1.

As an illustration of the 2 × 2 contingency table, let us use the survey mentioned earlier that assessed the number of hours studied per week by students classified as Type A or Type B. A total of 110 Type A or Type B students reported studying either 0 to 10 hours or 30 hours or more per week, and the results are recorded in Table 13.4.

The χ^2 of 12.81 is significant at the .001 level, so we conclude that there is a significant difference in the number of study hours reported by Type A and Type B students. By inspecting Table 13.4, we see that proportionately more Type A students reported studying more than 30 hours per week, while more Type B's said they studied less than 10 hours per week.

TABLE 13.4. **Number of Weekly Study Hours Reported by Type A and Type B Students**

		Study Hours per Week		
		0 to 10	30 or more	
Personality	Type A	25 (*A*)	26 (*B*)	51 (*A* + *B*)
	Type B	48 (*C*)	11 (*D*)	59 (*C* + *D*)
		73 (*A* + *C*)	37 (*B* + *D*)	110 (*N*)

$$\chi^2 = \frac{N(AD - BC)^2}{(A + B)(C + D)(A + C)(B + D)} = \frac{110(25(11) - 48(26))^2}{(51)(59)(73)(37)}$$

$$= \frac{110(275 - 1{,}248)^2}{8{,}127{,}309} = \frac{110(-973)^2}{8{,}127{,}309} = \frac{104{,}140{,}190}{8{,}127{,}309} = 12.81$$

$\chi^2_{.001}$, $df = 1$, is 10.83
Significant, $p < .001$

This computational formula for the 2 × 2 table eliminates the separate steps involved in calculating the expected values and performing the arithmetic operations for each step. However, the numbers do get unwieldy at times, and some pocket calculators may not be able to handle the large number of digits.

ASSUMPTIONS NECESSARY FOR CHI SQUARE. Even though a nonparametric statistic does not require assumptions regarding the population, there still are some restrictions regarding its use. Assumptions necessary for use of the chi square technique are

1. *The data must be in frequency form.* The entries in the cells indicate *how many* are in a given category and involve only a counting procedure. This is, basically, a technique for nominal data.
2. *The individual observations must be independent of each other.* This means, for example, that you cannot "inflate" χ^2 by asking each of 10 people to guess the suit of a playing card on four successive draws and then claim that you have an *N* of 40. Since a given person is making four guesses, we could expect that guesses on one occasion might be influenced by the guesses on previous occasions. Clearly, the four observations would be related and not independent of each other.
3. *Sample size must be adequate.* We have already noted that there were restrictions on the size of sample for the one-way classification χ^2. For the two-way classification we need to remember that:

a. In a 2 × 2 table, chi square should not be used if N is less than 20.
b. In a larger table, no cell should have an expected value of less than 1, and no more than 20% of the cells can have expected values of less than 5. For example, the data of Table 13.3 have one cell with an expected value of only 3.83. However, there are six cells, and since only one is short, $\frac{1}{6}$ = 16.7%, or less than 20%, as required.

There is not complete agreement among statisticians on this requirement, and your instructor may have a different view than that presented here. A correction is sometimes applied to the basic chi square formula, called a "correction for continuity," and your instructor or an advanced textbook may develop this concept further.†

4. *Distribution basis must be decided on before the data are collected.* In the grade distribution of Table 13.2, our psychology instructor was testing the deviation of his final grade distribution from a normal distribution. You can see that it would be inappropriate for him to "eyeball" his data and choose some kind of a distribution (i.e., rectangular, bimodal, or skewed) to compare with his distribution. Obviously, he could come up with a distribution that would yield nonsignificant or significant results to please his fancy. There must be a logical basis for his choosing the categories before he collects his data.
5. *The sum of the observed frequencies must equal the sum of the expected frequencies.* In order for the chi square formula to "work," the sums of the observed and expected frequencies have to be equal. Tables 13.1, 13.2, and 13.3 all show the expected and observed sums to be the same. Suppose we rolled a die 60 times and observed the frequency with which a "2" appeared. We could *not* use chi square to see whether there is a significant difference between the 14, for example, observed "2's" and the 10 expected by chance, because 14 and 10 are not the same. What is missing, of course, are the "non-2's." We can indeed run a chi square if we have 14 observed "2's" and 46 "non-2's," with expected values of 10 and 50, respectively, since now the sum of the observed frequencies is equal to the sum of the expected frequencies.

The Median Test

We can use this handy nonparametric test with two independent groups to see if they differ with respect to a combined median. The null hy-

†Camilli and Hopkins (1978) and others have reported that chi squares may be calculated on a 2 × 2 table where expected values are less than 5 without seriously affecting the usual probability levels.

pothesis would state that both groups are samples from populations with the same median. The scores (which must be ordinal or above) for both groups are used to calculate a combined median, and then the numbers of scores above and below the group median are tabulated. If there is no difference between the two groups, we would expect that about half of the scores in each group would be above this median and half below. On the other hand, if a majority of one group scores above the combined median while a majority of the other group scores below, we would have reason to believe that the two groups are different with respect to what is being measured. The following example should help clarify these preliminary statements.

A research psychologist was studying frustration in children and noticed that the maturity of a child's play activities decreased when the child was frustrated. She had a group of 30 nursery school children and divided them into two groups of 15 each. One group ate their noon lunch at the usual time (the satiated group), and the other group did not eat until mid-afternoon (the hungry group). Starting at 2:00 P.M. four nursery school teachers rated each child on "maturity of play activity" on a scale of 1 (very immature play) to 20 (extremely mature play). The scores are listed in the two columns at the top of Table 13.5. The psychologist hypothesized that the maturity scores would be higher for the satiated group than for the hungry group.

The first step in the median test is to calculate the median of *both groups combined*. The median for the group of 30 scores is 9.5. The next step in the median test is to tabulate the number from each group scoring above and below this combined median to see if there is a significant deviation from the 50-50 split that we would expect just by chance.

The numbers of each group scoring above or below the median are entered in a 2 × 2 contingency table, as shown in Table 13.5, and the computational formula for a 2 × 2 chi square is used as before. The calculated $\chi^2 = 1.20$ is not large enough for the difference to be significant at the .05 level, so we would conclude that the distribution of frequencies in the contingency table could have happened by chance. In terms of the experiment, not enough hungry children scored below the combined median and not enough satiated children scored above the combined median for us to attribute the difference to something other than chance.

The median test is easy to use, and you can modify it slightly to avoid troublesome complications. For example, if a number of scores are equal to the median, you can simply eliminate them from the analysis. This, of course, reduces the number of scores, which may not be desirable. If too many scores are at the median, you can avoid that problem by using the labels "number above the median" and "number

TABLE 13.5. **Median Test on the Maturity of Play Activity**

Hungry Group	Satiated Group
14	15
13	14
12	14
11	13
11	12
10	12
9	11
9	11
8	10
8	9
7	9
7	7
6	4
5	3
4	3

Combined *Med* = 9.5

	Hungry	Satiated	
Number above the median	A 6	B 9	15 (*A* + *B*)
Number below the median	C 9	D 6	15 (*C* + *D*)
	15 (*A* + *C*)	15(*B* + *D*)	30 (*N*)

$$\chi^2 = \frac{N(AD - BC)^2}{(A + B)(C + D)(A + C)(B + D)} = \frac{30(36 - 81)^2}{(15)(15)(15)(15)}$$

$$= \frac{30(2{,}025)}{50{,}625} = \frac{60{,}750}{50{,}625} = 1.20$$

$\chi^2_{.05}$, $df = 1$, is 3.84
Not significant, $p > .05$

not above the median" instead of the usual "above" and "below" dichotomy.

Since the 2 × 2 contingency table for chi square is used in the median test, the same restrictions apply to the median test as affected the 2 × 2 chi square. That is, the median test cannot be used when *N* is less than 20.

The Mann–Whitney *U* Test

The Mann–Whitney test is a very powerful nonparametric technique for determining whether two independent samples have been drawn from the same population. The measurements in both groups must be at least of an ordinal nature. The logic of the method is simple enough: If you have two populations from which you are drawing your two samples, the null hypothesis would state that both populations have the same distribution. That is, if you selected a score from one population, the probability that it would be larger than a score from another population would be $p = .50$. However, if we found that more scores from one population were larger than scores from the other population than would be expected by chance, we would reject the null hypothesis. We would conclude that there was a significant difference between the two populations.

SMALL SAMPLES. To see how the method works with small samples, let us consider the data of Table 13.6. An interviewer at a state political rally asked four people from community A and five people from community B to rate the efficiency of their local government on a scale from 1 (very inefficient) to 20 (extremely efficient). Was there a significant difference in the ratings of the two local governments?

TABLE 13.6. **Mann–Whitney *U* Test on Government Efficiency Data**

Community A	Community B
8	10
11	13
6	16
9	12
	18
$N_1 = 4$	$N_2 = 5$

6	8	9	10	11	12	13	16	18
A	A	A	B	A	B	B	B	B

A precedes B 19 times, $U' = 19$
B precedes A once, $U = 1$

$p = .016 \times 2 = .032$
Significant

The Mann–Whitney test requires only three bits of information:

1. N_1 = sample size of smaller group.
2. N_2 = sample size of larger group.
3. U = number of times a score in the larger group precedes a score in the smaller group when both groups are ranked together.

As shown in Table 13.6, we take the scores from the columns and rank them from smallest to largest, keeping their identities (A or B) intact. We then simply count the number of times that B (the largest group) precedes A. The only B preceding an A is the B of 10, which precedes an A of 11. Since this is the only B in front of an A, our $U = 1$.

We can check the accuracy of our counting by tabulating *the number of times A precedes B*. We see that the first B (10) is preceded by three A's, the second B (12) by four A's, and the remaining B's each by four A's. The total would be $3 + 4 + 4 + 4 + 4 = 19$, and this statistic is denoted U'. This can be a check on the accuracy of our tabulation of U, since there is a relationship between U and U', in that

$$U = N_1N_2 - U'$$

For the data of Table 13.6,

$$U = (4)(5) - 19$$
$$U = 1$$

It must be remembered that U is *always* smaller than U'. In fact, if we tabulate both, we can tell which value is U because it is the smaller of the two.

Once we have tabulated U, we again resort to a sampling distribution, and the probabilities or critical values are shown in Tables H and I in Appendix 4. Table H is to be used when neither N_1 nor N_2 is larger than 8, and Table I can be used when N_2 (the size of the larger group) is between 9 and 20.

Since the data of Table 13.6 have a small sample size, with $N_2 = 5$, we will use Table H to determine whether our $U = 1$ is statistically significant. The first page of Table H shows a table of $N_2 = 5$, and we go down the left-hand column to a U of 1. With $N_1 = 4$, the probability associated with values this small is $p = .016$. However, the probabilities given in Table H are for a *one-tailed test*. This means that whenever a two-tailed test is used, *these probabilities must be doubled*. Since we did not predict in advance which community would be rated more efficient, we have to make a two-tailed test, and the probability of obtaining a U as small as 1 in a two-tailed test is $.016 \times 2 = .032$. Since this probability is less than the usual .05, we conclude that there is a significant difference between the ratings of the two communities.

USING TABLE I WITH LARGER SAMPLES. When N_2 (the larger of the two samples) is between 9 and 20, we cannot use Table H to determine the exact probability of obtaining a value as small as U. With these larger samples, one of the tables from Table I must be used that shows the value of U needed for the difference to be significant at the .001, .01, .025, and .05 levels for a one-tailed test and the .002, .02, .05, and .10 levels for a two-tailed test.

As an example let us consider the data of Table 13.7, where a sample of 7 women drivers and 10 men drivers were assigned safe driving scores based on a complex formula involving accident rate and miles traveled. A higher score indicates a poorer driver. Was there a significant difference between the driving scores of men and women?

TABLE 13.7. **Mann–Whitney *U* Test on Safe Driving Scores**

Women	Men
37	17
52	72
67	62
27	75
32	69
22	85
13	42
	77
	71
	81
$N_1 = 7$	$N_2 = 10$

13	17	22	27	32	37	42	52	62	67	69	71	72	75	77	81	85
W	M	W	W	W	W	M	W	M	W	M	M	M	M	M	M	M

$$U = 1 + 1 + 1 + 1 + 2 + 3 = 9$$

Significant, $p < .02$ (two-tailed test)

Since U is the number of times scores from the larger sample precede scores from the smaller sample, we tabulate the number of times a woman's score (W) is preceded by a man's score (M). We see in Table 13.7 that the women's scores of 22, 27, 32, and 37 are each preceded by one man's score, a woman's score of 52 is preceded by two men's scores, and a woman's score of 67 is preceded by three men's scores. This gives the total of $U = 1 + 1 + 1 + 1 + 2 + 3 = 9$.

As a check on our calculations we tabulate U' and find that a man's score of 17 is preceded by one woman's score, a man's score of 42 is preceded by five women's scores, a man's score of 62 is preceded by six women's scores, and the remaining seven men's scores are each preceded by seven women's scores. This would give a total of $U' = 1 + 5 + 6 + 7 + 7 + 7 + 7 + 7 + 7 + 7 = 61$. Applying the check for accuracy, we obtain

$$U = N_1N_2 - U'$$
$$U = (7)(10) - 61$$
$$U = 9$$

Keeping in mind that we are running a two-tailed test, we now consult Table I in Appendix 4 and find the value of U for $N_2 = 10$ and $N_1 = 7$ in the four tables. We note that our value of $U = 9$ is between the tabled U of 5 in the .002 table and the U of 11 in the .02 table. Since our value of 9 is less than the tabled value of 11 in the .02 table, we say that our U is significant between the .02 and the .002 levels. (Note that with the Mann–Whitney test the smaller the value of U the greater the level of significance.) In terms of the experimental data, we conclude that the women's safe driving scores are significantly lower (indicating better drivers) than the men's scores.

The Mann–Whitney test can also be used for samples larger than 20, but space does not permit inclusion of the procedure in this introductory textbook. Other sources (e.g., Daniel, 1978) can provide you with this technique, as well as alternate ways of computing U and U' without going through the laborious (especially if N is large) counting procedure. It should be mentioned again that the Mann–Whitney test is one of the most powerful nonparametric tests, and it is strongly recommended when the assumptions cannot be met for the parametric t test.

The Kruskal–Wallis One-Way Analysis of Variance

The last nonparametric technique for independent samples to be discussed in this chapter is one that is markedly different from the procedures described earlier. Like the parametric analysis of variance, the Kruskal–Wallis test can be used *for more than just two samples.* This test assumes that the data are at least ordinal in nature so they can be converted to ranks.

The procedure requires converting the scores of the individual groups to one *overall* set of ranks. The data of Table 13.8 show the scores for

TABLE 13.8. **Kruskal–Wallis Test on Personal Concerns Inventory**

Freshman		Sophomore		Junior		Senior	
Score	Rank	Score	Rank	Score	Rank	Score	Rank
7	9.5	9	14	17	23	21	28
6	7.5	7	9.5	12	18.5	20	27
8	12	6	7.5	19	25.5	18	24
12	18.5	11	17	10	15.5	19	25.5
5	5.5	10	15.5	8	12	16	22
3	2.5	5	5.5	14	20	15	21
8	12	3	2.5				
2	1						
4	4						
$R_1 = 72.5$		$R_2 = 71.5$		$R_3 = 114.5$		$R_4 = 147.5$	
$N_1 = 9$		$N_2 = 7$		$N_3 = 6$		$N_4 = 6$	

$$H = \frac{12}{N(N+1)} \sum \frac{R_G^2}{N_G} - 3(N+1)$$

$$= \frac{12}{28(28+1)} \left[\frac{(72.5)^2}{9} + \frac{(71.5)^2}{7} + \frac{(114.5)^2}{6} + \frac{(147.5)^2}{6} \right] - 3(28+1)$$

$$= \frac{12}{812}(584.03 + 730.32 + 2{,}185.04 + 3{,}626.04) - 87$$

$$= 0.0148\,(7{,}125.43) - 87$$

$$H = 105.46 - 87 = 18.46$$

$\chi^2_{.001}$, $df = 3$, is 16.27
Significant, $p < .001$

four separate groups ($k = 4$), and their ranks in a single overall series. The smallest score is given a rank of 1, the next smallest a rank of 2, and so on. In Table 13.8, the freshman score of 2 has a rank of 1 and the senior score of 21 has a rank of 28. As with the Spearman method of correlation, from Chapter 8, scores that are tied are given the average ranking. For example, there are two scores of 3, which would be tied for second and third rank, so both scores of 3 are given a rank of 2.5, and the next score is given a rank of 4. Similarly, there are three scores of 8, tied for ranks of 11, 12, and 13, so each score of 8 has a rank of 12, and the next score is given a rank of 14.

The null hypothesis for the Kruskal–Wallis test should be obvious now that we have converted all the scores to ranks. If there were no differences between the groups, the sums of the ranks for each group would be about the same. That is, if only sampling error were responsible for differences in *scores* among the groups, the *sums of the ranks* for each group would only differ because of sampling error—and we would conclude that the samples had been drawn from the *same* population. On the other hand, if the sums of the ranks for each group differ from each other by more than we would expect by sampling error, we would conclude that these groups were samples from *different* populations, and we would have demonstrated that there is a significant difference between the populations.

The statistic to be evaluated in the Kruskal–Wallis test is H, and it is given by

$$H = \frac{12}{N(N + 1)} \sum \frac{R_G^2}{N_G} - 3(N + 1) \qquad (13.3)$$

where

N is the total numbers of scores
R_G^2 is the squared sum of the ranks in a group
N_G is the number of scores in a group

Although the formula for H looks somewhat imposing, it requires only that after ranking the scores in the manner described above, we proceed as follows.

1. Add the ranks for group 1 to obtain R_1.
2. Square this sum of ranks to obtain R_1^2.
3. Divide R_1^2 by N_1, the number of scores in group 1.
4. Since the above steps give us R^2/N for the first group only, we repeat these steps for all k groups (k = number of groups).
5. We sum the individual R^2/N for all groups, multiply by $12/N(N + 1)$, and subtract the quantity $3(N + 1)$.
6. If there are at least five cases in every group, the resultant statistic, H, is distributed as chi square with $df = k - 1$. H is also distributed as chi square if $k = 3$ and there are *more* than five cases in every group. For times when $k = 3$ and there are five cases or fewer in any group, Daniel (1978) has a special table, since the chi square distribution cannot be used.

Table 13.8 shows the results of a "Personal Concerns" questionnaire given to a sample of 28 college students. A personnel dean had predicted earlier that as the students got nearer to receiving their college degrees, they would check more and more items on the questionnaire that dealt with vocational and marital decisions. The sample consisted of 9 freshmen, 7 sophomores, 6 juniors, and 6 seniors. Was there a significant difference between the groups on the number of items checked relating to future decisions?

Since the value of H is greater than the tabled chi square value for $df = 3$ of 16.27, we conclude that our samples are from different populations, with $p < .001$ that such a result would happen by sampling error alone. In terms of the data, the seniors seem to have checked a significantly greater number of items relating to future vocational and marriage plans than have the freshmen and sophomores.

THE PROBLEM OF TIED RANKS. When the number of tied ranks becomes quite large, the value of H calculated by the formula is somewhat smaller than it should be. Daniel (1978) presents a correction for tied ranks that we will not consider here, since H is not seriously affected unless the tied ranks are unusually severe. In the data of Table 13.8, for example, 14 of the 28 scores are tied, and the corrected H is 18.52 instead of the 18.46 calculated in Table 13.8. Obviously, this did not affect our decision to reject the null hypothesis and conclude that the differences were significant. The Kruskal–Wallis test is a "conservative" test with respect to tied scores; that is, when the correction is made, it tends to make the test even more powerful in its rejection of a false null hypothesis.

In conclusion, we must note that the Kruskal–Wallis test is a reasonable alternative to the one-way analysis of variance when we cannot meet the assumptions of the parametric F test.[†]

Nonparametric Tests Using Related Samples

The nonparametric tests in this section are alike in that they all require *correlated data*. As you may remember from Chapter 10, we noted that one type of t test (the direct difference method) used correlated samples. We further noted that correlated data could involve matched samples (split litters, cotwin controls, or matched pairs) or repeated measurements of the same subjects. The nonparametric tests to be described in the following pages are designed for use with such correlated data.

The Sign Test

A very convenient test using matched pairs or repeated measurements is the sign test (so called because plus and minus signs are used to indicate changes). Each pair of measures is considered separately, and, if the second score is *greater* than the first, a minus sign is entered in the *Sign* column. If the second score is *smaller* than the first, a plus

[†]As with the parametric F test, there is also a procedure following a Kruskal–Wallis test to see which groups are significantly different. The technique is beyond the scope of this introductory textbook but is described in a text by Daniel (1978).

sign is entered. If both observations are the same, a 0 is entered. Since we must be able to tell which of the two scores is greater in a given pair, data must be at least ordinal. A typical sign test is shown in Table 13.9.

TABLE 13.9. **Sign Test on Number of Negative Statements Made During Therapy**

Client	First Session	Eighth Session	Sign
A	35	23	+
B	17	17	0
C	14	6	+
D	18	12	+
E	14	10	+
F	36	26	+
G	27	24	+
H	10	14	−
I	16	20	−
J	15	2	+
K	36	13	+
L	25	18	+
M	10	17	−
N	16	14	+

$N = 13$
$x = 3$

Significant, $p = .046$

It should be obvious from Table 13.9 that the null hypothesis would state that half the signs would be positive and half would be negative. In other words, if there are no significant differences between the pairs of measurements, about half of the changes should occur in a positive direction and half should be in a negative direction.

The example in Table 13.9 shows the results of a study involving the amount of negative, self-deprecating statements made by clients during psychotherapy. Tape recordings of the first therapy session for 14 clients at a regional mental health center were studied, and the number of negative statements made during therapy were tabulated. Eight sessions later a similar tabulation was made. Was there a significant reduction in the number of negative statements made?

The sign test makes use of two bits of information:

1. N is the number of changes. If there are any pairs that show a 0, they are subtracted from the total number of pairs. In the data of Table 13.9, there is one pair with a 0, so $N = 14 - 1 = 13$.

2. x is the frequency of occurrence of the sign that appears the least. There are 10 plus signs and 3 minus signs, so $x = 3$.

When N and x have been determined, we need to consult Table J in Appendix 4 to determine the probability of obtaining an x as small as our obtained value. Table J can be used for N values as large as 25. With $N = 13$ and $x = 3$, we see that the one-tailed probability of obtaining an x as small as 3 is .046. Since .046 is less than the conventional .05, we state that $p < .05$, and we conclude that there was a significant reduction in the number of negative statements made by clients in the therapy sessions.

We must remember that Table J shows *one-tailed* probabilities, meaning that the researcher is specifying in advance of collecting the data that the change will be in a specific direction. If the researcher is not specifying the direction of the difference, a two-tailed probability is needed, and the values in Table J must be *doubled*. Our researcher in the therapy survey *was* predicting a reduction in the number of negative statements, so she used the one-tailed probabilities in Table J.

The Wilcoxon Matched-Pairs Signed-Ranks Test

The sign test discussed in the previous section requires that we merely be able to tell whether one of a pair of scores is greater or less than the other. There may be times when our measuring scale is such that the difference between the two itself can be measured. For example, if pair A has a small difference between the two scores and pair B has a large difference, we would certainly attach more meaning to the larger difference. The sign test, of course, is not sensitive to the magnitude of the differences but only to the *direction* of the differences (plus or minus).

The Wilcoxon test uses the size of each difference. It requires that we determine the size of each difference, rank each difference without regard to its algebraic sign (smallest difference gets a rank of one), and then give each rank the sign of the original difference. Thus, if a rank is associated with a positive difference, it is given a plus sign. If the rank is associated with a negative difference, it is given a minus sign.

If the null hypothesis were true and the differences between paired values were due to sampling error, we would expect about as many small positive differences as small negative differences and about as many large positive differences as large negative differences. We would also expect the sum of the positive ranks to be about the same as the sum of the negative ranks. However, if we found one of these sums to be much smaller than the other, we would suspect that there was a significant difference between the two sets of scores.

Before we examine an example using the Wilcoxon test, we have to agree on how to handle the equivalent pairs and the tied values. If both scores in a single pair are the same, their difference is 0, and such pairs are eliminated from the data analysis, just as with the sign test. That is, N = the number of pairs minus the number of pairs whose difference is 0. In Table 13.10, pairs K and O were eliminated because both differences were 0, so $N = 20 - 2 = 18$.

To handle ties when we are ranking the individual differences, we again use the technique we used in the Spearman rank-difference method of correlation and in the Kruskal–Wallis test: Each tied difference is given the average rank. For example, in Table 13.10, differences of 6 and −6 are tied for fourth and fifth ranks, so each is given a rank of 4.5, and the next score is given a rank of 6. (Remember that in the Wilcoxon test the differences are ranked without regard to sign.)

TABLE 13.10. **The Wilcoxon Test on Knowledge-of-Results Data**

Pair	Control Group	Experimental Group	D	Rank of D	Rank with Less Frequent Sign
A	80	45	35	17	
B	24	62	−38	18	18
C	40	45	− 5	2.5	2.5
D	46	30	16	9	
E	85	56	29	13	
F	52	31	21	11	
G	55	35	20	10	
H	29	36	− 7	6	6
I	77	46	31	14	
J	59	49	10	7	
K	26	26	0	—	
L	46	35	11	8	
M	57	51	6	4.5	
N	46	52	− 6	4.5	4.5
O	69	69	0	—	
P	88	56	32	15	
Q	49	44	5	2.5	
R	49	25	24	12	
S	93	60	33	16	
T	35	39	− 4	1	1
					$T = 32.0$

$N = 18$
$T = 32$

Significant, $p < .01$

The data in Table 13.10 show the results of the project of an educational psychologist who was studying the effect that knowledge of results has on the learning process. Forty high school sophomores were used to form 20 pairs that were matched for sex, grade-point average, and manual dexterity. All students were tested on a tracking device, called a pursuit rotor, that tested motor coordination skills. Each member of the experimental group heard a click and saw an indicator light flash every time he or she made an error, while members of the control group had no extra information when they strayed off-target. Was there a significant difference between the number of errors made by the students who had immediate knowledge of results (experimental group) and the number made by those who did not receive additional knowledge of results (control group)?

The paired scores are first listed in columns, as shown in Table 13.10. The differences (D) are then found, the second score being subtracted from the first score in each pair, and the difference, with its algebraic sign, is placed in the D column. These differences are now ranked, *without regard to sign*, and each rank is placed in the appropriate column. As mentioned earlier, zero differences are eliminated from the analysis and tied differences are given the average of the ranks for which they are tied.

After the ranking is completed, we look at the D column again and note the differences that have the less frequent sign. In Table 13.10, these are the differences with the minus sign, *and the ranks corresponding to these differences* are then placed in the last column on the right. There are only five minus differences in the D column; their ranks are 18, 2.5, 6, 4.5, and 1, and the total of these ranks with the less frequent sign is 32.0. The total of this column is T, the statistic to be evaluated in the Wilcoxon test.

Table K in Appendix 4 contains the critical values of T necessary for significance at the customary significance levels for values of N up to 25. Since our researcher specified in advance that the knowledge-of-results group would have fewer errors, we use the values for the one-tailed test. We see that for $N = 18$, a T of 33 is significant at the .01 level. Since values smaller than the tabled values of T have a smaller probability of occurrence, our T of 32 is significant beyond the .01 level, and $p < .01$.

We noted earlier that the Wilcoxon test is preferred to the sign test when the magnitude of the differences between paired scores can be measured. If we were to apply the sign test to the data of Table 13.10, with $N = 18$ and $x = 5$, we would see from Table J that the one-tailed probability is .048. This is considerably larger than the probability reported by the Wilcoxon test in Table 13.10, reflecting the fact that the

Wilcoxon test is a more powerful test when the magnitude as well as the direction of the differences can be measured.

The Friedman Two-Way Analysis of Variance by Ranks

The Friedman test is a useful nonparametric test when we have *more* than two related samples. If our data are at least ordinal in nature, we can use the Friedman test to test the null hypothesis that our samples have been drawn from the same population. Although this procedure can be used for matched samples, it probably has its greatest application in situations where repeated measurements have been made on the same individual.

The rationale for the analysis is reasonably straightforward. If a table is constructed with the usual columns and rows, the rows correspond to individuals, and the columns are the experimental conditions. In the Friedman test, N is the number of rows, and k is the number of columns. The scores in the different conditions for each individual are entered in the appropriate columns in the row for that individual. Another table is now constructed which shows these scores converted to ranks, and the scores for *each* individual are ranked from 1 to k, where k is the number of conditions. That is, the ranks in each row should go from 1 to k. For example, in Table 13.11, student A has scores of 82, 57, 67, and 60 on four tests ($k = 4$), and these are ranked as 4, 1, 3, and 2 in the table of ranks. Similar rankings have been done for the other nine students.

Under the null hypothesis, the sums of the ranks should be about the same for each column. That is, if there are no differences in the experimental conditions, the ranks should be distributed evenly over the k conditions. However, if one of the conditions, let us say, has consistently lower scores, there will obviously be more ranks of 1 in the column corresponding to that condition, resulting in a smaller sum of the ranks for that column.

Let us consider an example to clarify some of these points. A freshman adviser at a small liberal arts college was curious about the special academic abilities of students who had decided to major in psychology. He wanted to know if beginning freshmen who chose psychology as a major had strong abilities in some special area such as mathematics or the natural sciences. To help answer this question, he chose 10 freshman psychology majors and examined their college entrance examination scores. This particular exam yielded a student's percentile rank in the areas of English, mathematics, social science, and natural science. The data for the 10 students are shown in Table 13.11. The first part of the table shows the percentile ranks in each of the four ability areas,

TABLE 13.11. **The Friedman Analysis of Variance on Special Abilities of Psychology Majors**

College Entrance Exam Percentile Ranks

Student	English	Mathematics	Social Science	Natural Science
A	82	57	67	60
B	61	95	79	92
C	88	82	99	70
D	86	98	77	97
E	77	89	72	97
F	99	85	95	98
G	77	46	95	87
H	61	71	79	56
I	10	30	28	19
J	61	34	54	57

Ranks

Student	E	M	SS	NS
A	4	1	3	2
B	1	4	2	3
C	3	2	4	1
D	2	4	1	3
E	2	3	1	4
F	4	1	2	3
G	2	1	4	3
H	2	3	4	1
I	1	4	3	2
J	4	1	2	3
	$R_1 = 25$	$R_2 = 24$	$R_3 = 26$	$R_4 = 25$

$$\chi_r^2 = \frac{12}{Nk(k+1)} \Sigma (R_G)^2 - 3N(k+1)$$

$$= \frac{12}{10(4)(4+1)} [(25)^2 + (24)^2 + (26)^2 + (25)^2] - 3(10)(4+1)$$

$$= \frac{12}{200} (625 + 576 + 676 + 625) - 150$$

$$= 0.06\ (2{,}502) - 150$$

$$\chi_r^2 = 0.12$$

$\chi^2_{.05}$, $df = 3$, is 7.82
Not significant, $p > .05$

and the second part of the table shows the four percentiles ranked for each student. Note that the null hypothesis would state that if there are no differences between the four ability areas the ranks will be distributed in such a way that the sums of the columns are about the same.

The ranks are assigned for each student's scores, with the smallest score being given a rank of 1. If there are tied scores, the familiar procedure of averaging the tied ranks is used. The ranks in the first column are then summed, yielding R_1. The sums are also calculated for the remaining columns. These values of R, along with N (the number of rows) and k (the number of columns) are then substituted into Friedman's formula:

$$\chi_r^2 = \frac{12}{Nk(k+1)} \sum (R_G)^2 - 3N(k+1) \qquad (13.4)$$

Note that the sum of ranks for each column is squared, these squares are added, this sum is multiplied by $12/Nk(k + 1)$, and, finally, the quantity $3N(k + 1)$ is subtracted from the above.

NOTE 13.1

NONPARAMETRIC STATISTICAL TESTS AND HOW THEY GREW

Compared to the parametric techniques, which evolved early in this century, nonparametric statistical methods have been developed in recent years. Frank Wilcoxon (1892–1965), an industrial chemist, proposed the test bearing his name in 1945, which helped to stimulate a variety of researchers in the development of other nonparametric methods. These methods found extensive use in education and the behavioral sciences, because much of the data of rating scales, surveys, and other scaling techniques yielded data that was on an ordinal scale and did not meet the assumptions of the usual parametric tests. This interest in small-group interactions and measurements of attributes was a primary factor in the immediate popularity of the techniques.

The resulting statistic is χ_r^2 (read "chi square for ranks"), and it is distributed approximately as chi square with $k - 1$ degrees of freedom. From Table G we see that, for $df = 3$, a chi square of 7.82 is needed for the difference to be significant at the .05 level. Our $\chi_r^2 = 0.12$ does not come close to being significant, so we must accept the null hypothesis that there are no differences between the conditions. This does not surprise us, since a glance at the first part of Table 13.11 shows that the ranks for the ability levels are spread out among the 10 students and that there are an approximately equal number of ranks of 1, 2, 3, and 4 in all four conditions. So, we conclude that, at least for this group of students, no ability area seemed to rank consistently high or low among those students choosing a psychology major.

It turns out that the statistic χ_r^2 is distributed as chi square with $df = k - 1$ only if the number of columns or the number of rows is not too small. If k is 3 and N is less than 10, or if k is 4 and N is less than 5, you cannot use the usual chi square table to determine significance levels. For these smaller samples, a special table of critical values of χ_r^2 is available (Daniel, 1978).

Concluding Remarks

At the beginning of the chapter we noted that the nonparametric tests were necessary for situations in which (1) the parametric assumptions regarding normality and homogeneity of variance of populations could not be met or (2) data were from a nominal or ordinal scale. We have seen in this chapter a wide variety of statistical tests that are appropriate for just these situations. A convenient way to remember the various nonparametric tests is to study them in tabular form according to the level of measurement (nominal or ordinal) and type of situation (independent samples or related samples). Table 13.12 summarizes these characteristics in a convenient form.

TABLE 13.12. **Level of Measurement and Application for Selected Nonparametric Tests**

	Independent Samples	**Related Samples**
Nominal	Chi square	
Ordinal	Median test Mann–Whitney U test Kruskal–Wallis ANOVA	Sign test Wilcoxon matched-pairs test Friedman ANOVA

The main weakness of nonparametric tests is that they are less powerful than parametric tests; that is, they are less likely to reject the null hypothesis when it is false. We must remember that when the assumptions of parametric tests can be met, parametric tests should definitely be used, because they are the most powerful tests available. However, if these assumptions cannot be met, we simply do not have any choice in the matter. We are left with nonparametric tests alone, and we must do the best we can.

But this is a rather negative approach, and it has led to the feeling that nonparametric tests are not quite legitimate and should be avoided whenever possible. This feeling is entirely unwarranted because nonparametric tests can be just as powerful as parametric tests if *the size of the samples* is increased. A researcher who had planned initially on using 40 subjects in her project might increase her sample size to 50 if she knew that her data could not meet the assumptions of a parametric test. The precise increases in sample size needed to achieve similar power in nonparametrics are discussed by Daniel (1978).

In some of the tests described in this chapter, there have been restrictions placed on the size of samples that could be used. In an introductory textbook it is just impossible to list all the alternative ways of handling very small samples or samples with an N of more than 25. The text by Daniel (1978) is an excellent resource for these special situations, and the researcher can find answers to just about any question on statistical analysis by nonparametric procedures.

SAMPLE PROBLEM

In a pilot study that established the link between Reye's (rhymes with "wise") syndrome and aspirin, investigators at the Centers for Disease Control in Atlanta noted that of 29 children contracting the disease, 28 had been given aspirin during the chicken pox or influenza that preceded the disease. For control groups, the researchers also contacted (1) 62 other children who were in the same school or hospital as the children with Reye's syndrome, and (2) 81 children whose parents responded to randomly placed telephone calls. Sixteen of the 62 children in the first control group and 45 of the 81 children in the second control group had been given aspirin during their chicken pox or influenza.

Let us construct a 2 × 3 contingency table and use a two-way chi square to see if there was a significantly larger proportion of children with Reye's syndrome that had been treated with aspirin.

		Group Category			
		Reye's Syndrome	School or Hospital	Telephone Calls	
Treatment	Aspirin	28 (15.0)	16 (32.1)	45 (41.9)	89
	No Aspirin	1 (14.0)	46 (29.9)	36 (39.1)	83
		29	62	81	172

Independence values:

$$\frac{89 \times 29}{172} = 15.0 \qquad \frac{89 \times 62}{172} = 32.1 \qquad \frac{89 \times 81}{172} = 41.9$$

$$\frac{83 \times 29}{172} = 14.0 \qquad \frac{83 \times 62}{172} = 29.9 \qquad \frac{83 \times 81}{172} = 39.1$$

Chi square:

$$\frac{(13.0)^2}{15} = 11.27 \qquad \frac{(16.1)^2}{32.1} = 8.08 \qquad \frac{(3.1)^2}{41.9} = 0.23$$

$$\frac{(13.0)^2}{14} = 12.07 \qquad \frac{(16.1)^2}{29.9} = 8.67 \qquad \frac{(3.1)^2}{39.1} = 0.25$$

$$\chi^2 = 40.57$$

$\chi^2_{.001}$, $df = 2$, $= 13.82$

Significant, $p < .001$

With $\chi^2 = 40.57$, $p < .001$, we would conclude that there is a significant difference between the observed and expected frequencies in the contingency table. In inspecting the table, we note that the large chi square value is due to the large proportion of children with Reye's syndrome that had taken aspirin.

STUDY QUESTIONS

1. Under what conditions would a nonparametric statistic be used instead of a parametric statistic?
2. What is the difference between *independent* samples and *correlated* samples?
3. Describe how expected frequencies are determined for the one-way chi square.
4. How does sample size affect the use of the one-way chi square?
5. How are expected frequencies calculated for the two-way chi square?
6. How are the degrees of freedom (*df*) determined for a one-way chi square?
7. Describe how the degrees of freedom for a two-way chi square are calculated.
8. List the assumptions for the chi square.
9. How would you state the null hypothesis for the median test?

10. State the null hypothesis for the Mann–Whitney test.
11. In the Kruskal–Wallis test, observations in each category are given an overall rank. Why should the rank sums of each category be the same under the null hypothesis?
12. In what ways are the sign test and the Wilcoxon test alike? How are they different?
13. How would you state the null hypothesis in the Friedman test?
14. What is the main weakness of the nonparametric tests? How might this be overcome?

EXERCISES

1. One retail store of a national discount chain caught 85 adults in the act of shoplifting during a 7-day period. There were 56 women and 29 men. According to FBI statistics, approximately 60% of shoplifters apprehended are women. Use a one-way chi square to see if this store's data agree with the 60%-40% split.
2. A recent study on the use of relaxation therapy and biofeedback to help control high blood pressure showed that of 21 patients, 18 could reduce blood pressures to normal levels using these behavioral techniques. Since, by chance, it should be as likely for blood pressures to increase as decrease, use a one-way chi square to see if this is a significant departure from a 50-50 split. (*Hint:* Assume two categories: blood pressure "decreased" and "did not decrease.")
3. In the study on drivers' behavior when a marked police car was present or absent referred to in Chapter 2, Nelson and Nilsen (1984) assessed the degree of "stopping" at a four-way stop sign. Use the two-way chi square to see if there was a significant difference in the stopping behavior of the 356 drivers when the police car was present and when it was absent.

		Police Car Condition	
		Present	Absent
Stopping Behavior	Complete Stop	132	39
	Rolling Stop	38	82
	Slight or No Speed Change	10	55

4. An educational psychology instructor kept a record of the number of times students requested a make-up of a chapter quiz during the semester. Her policy was to give a make-up only under "unusual circumstances," so a make-up request was a rare occurrence, which she noted in her grade book. At the end of the semester she saw that 49 different students had missed a quiz, and 19 of them had asked if they could take a make-up. She also noted their final course grade, and concluded that the superior students did not often ask for a make-up quiz, but students of lesser ability did. Use a two-way chi square to see if you would agree with her conclusions.

		Course Grade		
		A	*B*	*C*
Make-up Exam	Requested	1	9	9
	Not Requested	11	15	4

5. In the Pennsylvania State University study mentioned earlier, resting heart rates were measured for a group of 92 men and women students. The number of men and women above the combined median of 71 BPM is shown below. Use the median test to see if there is a significant difference in the heart rates of the men and women.

	Men	Women
Above Median	23	23
Below Median	34	12

6. In a study referred to earlier the College Life Questionnaire was administered to Type A and Type B students. The researchers hypothesized that for those students holding part-time jobs, Type A's would be more likely to work more hours per week. The number of Type A's and Type B's scoring above and below the combined median of 6.4 hours/week is shown below. Use the median test to see if there is a significant difference in the number of hours worked per week by Type A's and Type B's.

	Type A	Type B
Above Median	24	14
Below Median	16	22

7. The Veterans Administration long-term Normative Aging study has shown that men between the ages of 55 and 70 who were retired were just as healthy as a control group of similar age who continued working. One of the reasons for the study was to put to rest the folklore that a disproportionate number of seemingly healthy men become ill or die shortly after retirement. In a follow-up study, 18 male volunteers in good health at age 62 were tested at age 67. Eight of the men had retired at age 65 while the remaining 10 were still employed. Each volunteer filled out the Physical Symptoms Checklist (PSC), and the scores (higher scores = more physical symptoms and/or more frequent symptoms) are shown below. Use the Mann–Whitney test to see if there was a significant difference between the PSC scores reported by retired and working men.

Retired	Working
37	40
30	35
48	25
45	39
29	43
33	26
16	20
44	36
	42
	32

8. A zoologist was comparing island species with closely related species on the U.S. mainland, and noticed that a bird, the island jay, seemed more inquisitive than its mainland counterpart, the California jay (Haemig, 1984). In one experiment the researcher sounded a novel noise ("squeaking" a balloon), and when a jay arrived in response to the sound, timed how long the bird or birds remained in the vicinity. The researcher ran this trial 16 times on the island and 10 times on the mainland. The length of time the birds remained near the balloon (in seconds) is shown below for the 26 trials. Use the Mann–Whitney test to see if there was a significant difference in the time that the birds spent near the object.

Island		Mainland	
321	110	16	143
370	240	82	75
352	282	40	37
326	298	43	51
400	183	36	20
316	305		
331	300		
376	221		

9. The National Institutes of Health study on improving mental functioning in the elderly through aerobic exercise was noted in Exercise 14 in Chapter 11. "Mental dexterity" scores of 30 volunteers who had participated for 4 months in one of three exercise programs are again shown below. Use the Kruskal–Wallis test to see if there was a significant difference between the mental dexterity scores of the three groups.

Aerobic Exercise	Nonaerobic Exercise	No Exercise
69	51	59
74	41	56
72	34	54
78	56	39
61	61	44
64	61	49
49	56	34
54	46	64
59	51	34
59	36	69

10. An instructor in business statistics had 8 sophomores, 15 juniors, and 10 seniors in his class and was curious about the relative performance of the three classes. At the end of the term, he compared the final exam scores of the three classes, with the following results. Use the Kruskal–Wallis test to see if there was a significant difference between the classes.

Sophomores	Juniors	Seniors
20	28	35
17	35	37
27	30	18
29	36	32
19	20	19
18	36	27
26	35	37
28	34	17
	27	36
	22	29
	35	
	36	
	26	
	35	
	33	

11. In a test of a new drug intended to increase the attention span of hyperactive children, a neurologist asked the mothers or fathers of 16 children to observe their child's TV viewing for a period of 1 week prior to the onset of medication. After each child had been on medication for 10 days, the parents were again asked to indicate to the neurologist whether they felt that the child's attention span was shorter than (−), the same as (0), or longer than (+) when the child received no medication. Use the sign test (one-tailed probability) to see if there was an increased attention span after medication in a significant number of children.

Child	Span After Medication	Child	Span After Medication
A	+	I	−
B	0	J	+
C	+	K	0
D	+	L	+
E	−	M	0
F	0	N	+
G	0	O	+
H	+	P	+

12. Students in a psychological measurements course thought that their first 15 point quiz was very hard, while the instructor thought the last quiz was more difficult. The two quiz scores for the 20 students are shown below. Use the sign test to see if the students tended to score significantly lower on the last quiz.

Student	First	Last	Student	First	Last
1	11	10	11	14	11
2	12	7	12	11	11
3	13	12	13	15	10
4	14	10	14	13	11
5	13	11	15	13	11
6	11	12	16	12	13
7	11	10	17	12	14
8	13	7	18	13	8
9	10	11	19	14	13
10	14	14	20	12	12

13. In local golf tournaments, many participants have complained about "cup grabbers" who will score high on the first 18 holes; these higher scores are used to place them in their respective flights. Then they shoot better golf and score lower in order to win the flight in which they have been placed.

Such is not the case with the best golfers, since they must shoot a low score on the first 18 holes to be placed in the championship flight. To check on this bit of golfing folklore, a statistically minded sportswriter examined the scores for the first and second rounds for each golfer in the championship flight. Use the Wilcoxon method to see if there was a significant decrease from the first to the second round for the 16 golfers in the championship flight.

Championship Flight					
Golfer	**First 18**	**Second 18**	**Golfer**	**First 18**	**Second 18**
CS	73	78	JL	76	76
JM	74	76	JG	77	75
SM	74	79	RE	77	80
MO	75	73	BM	79	81
LD	75	74	MJ	80	72
MM	75	75	JK	80	79
MG	75	76	DS	80	83
MC	76	74	CB	80	77

14. Now use the Wilcoxon method to see if there was a significant decrease from the first to the second round for the 16 golfers in the last flight.

Last Flight					
Golfer	**First 18**	**Second 18**	**Golfer**	**First 18**	**Second 18**
ND	103	92	CL	109	107
PO	105	87	BL	109	101
AN	105	90	RE	110	111
CR	105	109	CP	111	95
LB	106	100	GE	111	98
TO	107	101	CA	112	104
FM	108	109	CT	112	102
NY	108	107	RC	115	108

15. A promotion for a popular soft drink invited customers in a supermarket to sample three brands and rank them according to taste. Twelve customers yielded the following data. Use the Friedman analysis of variance to determine if there was a significant difference in the ranks given the three brands.

Customer	Brand A	Brand B	Brand C
1	2	3	1
2	3	2	1
3	2	1	3
4	1	3	2
5	2	3	1
6	1	3	2
7	3	2	1
8	1	3	2
9	3	1	2
10	3	2	1
11	1	2	3
12	3	2	1

16. A TV sportscaster had 20 viewers call in, ranking their preferences for being a spectator at a World Series game, a Super Bowl, or a heavyweight championship boxing match. Their rankings are shown below. Use the Friedman analysis of variance to see if there was a significant difference in preferences for sporting events.

Fan	World Series	Super Bowl	Boxing Match	Fan	World Series	Super Bowl	Boxing Match
A	1	2	3	K	1	3	2
B	1	3	2	L	2	1	3
C	2	1	3	M	1	2	3
D	3	1	2	N	1	2	3
E	1	2	3	O	1	3	2
F	1	3	2	P	2	1	3
G	2	1	3	Q	2	3	1
H	1	2	3	R	1	3	2
I	1	3	2	S	1	3	2
J	2	1	3	T	2	1	3

CHAPTER 14

Statistical Methods in Test Construction

The inclusion of the following material in this introductory textbook in statistics is somewhat arbitrary—and also a little strange, since it is the only chapter devoted entirely to an *application* of statistical methods to a content field. But, since we are bombarded with tests almost from birth to death, any self-respecting student of education and the behavioral sciences should be acquainted with some of the statistical concepts that are important in test construction.

The application of statistics to testing is a broad subject, its topics ranging from construction of an individual test item all the way to mathematical theory about the accuracy of a test score. We can by no means cover all the statistical applications in a single chapter, or in a single volume for that matter. If you are interested in the subject, a textbook on tests and measurements (e.g., Aiken, 1985; Anastasi, 1982) may be helpful. Moreover, these may be able to point out additional sources for more specialized topics.

Our main interest in this chapter will be to examine two characteristics of a measuring instrument and to see how a statistical approach can be used to assess these characteristics. Specifically, we are interested in the *reliability* and the *validity* of a test.

Reliability

One important characteristic of any measuring device, be it a bathroom scale or a reading readiness test, is its *reliability*. That is, we expect the measurements yielded by any instrument to be *consistent* or *repeatable*. If Tommy's weight as taken at 1-hour intervals on the same scale were 75, 98, and 36 pounds, or his IQ scores as measured at 1-month intervals by the same test were 126, 180, and 92, we would seriously doubt the reliability of the bathroom scale and the particular IQ test. We expect a reliable instrument to give us consistent results. We do not, of course, expect to get *identical* results each time we measure some human characteristic more than once, because most instruments are not perfectly reliable, but we do expect a certain degree of "sameness" in our measurements.

It does not take a great deal of imagination to see that the methods of correlation discussed in Chapter 8 could be used to determine the reliability of a test. If we gave the same test twice to the same group of students, we would expect a high positive correlation if the test were reliable. If the measurements were consistent, those pupils who scored high on one administration of the test would tend to score high on the second administration, and those making low scores the first time should also score low the second time. The coefficient of correlation, the Pearson r, for example, would be a direct indicator of the amount of reliability. If r were high, it would indicate a great degree of consistency between the two administrations of the test. If some students who scored high on the first testing scored low on the second, the value of r would be lower, indicating a lower degree of reliability.

However, determining the reliability of a test is not quite as simple as it may first appear, and a number of special methods have been developed. We will consider three general methods often used by test writers and publishers to determine the reliability of their tests: the test–retest, the parallel forms, and the internal consistency methods.

Test–Retest Method

The simplest and most straightfoward method for determining reliability would be to give the test twice to the same group of individuals. We could then use the Pearson r to determine the correlation between the two administrations of the test. As we noted earlier, the coefficient of correlation yielded by this method would indicate the amount of relationship. If r were high, the measurements would be consistent and the test would be considered reliable.

Despite its simplicity, there are a number of problems involved in the test-retest method. If the test is repeated within a short time interval, many individuals might be able to recall answers that they had given previously and could spend more time on the more difficult material. Not all individuals would do this, of course, so this would increase some scores and not others and would lower the coefficient of correlation.

On the other hand, if the time interval between the initial administration of the test and its repetition is too long, growth and maturity (especially if the examinees are children) would affect performance on the second administration. Certain experiences by different individuals during this interval (e.g., peer tutoring or an extended illness) might influence their performance also—again tending to lower the correlation coefficient.

Because of the difficulty in controlling these conditions, other methods for determining test reliability are often used. However, the test–retest method is always included in a discussion of reliability, since it *defines* the term. Theoretically, a reliable test should always give consistent results from one administration to the next.

Parallel Forms Method

One obvious way to avoid the practice effect of the test–retest method for short time intervals would be to use another form of the same test. This parallel forms method (also called alternate forms or comparable forms) means that a test author has to write twice as many items as would be needed for just a single test. Two separate tests, for example, Form A and Form B, are then constructed and both forms of the test would be given to a sample of individuals. A time interval of 5 days to 2 weeks might separate the two administrations, but there should be very little practice effect, since the two forms contain completely different items. The reliability of the test is given by the Pearson r calculated between the two sets of test scores.

When accurate parallel forms have been constructed (such as Form L and Form M of one of the earlier editions of the Stanford-Binet IQ test), they have provided a very useful research tool and a defensible definition of reliability. Nevertheless, it is imperative to keep in mind that the use of parallel forms does not really eliminate the problem of growth or maturity over longer time intervals that we encountered in the test–retest method. As individuals change, so do their responses to an identical item or to a completely different item that is testing the same content. What the use of parallel forms does do is eliminate the *need* for long time intervals between two administrations of a test.

One obvious drawback to the parallel forms method lies in the sheer amount of labor required for an author to construct the parallel form. Writing good items for one test is a difficult task (as you may have learned in a tests-and-measurements course), and writing twice as many is a real problem. There may be a number of occasions when we would like to determine the reliability of a test we have constructed for classroom purposes, but when the extra time and effort involved would prevent us from using the parallel forms approach.

The two methods described above are similar in that they both involve *two* administrations: either of the same test or of parallel forms. However, methods involving *internal consistency* are based on a *single* administration of the test. For example, in the *split-half method,* the test is divided into two parts (such as two subsections, with random assignment of items into the two parts), and a correlation coefficient is calculated between the two portions of the test. Obviously, this correlation coefficient has a different interpretation from the coefficient for the test–retest and parallel forms methods: it describes *internal consistency,* the degree to which the different parts of the *same* test are in agreement.

One commonly used technique for obtaining split halves in order to determine internal consistency is the *odd–even method.* With this method, each individual has two scores: a score on the odd-numbered items in a test and a score on the even-numbered items. It is necessary to go through the answer sheet for each individual and tabulate the number of items correct on the odd items and the number correct on the even items. The scores are then placed in the familiar X and Y columns of the Pearson correlation method, as shown in Table 14.1. For example, suppose that we gave a 30-item multiple choice test to 10 students. (The examples in this chapter will, by necessity, involve small samples. Much larger numbers of examinees are used in actual situations.) Student A received a score of 22 on the test, so we go through the answer sheet and find that he got 12 of the odd-numbered items correct and 10 of the even-numbered items correct. These scores, and the odd and even scores for the remaining students, are shown in Table 14.1.

As you see from Table 14.1, the computational steps are identical to those for the correlation coefficients that you calculated in Chapter 8. However, $r_{oe} = .53$ is *not* the reliability coefficient but only the correlation between the two halves of the test. It has been demonstrated that reliability is directly related to the *length* of the test, and we have, in effect, cut our test in half by separating odd and even items. Our r_{oe} of .53 is artificially low as an estimate of the reliability of the original test.

For this reason it is necessary to correct for the effective length of the test by using the Spearman–Brown formula:

$$r_{tt} = \frac{2r_{oe}}{1 + r_{oe}} \tag{14.1}$$

Student	X (Odd)	Y (Even)	X^2	Y^2	XY
A	12	10	144	100	120
B	10	8	100	64	80
C	9	11	81	121	99
D	14	11	196	121	154
E	13	10	169	100	130
F	8	8	64	64	64
G	12	11	144	121	132
H	11	10	121	100	110
I	11	11	121	121	121
J	10	10	100	100	100
	110	100	1240	1012	1110

Correlation Coefficient:

$$r_{oe} = \frac{N\Sigma XY - \Sigma X \Sigma Y}{\sqrt{N\Sigma X^2 - (\Sigma X)^2}\sqrt{N\Sigma Y^2 - (\Sigma Y)^2}}$$

$$= \frac{10(1110) - (110)(100)}{\sqrt{10(1240) - (110)^2}\sqrt{10(1012) - (100)^2}}$$

$$r_{oe} = \frac{100}{189.65} = .53$$

Spearman–Brown:

$$r_{tt} = \frac{2r_{oe}}{1 + r_{oe}} = \frac{2(.53)}{1 + .53} = \frac{1.06}{1.53} = .69$$

where
r_{tt} is the reliability coefficient of the entire test
r_{oe} is the coefficient of correlation between the two halves

Using this formula with the data of Table 14.1, we see that the reliability of the entire test, r_{tt}, is .69. We will have occasion later in this chapter to discuss the relative sizes of reliability coefficients.

But what is the rationale for the odd–even method of determining reliability? The test–retest and parallel forms methods were straightforward: If the test were reliable, an individual would make comparable scores relative to the rest of the group on both administrations. Is there similar reasoning behind the odd–even method (and other internal consistency methods)? When we separate an individual's total score into odd and even correct, we expect to see similar results on the odd and even items (if odd and even items are of similar difficulty and are covering the same content) relative to the rest of the group. If this is the case, the test is reliable; that is, its measurements are consistent. If the individual does not do equally well on the odd and even halves, the test is not measuring consistently and, therefore, is not reliable. However, this "split-half" method for determining reliability is not measuring the same concept as the test–retest and parallel forms methods. It measures *internal* consistency and not consistency from administration to administration.

The split-half method should not be used on what are termed "speed tests," where every examinee would get *all* the items correct if enough time were allowed. Obviously, this would lead to problems with internal consistency, since every examinee would have the same score (i.e., 100% correct) on each half. On the other hand, "power tests," where items may vary in difficulty from very easy to very difficult, result in a variety of scores, and lend themselves to reliability estimates by the split-half technique.

In our discussion of internal consistency we have emphasized and illustrated the split-half method. Actually, there are many different methods for analyzing internal consistency, and our odd–even illustration is just one of these methods. There are several methods employing *item statistics,* which do not require that the test be split into separate halves. Some of these methods require information on how each student did on every test item, but this approach generally requires electronic scoring by a computer. You may find references to such techniques as Kuder–Richardson 20, Kuder–Richardson 21, and coefficient alpha. You will find information on these methods in most tests-and-measurements textbooks, such as Aiken (1985), Anastasi (1982), or Mehrens and Lehmann (1984).

Comparing the Three Methods

The test–retest, parallel forms, and internal consistency methods for estimating the reliability of a test will result in different estimates for the same test. This does not surprise us, since we know that there are specific characteristics that are unique to each method. For example, we have already noted that the test–retest method can be affected by practice and lengthy time intervals, the parallel forms method is sensitive to lengthy time intervals, and internal consistency methods are different from stability (test–retest) or equivalency (parallel forms) methods. In fact, split-half reliability coefficients tend to be higher than those obtained by either of the other two methods. For these reasons, it is important that the type of method used be cited in test manuals and journal articles, so the reader can interpret the results correctly.

Since all the methods have their shortcomings, is there any one method that is preferred? The answer to this question lies in the purpose of the investigator. If the researcher is involved in test theory—for example, the development of alternative methods for estimating reliability, or the effect of test length or item difficulty on reliability—he or she will probably choose one of the internal consistency methods. On the other hand, the test *user* (clinical psychologist, educational consultant) might be more interested in a parallel forms approach, since he or she may be running *before and after* studies where the same test is given a second time but in a different form. And the test–retest method is especially good where it is difficult to identify specific "items' in a test, such as might occur with the block design or picture arrangement subtests of a performance test. We are likely to see all three methods cited frequently in the literature; however, the fact that internal consistency methods are well-suited for computer analysis makes them very desirable.

One last word about reliability: It is not something "possessed" by a test, like an answer sheet or a title or a scoring key. There is nothing about the test itself that has the quality "reliability," because reliability is partly a function of *how* a test is used and on *whom* it is used. We certainly would not expect identical reliability estimates from a motor skills test given to seventh-graders in a health class and to retirement home residents in an exercise class.

We would be more accurate in our discussion if we talked about a test yielding "reliable results" or said. "The scores on such and such a test were reliable." Technically, we can talk of a "reliable test" only when we specify the conditions under which it was administered and the sample to whom it was given.

Standard Error of Measurement

Correlation coefficients and their interpretation may be difficult to apply in specific settings, especially for the person who has relatively little training in statistics. An alternative approach to using the reliability coefficient, r_{tt}, as an index of accuracy is the *standard error of measurement*. Although r_{tt} is a statistic describing *all* the scores in the distribution, the standard error of measurement is applied to a *single* score, and is used, basically, to set up a confidence interval that allows us to interpret an individual's score with a certain degree of precision.

Suppose, for example, that a student scores 79 on an examination. How much confidence do we have in this score? How much confidence do we have that our test is measuring accurately and that 79 is the individual's *true* score? (By a *true score,* we mean one in whose determination there are no errors of measurement present). We assume that if we could somehow measure this student over and over again, the mean of the student's scores would be the *true* score. The standard deviation of the distribution of scores would be called the *standard error of measurement.*

It is impractical to administer a test enough times to find a mean of the distribution to identify the student's true score. It is also unnecessary, for we can *estimate* the standard deviation of this distribution of repeated measurements and use a confidence interval approach to make a probability statement about the individual's true score. This estimate is the standard error of measurement, and it is given by

$$S_{EM} = S_T\sqrt{1 - r_{tt}} \tag{14.2}$$

where

S_T is the standard deviation of the original test scores
r_{tt} is the reliability coefficient of the test

For example, suppose that, in the situation above, the standard deviation of the test scores was 6.45 and the reliability coefficient of the test was .87. The standard error of measurement would then be

$$\begin{aligned} S_{EM} &= S_T\sqrt{1 - r_{tt}} \\ &= 6.45\sqrt{1 - .87} \\ &= 6.45\sqrt{.13} \\ &= 6.45(.36) \\ S_{EM} &= 2.33 \end{aligned}$$

We can now use S_{EM} just like any other standard error to determine a 95% confidence interval in which we expect the true mean (the individual's true score) to fall. That is, we could mark off $X \pm 1.96S_{EM}$ or $X \pm 2.58S_{EM}$, and these intervals would give probabilities of .95 and .99 that the individual's true score is in the interval. However, it has been conventional in test theory to mark off 1 S_{EM} above and below the individual's observed score to obtain $X \pm S_{EM}$. This, of course, gives the 68% confidence interval. For the situation above, with the individual's score of 79 and $S_{EM} = 2.33$, the confidence interval would be

$$\begin{aligned} X \pm S_{EM} &= 79 \pm 2.33 \\ &= 76.67 \text{ to } 81.33 \end{aligned}$$

We interpret this by saying that we are 68% certain that the true score is in the interval 76.67 to 81.33 or that the chances are 2 out of 3 that the true score is in that interval. Table 14.2 shows the standard error of measurement based on the data of Table 14.1. As an example, student H had a total score of 21 (11 on the odd items and 10 on the even items). What does the standard error of measurement tell us about the accuracy of this score?

TABLE 14.2. **Calculation of the True Score Interval for Student H**

Test Data:

$$\Sigma X_T = \Sigma X + \Sigma Y = 110 + 100 = 210$$

$$\Sigma X_T^2 = \Sigma X^2 + 2\Sigma XY + \Sigma Y^2 = 1240 + 2(1110) + 1012 = 4472$$

$$S_T = \frac{1}{N}\sqrt{N\Sigma X_T^2 - (\Sigma X_T)^2} = \frac{1}{10}\sqrt{10(4472) - (210)^2}$$

$$= \frac{\sqrt{620}}{10} = 2.49$$

$$r_{tt} = .69 \text{ (calculated previously)}$$

Standard Error of Measurement:

$$S_{EM} = S_T\sqrt{1 - r_{tt}} = 2.49\sqrt{1 - .69} = 2.49(0.56) = 1.39$$

Confidence Interval for Student H:

$$X \pm S_{EM} = 21 \pm 1.39 = 19.61 \text{ to } 22.39$$

Note in Table 14.2 that, before we can calculate S_{EM}, we have to determine S_T, the standard deviation of the original test scores. (The subscript, T, denotes the original test scores, so S_T is the standard deviation of the test scores, *not* of the odd or even scores.) We first calculate ΣX_T,the sum of the test scores, by simply adding ΣX and ΣY, where X and Y denote the odd and even data from Table 14.1. We then determine ΣX_T^2, the sum of the squared test scores, by finding $\Sigma X^2 + 2\Sigma XY + \Sigma Y^2$, where X and Y again refer to the odd and even values in Table 14.1. We then use the machine formua to obtain $S_T = 2.49$. The remaining steps to determine S_{EM} and the confidence interval are shown in Table 14.2.

We would conclude that the chances are 2 out of 3 that the student's true score lies between 19.61 and 22.39. We have done what we set out to do: Even though we cannot know what the true score may be for any examinee, we can state the probability that the true score is included in a certain interval.

The confidence interval for the true score has been used by some test publishers in constructing norms, and we noted in Chapter 4 that the use of percentile bands aided us in interpreting an individual's score. For example, we noted that a test score of 157 on the Cooperative English Test had a percentile band of from 29 to 52. This means that we are 68% certain that the individual's true score would lie in such a way that from 29% to 52% of the norm distribution was below the true score.

It should be obvious that errors of measurement are related to the size of the reliability coefficient. The greater the value of r_{tt}, the smaller the standard error of measurement and the greater the accuracy of the score. This is illustrated quite simply by the formula for S_{EM}. The size of the confidence interval indicates the amount of error, which is directly dependent upon $S_T\sqrt{1 - r_{tt}}$. If there is perfect reliability, $r_{tt} = 1.0$, S_{EM} is 0, and there is no error at all. On the other hand, if $r_{tt} = 0$, S_{EM} becomes the same as the standard deviation of the test. We can see that the higher the reliability of the test, the smaller the standard error of measurement.

Validity

When we ask the question "Are these test results consistent and repeatable?" we are talking about reliability. But when we ask, "Is this test testing what it is supposed to test?" we are concerned with *validity*, or how well the test is meeting its purpose. Various tests may yield valid measures of intelligence, manual dexterity, vocational interest, and a host of other traits that are amenable to measurement. Basically, we say that a test is valid if it measures what it purports to measure.

A test can be highly reliable without being valid. With a primitive IQ test that consisted of measuring the circumference of the head with a tape measure, we might get very *reliable* (i.e., consistent) measurements. However, such measurements would not be *valid* for estimating intelligence.

Test theorists have identified three basic types of validity: *content validity, construct validity,* and *criterion-related validity*. These terms all represent different ways of determining whether a test is "testing what it is supposed to test." Since the purpose of this chapter is to illustrate some of the statistical methods in test construction, we will confine our discussion to the third type, the criterion-related validity, because statistical analysis is a major factor in this method.

In order to determine whether a test is testing what it is supposed to test, we need to have some other measurement that is totally independent with which to compare our test. This independent "other measurement'' is called the *criterion*, and the proper selection of a criterion is a vital part of criterion-related validity. The criterion must be a valid and reliable measure of what we are interested in measuring with our own test. For example, if we are constructing a new intelligence test, we would probably choose a test like the Stanford-Binet or one of the Wechsler tests as our criterion. These tests have been shown previously to be valid and reliable measures of intellectual ability. If our new test compares favorably with the criterion, we can say that our test is also a valid measure of intelligence.

But a criterion need not necessarily be another test. If our task was to construct a test that would predict success in soldering electronic parts in an assembly line procedure, we would need some independent measure of job success as a criterion. Since job ''success'' is quite often defined in terms of production, we could use the number of parts produced by a worker per hour or per day as the criterion.

The criterion may take many forms in validity studies. For a college entrance examination the criterion might be GPA at the end of the freshman year; for a personality inventory the criterion might be a clinical evaluation by a psychiatrist or clinical psychologist; and for a mechanical aptitude test the criterion might be supervisor ratings of a group of diesel mechanics.

The precise *amount* of validity shown by a test in a given situation is the *statistical correlation* between the test and its criterion. If the coefficient is positive and relatively high, we would conclude that both the test and the criterion are yielding similar results, so *both must be measuring the same traits*. In other words, a test is valid in a given situation to the extent that it is yielding the same results as the criterion. The logical tool for measuring how pairs of scores are related is the Pearson r, and we call the result a validity coefficient.

There are two ways of looking at criterion-related valdity: we can speak of *predictive validity* or *concurrent validity*. These two methods are distinguished by the difference in the amount of time elapsing between the test administration and the measurement of the criterion.

Predictive Validity

Establishment of the *predictive* validity for a test in a given situation means that there will be a significant period of time between the test administration and the criterion measurement. This time interval must be of sufficient length that progress in the trait that is being predicted can be evaluated. Suppose, for example, that a college entrance exam is to be validated. Most exams are given to high school seniors during the fall of their senior year, so if the criterion were college GPA at the end of the freshman year there would be a time interval of at least a year and a half between the test administration and the correlation of the test scores with GPA.

We have already discussed an example of predictive validity back in Chapter 9, again using college entrance exams and GPA. The data are reproduced in Figure 14.1, and the validity coefficient is $r = .53$. By inspecting the scattergram, we see that, *in general*, those with higher entrance scores tended to make higher grades and those with the lower entrance scores tended to make lower grades.

Time intervals in predictive validity studies can be quite lengthy. It is not unusual to see validity coefficients reported on performance tests

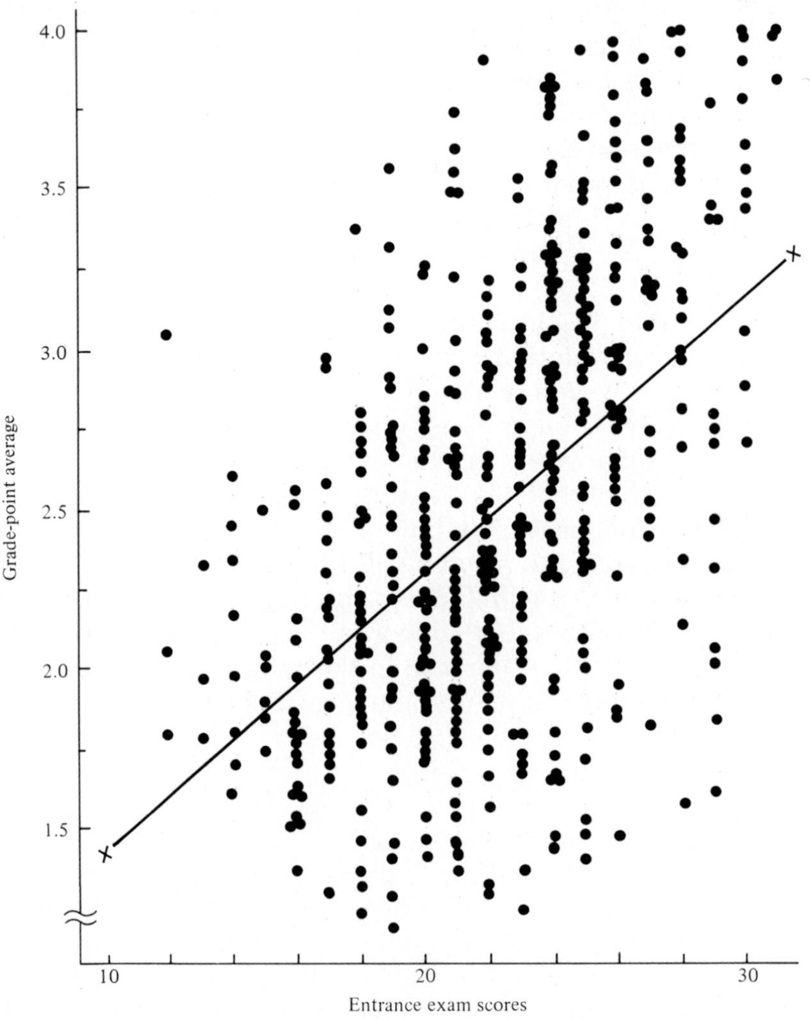

FIGURE 14.1. **Scattergram for examining the validity of a college entrance exam.**

of intelligence given to 4-year-olds with the criterion of high school or college grades. Of course, any longitudinal study of this sort has problems—not the least of which is the dropout rate of the examinees, whose families may move away before the study is completed.

Concurrent Validity

There are times when the investigator cannot wait the months or years between the administration of the test and the measurement of criterion performance. It might be too expensive to gather data and then wait until the examinees reach the criterion, or maybe the results are needed immediately and a lengthy time interval is just not a possibility.

In such cases, *concurrent* validity is the procedure of choice; it involves administering the test and measuring the criterion within a very short time interval. For example, a personnel director of a business firm may want to develop a clerical aptitude test that will predict how well an applicant will do as a data entry operator in the business office. If she were using a predictive validity approach, she would administer the test to applicants for typing jobs and then wait months (or even longer, if other specialized skills are involved) before measuring their criterion performance, such as on-the-job ratings by a supervisor. However, using a concurrent validity approach, she administers the test to typists *already* employed by the company and has a supervisor rate the typists on their job performance. The Pearson method of correlation then yields a validity coefficient for the test.

Concluding the topic of validity, we should note that it is somewhat misleading to say that "such and such a test is valid." A test is not valid in itself but is valid *for* a particular purpose. A given set of items may be valid for assessing English vocabulary, but *not* valid for assessing clerical skills or knowledge of internal combustion engines. In summary, we would say that validity is a *specific* characteristic of a test rather than a general one. For a good discussion of validity, as well as other topics in tests and measurements, see the excellent paperback by Tyler and Walsh (1979).

Concluding Remarks

Test theorists have shown that the size of the validity coefficient depends on the reliability of *both* the test *and* the criterion. This is an important point to remember, since many test constructors take great pains to devise a reliable test but then choose a criterion with questionable reliability. To obtain maximum predictive efficiency with a validity coefficient, you must have high reliability in both your test and the criterion.

Nothing has been said so far about the relative sizes of reliability and validity coefficients, and a complete discussion of this topic more properly belongs in a textbook on tests and measurements. However, you probably can get an idea of relative sizes by referring to the data of some representative tests shown in Table 14.3.

TABLE 14.3. **Some Reliability and Validity Coefficients for Selected Tests**

Test	**Reliability**	**Validity**
Differential Aptitude Test, Form L, Verbal Reasoning	Split-Half $r = .92$	With high school English grades $r = .63$
Minnesota Paper Form Board	Parallel forms $r = .78$	With freshman mechanical drawing grades $r = .49$
Peabody Picture Vocabulary Test, Age Level 18	Parallel forms $r = .84$	With Stanford-Binet $r = .71$
Purdue Pegboard	Test-retest $r = .84$	With radio assemblers $r = .64$
Stanford-Binet IQ, Form L-M	Test–retest (Age 3 and 1 year later) $r = .83$	Subtest with full test IQ $r = .61$

Reliability coefficients are usually higher than validity coefficients, and it is not unusual to find split-half coefficients in the .90 to .95 range. As we noted earlier, these internal consistency coefficients are generally higher than those found by either the parallel forms or test–retest methods. Not all reliability coefficients are as high as those shown in Table 14.3, however, and the reliability of such measures as supervisor or interview ratings are much lower.

Validity coefficients may range anywhere from .20 and up, with the .60 to .70 range considered quite high. Since the predictive validity coefficient is an index of how well the test predicts future performance, we are naturally interested in having r as large as possible. We must remember, though, that *any* significant validity coefficient indicates that the test is better than nothing. If a manual dexterity test correlates only .35 with some industrial job, it is still better to use the test as a means of screening new workers than just to give the job to any warm body that happens to wander in. If the turnover rate for a particular job was 19% before the test was used for screening and drops to 17% after the test is adopted, the difference of 2% may save the company thousands of dollars in supervisor time, reduce the waste of materials, and increase production during training.

We should also call attention to the fact that in many statistical studies of reliability and validity, relatively large numbers of examinees are used. The illustrative examples and problems in this chapter have used a very small N to prevent us from getting bogged down in computational procedures, but real-life situations demand as large a sample as possible to reduce measurement error. Most popular standardized tests have been evaluated on the basis of thousands of examinees.

We started this chapter with the reservation that we would be discussing only a small sample of statistical applications in test construction. If these selected topics have aroused your curiosity about the test construction field, you are referred to traditional textbooks on tests and measurements or educational statistics.

SAMPLE PROBLEM 1

A graduate class in educational testing constructed a short test that was designed to measure arithmetic skills. The test was given to 21 eighth-graders, and their scores on odd and even items follow. Calculate r_{tt}, S_{EM}, and the confidence interval for student L.

Student	*X*(Odd)	*Y*(Even)	X^2	Y^2	*XY*
A	15	14	225	196	210
B	15	16	225	256	240
C	16	16	256	256	256
D	17	18	289	324	306
E	13	12	169	144	156
F	15	16	225	256	240
G	16	16	256	256	256
H	16	15	256	225	240
I	16	18	256	324	288
J	19	18	361	324	342
K	16	11	256	121	176
L	13	13	169	169	169
M	15	15	225	225	225
N	12	11	144	121	132
O	17	15	289	225	255
P	14	12	196	144	168
Q	12	11	144	121	132
R	15	14	225	196	210
S	11	7	121	49	77
T	12	12	144	144	144
U	9	5	81	25	45
N = 21	304	285	4512	4101	4267

Correlation Coefficient:

$$r_{oe} = \frac{N\Sigma XY - \Sigma X \Sigma Y}{\sqrt{N\Sigma X^2 - (\Sigma X)^2}\sqrt{N\Sigma Y^2 - (\Sigma Y)^2}}$$

$$= \frac{21(4267) - (304)(285)}{\sqrt{21(4512) - (304)^2}\sqrt{21(4101) - (285)^2}}$$

$$= \frac{2967}{3381.65} = .88$$

Spearman–Brown:

$$r_{tt} = \frac{2r_{oe}}{1 + r_{oe}} = \frac{2(.88)}{1 + .88} = \frac{1.76}{1.88} = .94$$

Whole Test Data:

$$\Sigma X_T = \Sigma X + \Sigma Y = 304 + 285 = 589$$

$$\Sigma X_T^2 = \Sigma X^2 + 2\Sigma XY + \Sigma Y^2 = 4512 + 2(4267) + 4101 = 17{,}147$$

$$S_T = \frac{1}{N}\sqrt{N\Sigma X_T^2 - (\Sigma X_T)^2} = \frac{1}{21}\sqrt{21(17{,}147) - (589)^2}$$

$$= \frac{\sqrt{13{,}166}}{21} = 5.46$$

Standard Error of Measurement:

$$S_{EM} = S_T\sqrt{1 - r_{tt}} = 5.46\sqrt{1 - .94} = 5.46(.24) = 1.31$$

Confidence Interval for Student L:

$$X \pm S_{EM} = 26 \pm 1.31 = 24.69 \text{ to } 27.31$$

We conclude that the chances are 2 out of 3 that L's true score is included in the interval 24.69 to 27.31.

SAMPLE PROBLEM 2

The arithmetic skills test described above was found to be highly reliable, with $r_{tt} = .94$. The class then proceeded to determine the validity of this test by comparing it with a nationally standardized arithmetic achievement test. Both tests were administered to another group of 20 eighth-graders. Use the Pearson method of correlation to determine the validity of the class-made test.

Student	Class Test X	National Y	X^2	Y^2	XY
A	31	24	961	576	744
B	32	18	1,024	324	576

SAMPLE PROBLEM 2 (*cont.*)

C	33	29	1,089	841	957
D	34	23	1,156	529	782
E	36	26	1,296	676	936
F	35	32	1,225	1,024	1,120
G	34	23	1,156	529	782
H	34	30	1,156	900	1,020
I	34	26	1,156	676	884
J	37	27	1,369	729	999
K	35	37	1,225	1,369	1,295
L	36	34	1,296	1,156	1,224
M	34	31	1,156	961	1,054
N	37	32	1,369	1,024	1,184
O	37	31	1,369	961	1,147
P	37	25	1,369	625	925
Q	37	35	1,369	1,225	1,295
R	37	32	1,369	1,024	1,184
S	38	31	1,444	961	1,178
T	38	29	1,444	841	1,102
N = 20	706	575	24,998	16,951	20,388

Validity Coefficient:

$$r = \frac{N\Sigma XY - \Sigma X\Sigma Y}{\sqrt{N\Sigma X^2 - (\Sigma X)^2}\sqrt{N\Sigma Y^2 - (\Sigma Y)^2}}$$

$$= \frac{20(20{,}388) - (706)(575)}{\sqrt{20(24{,}998) - (706)^2}\sqrt{20(16{,}951) - (575)^2}}$$

$$= \frac{1{,}810}{(39.04)(91.62)} = .51$$

With a validity coefficient of .51, we would conclude that the class-constructed test is measuring to a moderate extent what the national test is measuring.

STUDY QUESTIONS

1. Many of the longer standardized tests are highly reliable. What does *reliable* mean?
2. Briefly summarize the three methods for determining the reliability of a test.
3. What is the basic difference between the internal consistency methods for determining reliability and the other two?
4. Why is the Spearman–Brown formula necessary when the split-half method is used?

5. Why should the split-half method not be used on speed tests?

6. For what kinds of test items is the test–retest method especially well-suited?

7. How are r_{tt} and S_{EM} different in the way in which they describe test consistency?

8. What does the interval $X \pm S_{EM}$ represent?

9. Many individuals confuse the terms *reliability* and *validity*. How would you answer a friend who asks "What is the difference between reliability and validity?"

10. How would you distinguish between *predictive* and *concurrent* validity?

11. How can a test be reliable without being valid?

EXERCISES

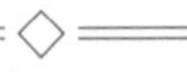

The majority of the following exercises have a small sample size (N) for ease of calculation. In actual practice, many studies on reliability and validity will have sample sizes in the thousands.

1. An instructor in health and recreation wanted to determine the reliability of the midterm examination for a wellness-fitness course he was teaching. He separated the odd (X) and even (Y) scores for each of the 50 students, with the following results.
   Caluclate r_{tt} for this examination.

Odd	**Even**
$\Sigma X = 911$	$\Sigma Y = 990$
$\Sigma X^2 = 20{,}950$	$\Sigma Y^2 = 23{,}850$

$$\Sigma XY = 21{,}700$$
$$N = 50$$

2. One of the students in the wellness-fitness class scored 29 on the midterm exam described in Exercise 1. Determine the 68% confidence interval for her score.

3. A communications instructor wanted to determine the reliability of her final examination. She broke down each student's score into odd-numbered answers (X) and even-numbered answers (Y) correct. The results for her 60 students follow. Determine r_{tt} for her exam.

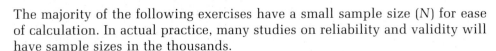

$$\Sigma X = 1{,}161 \qquad \Sigma Y = 1{,}290$$
$$\Sigma X^2 = 23{,}499 \qquad \Sigma Y^2 = 28{,}860$$
$$\Sigma XY = 25{,}778$$
$$N = 60$$

4. One of the students scored 41 on the communications exam described in Exercise 3. Determine the 68% confidence interval for his score.

5. A vocabulary test is given to 15 eighth-graders and their scores are separated into odd (X) and even (Y) answers correct. Calculate r_{tt} for this vocabulary test.

Student	X(Odd)	Y(Even)	Student	X(Odd)	Y(Even)
A	10	9	I	13	14
B	9	9	J	16	15
C	9	10	K	14	15
D	10	11	L	12	13
E	9	10	M	12	12
F	11	12	N	5	5
G	5	8	O	8	9
H	7	13			

6. One eighth-grader scored 31 on the vocabulary test in Exercise 5. Calculate the 68% confidence interval for his score.

7. An educational psychologist is constructing a college entrance exam that would predict a student's success in college. She administers her exam (X) to 15 college-bound high school seniors and, one year later, tabulates their college GPAs (Y). Calculate the Pearson r to determine the predictive validity of the exam.

Student	X	Y	Student	X	Y
A	21	2.7	I	15	2.7
B	20	2.5	J	14	2.7
C	19	3.1	K	13	2.6
D	17	3.1	L	13	2.3
E	16	3.1	M	12	2.5
F	15	2.7	N	10	2.9
G	15	2.2	O	10	2.8
H	15	2.6			

8. The personnel manager of a large insurance company is developing a new clerical aptitude test and wants to validate the test against the well-known Minnesota Clerical Aptitude Test. He administers both tests to a group of 20 clerical employees, and the scores (number of errors) on both tests follow. Use the Pearson r to determine the validity of this new test.

Employee	Local Test	Minnesota Clerical	Employee	Local Test	Minnesota Clerical
A	14	9	K	14	8
B	14	9	L	12	7
C	13	9	M	12	7
D	14	8	N	11	8
E	14	8	O	13	7
F	13	9	P	13	8
G	13	9	Q	12	7
H	12	7	R	13	5
I	14	7	S	10	7
J	13	7	T	10	6

The remaining exercises for this chapter are based on the research data shown below. A clinical psychologist is developing a culture-fair test that she hopes will be a true measure of intelligence unaffected by race, culture, religion, and so on. She administers the test to 16 college students; the following data show the test score, number of odd (X) and even (Y) correct, and the students' grade-point-average at the end of their freshman year.

Student	Score	*X*(Odd)	*Y*(Even)	GPA
A	56	27	29	2.90
B	73	37	36	3.67
C	71	35	36	3.52
D	59	28	31	3.50
E	59	29	30	3.00
F	52	28	24	3.10
G	57	29	28	3.00
H	71	35	36	3.25
I	54	28	26	3.30
J	68	38	30	2.30
K	53	28	25	3.10
L	66	35	31	2.97
M	67	35	32	3.50
N	52	27	25	3.33
O	64	34	30	2.50
P	64	34	30	3.37

9. Calculate the split-half reliability coefficient, r_{tt}, for this test.

10. One of the students scored 66 on the culture-fair test. Calculate the 68% confidence interval for her score.

11. The researcher was interested in how well the culture-fair test predicted success in college. Calculate the validity coefficient for the test scores with grade-point average.

References

Aiken, L. (1985). *Psychological Testing and Assessment*, 5th ed. Boston: Allyn & Bacon, Inc.

Anastasi, A. (1982). *Psychological Testing*, 5th ed. New York: Macmillan Publishing Company.

Bartz, A. (1976). Peripheral detection and central task complexity. *Human Factors, 18*, 63–70.

Blaske, D. M. (1984). Occupational sex-typing by kindergarten and fourth-grade children. *Psychological Reports, 54*, 795–801.

Camilli, G., and K. D. Hopkins (1978). Applicability of chi square to 2 × 2 contingency tables with small expected cell frequencies. *Psychological Bulletin, 85*, 163–167.

Daniel, W. (1978). *Applied Nonparametric Statistics*. Boston: Houghton Mifflin Company.

Follman, J. (1984). Cornucopia of correlations. *American Psychologist, 39*, 701–702.

Haemig, P. (1984). Enhanced exploratory behavior in island jays. Unpublished manuscript, Concordia College.

McCall, R. (1986). *Fundamental statistics for the behavioral sciences*, 4th ed. Orlando, Fla.: Harcourt Brace Jovanovich.

Mehrens, W., and I. Lehmann (1984). *Measurement and Evaluation*, 3rd ed. New York: Holt, Rinehart and Winston.

Nelson, M., and K. Nilsen (1984). Conformity in the presence and absence of a marked police vehicle. Unpublished manuscript, Concordia College.

Reynolds, B., and L. McDuffy (1985). Interviewer sex difference as a function in male self-disclosure. Unpublished manuscript, Concordia College.

Spielberger, C. D. (1983). *Manual for the State-Trait Anxiety Inventory*. Palo Alto: Consulting Psychologists Press, Inc.

Tifft, S., J. O'Reilly, and M. Vollers (1985). The triumphant spirit of Nairobi. *Time, 126*, 38–40.

Tyler, L., and W. Walsh (1979). *Tests and Measurements*. Englewood Cliffs, N.J.: Prentice-Hall, Inc.

Woods, P., and J. Burns (1984). Type A behavior and illness in general. *Journal of Behavioral Medicine, 7*, 411–415.

Appendixes

APPENDIX 1

Formulas

Name	Formula	First appears on page
Arithmetic mean (3.1)	$\overline{X} = \frac{\Sigma X}{N}$	53
Median (3.2)	$Med = L + \left(\frac{0.5N - cum\ f}{f_{MED}}\right)$	58
Population mean (3.3)	$\mu = \frac{\Sigma X}{N}$	64
Combined mean (3.4)	$\overline{X} = \frac{N_1\overline{X}_1 + N_2\overline{X}_2 + N_3\overline{X}_3 + \cdots}{N_1 + N_2 + N_3 + \cdots}$	65
Average deviation (5.1)	$AD = \frac{\Sigma \|(X - \overline{X})\|}{N}$	96
Standard deviation—deviation formula (5.2)	$S = \sqrt{\frac{\Sigma(X - \overline{X})^2}{N}}$	97
Standard deviation—mean formula (5.3)	$S = \sqrt{\frac{\Sigma X^2}{N} - \overline{X}^2}$	99
Standard deviation—machine formula (5.4)	$S = \frac{1}{N}\sqrt{N\Sigma X^2 - (\Sigma X)^2}$	100
z score—sample (5.5)	$z = \frac{X - \overline{X}}{S}$	107
z score—population (5.6)	$z = \frac{X - \mu}{\sigma}$	108
Semi-interquartile range (5.7)	$Q = \frac{P_{75} - P_{25}}{2}$	113
Standard error of the mean—with population standard deviation (7.1)	$\sigma_{\overline{X}} = \frac{\sigma}{\sqrt{N}}$	153
z score—sampling distribution of means (7.2)	$z = \frac{\overline{X} - \overline{X}_{\overline{X}}}{\sigma_{\overline{X}}}$	154

(continued)

Name	Formula	First appears on page
95% confidence interval for μ using population σ (7.3)	$\overline{X} \pm 1.96\sigma_{\overline{X}}$	161
99% confidence interval for μ using population σ (7.4)	$\overline{X} \pm 2.58\sigma_{\overline{X}}$	162
Biased estimate of population variance (7.5)	$S^2 = \frac{\Sigma(X - \overline{X})^2}{N}$	171
Unbiased estimate of population variance (7.6)	$s^2 = \frac{\Sigma(X - \overline{X})^2}{N - 1}$	171
Pearson r—z score method (8.1)	$r = \frac{\Sigma z_X z_Y}{N}$	184
Pearson r—mean and S formula (8.2)	$r = \frac{\frac{\Sigma XY}{N} - \overline{X}\overline{Y}}{S_X S_Y}$	186
Pearson r—machine formula (8.3)	$r = \frac{N\Sigma XY - \Sigma X \Sigma Y}{\sqrt{N\Sigma X^2 - (\Sigma X)^2}\sqrt{N\Sigma Y^2 - (\Sigma Y)^2}}$	187
Standard error of Pearson r (8.4)	$s_r = \frac{1}{\sqrt{N - 1}}$	191
Spearman rank-difference correlation coefficient (8.5)	$r_s = 1 - \frac{6\Sigma D^2}{N(N^2 - 1)}$	195
Regression equation (9.1)	$Y' = \left(\frac{rS_Y}{S_X}\right) X - \left(\frac{rS_Y}{S_X}\right) \overline{X} + \overline{Y}$	213
Standard error of estimate (9.2)	$s_E = S_Y\sqrt{1 - r^2}$	220
68% confidence interval for Y' (9.3)	$Y' \pm s_E\sqrt{1 + \frac{1}{N} + \frac{(X - \overline{X})^2}{NS_X^2}}$	221
z score—sampling distribution of differences (10.1)	$z = \frac{(\overline{X}_1 - \overline{X}_2) - 0}{\sigma_{D_{\overline{X}}}}$	243
Standard error of the difference between means—population standard deviation (10.2)	$\sigma_{D_{\overline{X}}} = \sqrt{\sigma_{\overline{X}1}^2 + \sigma_{\overline{X}2}^2}$	247
Standard error of the difference between means—sample standard deviation (10.3)	$s_{D_{\overline{X}}} = \sqrt{\frac{N_1 S_1^2 + N_2 S_2^2}{N_1 + N_2 - 2}\left(\frac{1}{N_1} + \frac{1}{N_2}\right)}$	247
t test—independent samples (10.4)	$t = \frac{(\overline{X}_1 - \overline{X}_2) - 0}{s_{D_{\overline{X}}}}$	248
Mean of the differences—direct difference method for correlated t test (10.5)	$\overline{D} = \frac{\Sigma D}{N}$	258

Name	Formula	First appears on page
Standard deviation—direct difference method for correlated *t* test (10.6)	$S_D = \sqrt{\frac{\Sigma D^2}{N} - \overline{D}^2}$	260
Standard error—direct difference method for correlated *t* test (10.7)	$s_{\overline{X}_D} = \frac{S_D}{\sqrt{N-1}}$	260
t test—correlated samples (10.8)	$t = \frac{\overline{D}}{s_{\overline{X}_D}}$	260
Standard error of the mean—with sample standard deviation (10.9)	$s_{\overline{X}} = \frac{S}{\sqrt{N-1}}$	269
95% confidence interval for μ using sample S (10.10)	$\overline{X} \pm t_{05}s_{\overline{X}}$	270
99% confidence interval for μ using sample S (10.11)	$\overline{X} \pm t_{01}s_{\overline{X}}$	270
Total sum of squares—deviation formula (11.2)	$SS_T = \sum^{k}\sum^{N_G}(X - \overline{X}_T)^2$	284
Within-groups sum of squares—deviation formula (11.3)	$SS_{WG} = \sum^{k}\sum^{N_G}(X - \overline{X}_G)^2$	284
Between-groups sum of squares—deviation formula (11.4)	$SS_{BG} = \sum^{k}N_G(\overline{X}_G - \overline{X}_T)^2$	285
Mean square between groups (11.5)	$MS_{BG} = \frac{SS_{BG}}{k-1}$	287
Mean square within groups (11.6)	$MS_{WG} = \frac{SS_{WG}}{N_T - k}$	287
F test (11.7)	$F = \frac{MS_{BG}}{MS_{WG}}$	289
Total sum of squares—computational formula (11.8)	$SS_T = \sum^{N_T}X^2 - N_T\overline{X}_T^2$	290
Between-groups sum of squares—computational formula (11.9)	$SS_{BG} = N_1\overline{X}_1^2 + N_2\overline{X}_2^2 + N_3\overline{X}_3^2 + \cdots - N_T\overline{X}_T^2$	290
Within-groups sum of squares—computational formula (11.10)	$SS_{WG} = \left(\sum^{N_1}X^2 - N_1\overline{X}_1^2\right) + \left(\sum^{N_2}X^2 - N_2\overline{X}_2^2\right) + \left(\sum^{N_3}X^2 - N_3\overline{X}_3^2\right) + \cdots$	291
Tukey's method—unequal N's (11.11)	$q = \frac{\overline{X}_L - \overline{X}_S}{\sqrt{\frac{MS_{WG}}{2}\left(\frac{1}{N_L} + \frac{1}{N_S}\right)}}$	294

(continued)

Name	Formula	First appears on page
Tukey's method—equal N's (11.12)	$q = \dfrac{\overline{X}_L - \overline{X}_S}{\sqrt{\dfrac{MS_{WG}}{N_G}}}$	296
Sum of squares for row variable (12.1)	$SS_R = N_{R_1}\overline{X}_{R1}^2 + N_{R_2}\overline{X}_{R2}^2 + \cdots - N_T\overline{X}_T^2$	313
Sum of squares for column variable (12.2)	$SS_C = N_{C_1}\overline{X}_{C1}^2 + N_{C_2}\overline{X}_{C2}^2 + \cdots - N_T\overline{X}_T^2$	313
Sum of squares for interaction (12.3)	$SS_{R\times C} = SS_{BG} - SS_R - SS_C$	313
df for rows (12.4)	$df_R = r - 1$	313
df for columns (12.5)	$df_C = c - 1$	314
df for interaction (12.6)	$df_{R\times C} = (r - 1)(c - 1)$	314
Within-groups *df* (12.7)	$df_{WG} = N_T - rc$	314
Mean square for rows (12.8)	$MS_R = \dfrac{SS_R}{r - 1}$	315
Mean square for columns (12.9)	$MS_C = \dfrac{SS_C}{c - 1}$	315
Mean square for interaction (12.10)	$MS_{R\times C} = \dfrac{SS_{R\times C}}{(r - 1)(c - 1)}$	315
Within-groups mean square (12.11)	$MS_{WG} = \dfrac{SS_{WG}}{N_T - rc}$	315
F for row effect (12.12)	$F_R = \dfrac{MS_R}{MS_{WG}}$	318
F for column effect (12.13)	$F_C = \dfrac{MS_C}{MS_{WG}}$	318
F for interaction effect (12.14)	$F_{R\times C} = \dfrac{MS_{R\times C}}{MS_{WG}}$	318
Chi square (13.1)	$\chi^2 = \sum \dfrac{(O - E)^2}{E}$	332
Chi square—2 × 2 table (13.2)	$\chi^2 = \dfrac{N(AD - BC)^2}{(A + B)(C + D)(A + C)(B + D)}$	338
Kruskal–Wallis test (13.3)	$H = \dfrac{12}{N(N + 1)} \sum \dfrac{R_G^2}{N_G} - 3(N + 1)$	348
Friedman analysis of variance (13.4)	$\chi_r^2 = \dfrac{12}{Nk(k + 1)} \sum (R_G)^2 - 3N(k + 1)$	356
Spearman–Brown formula (14.1)	$r_{tt} = \dfrac{2r_{oe}}{1 + r_{oe}}$	370
Standard error of measurement (14.2)	$S_{EM} = S_T\sqrt{1 - r_{tt}}$	374

APPENDIX 2

Basic Mathematics Refresher

Overview

Statisticians use a fair amount of mathematics in carrying out their work. However, many of the basic concepts in statistics can be learned with only a working knowledge of a first-year high school algebra course. It would be impractical to present a comprehensive survey of your mathematics training up to high school algebra (and there's little chance that you would have the time or energy to plow through it). So this refresher focuses only on those mathematical skills and ideas used in this book.

Content

Introductory sections on Estimation and Terms are followed by sections on seven major topics: fractions, decimals, order of operations, percent, signed numbers and absolute values, exponents and square roots, and algebra. Depending on your confidence and preparation, you may reference only a single topic or two or work through the entire review.

Format

Discussions on each major topic start with actual problems you will encounter in your text reading. The skills you'll need to handle those and similar problems in the textbook are summarized in the section titled, "You'll Need to Know." The next section, "Here's How," provides instruction on methods for solving the given problems. It's critical that you work out some of the examples and exercises to ensure that you gain the needed mathematical tools and have them readily available. Exercises appear at the end of each section to sharpen your skills. The solutions are available at the end of the appendix.

Statistics can be a fascinating subject. With a minimum amount of review, your mathematic skills can facilitate, rather than hinder, your studies.

Some Helpful Hints

If you feel your background in mathematics is weak, don't panic. This textbook is "mathematics friendly." A large number of problems are worked out, each complete with many steps to guide even the mathematically timid right to the solution. A very strong feature of this book is its abundance of helpful hints. A lot of time and effort can be saved by making full use of them.

Two hints are worthy of note.

1. Make a rough guess, that is, estimate your answer before attempting any calculation, and then . . .
2. Check your answer and see how reasonable it appears. This is particularly beneficial in your work in statistics. There are numerous ways to assess the plausibility of your answer. Consider the following examples:
   (a) A variance is the sum of squared numbers, so it's always nonnegative.
   (b) A correlation is never less than -1 nor greater than 1.
   (c) The size of a standard deviation can be approximated—it will never be greater than the range.

Although these hints may not be very enlightening at this stage, they are presented here so you are aware that there exist many ways you can evaluate your own answers. You will probably come up with many more on your own, as you become comfortable with statistics!

Terms

Knowing some of the nomenclature of mathematics will make statistics jargon more meaningful for you and formulas easier to remember.

Operations

Operations represent action in mathematics. Familiar operations include addition, subtraction, multiplication, division, and exponentiation.

Sum

A *sum* is a number found by *adding* two or more numbers. In Chapter 1, you use *summation notation* for problems requiring addition. In Chapter 5, you will calculate the "sum of squared deviations," which implies that you will be doing some adding.

Difference

A *difference* is a number found by *subtracting* one number from another. All of Chapter 10 is concerned with examining the difference between means. This difference can, of course, be found by subtracting one mean from another. (You will study more about means in Chapter 3.) The *deviation* is a related concept. The deviation from the mean is also a difference; the difference between an individual score and the mean score. Suppose the mean score for a class on a test is 75. A score of 83 would yield a deviation of $83 - 75$ or 8. You'll use this idea repeatedly in statistics.

Product

A *product* is a number found by *multiplying* two numbers. Each of the numbers being multiplied is referred to as a *factor*. Throughout most of this book, you need to multiply when you encounter the symbol "×" or two numbers written like this 10(17) or (10)(17). Often when variables are used, no symbol is used at all. For example, in Table 3.7, $N_1\overline{X}_1$ means multiply N_1 by $\overline{X}_1$.

Quotient

A *quotient* is a number found by *dividing* two numbers. You will need to divide when you encounter the familiar symbol "÷" or a quotient bar as in the problem $\frac{7.5}{3}$ or $\frac{1}{4}$. When a quotient bar is used, remember to divide the bottom number, sometimes called the *divisor*, into the top number. So, in our examples, $\frac{7.5}{3}$ is the same as $7.5 \div 3$ or 2.5 and $\frac{1}{4}$ is the same as 1 ÷ 4 or .25. A *ratio* is used to compare two numbers. It may be expressed in words such as 3 to 5, or as a quotient, such as $\frac{3}{5}$. You'll see ratios used in your study of probability and *F*-ratios used to compare numbers in your study of ANOVA in Chapter 11.

Power

The *power*, a^n, is the product of *n* factors of *a*. For example, the *power*, 5^2, is the product of 2 factors of 5. It can be found by multiplying 5×5 yielding 25. In this example, 5 is the base and 2 is the exponent. Throughout this book, 2 is the exponent you will see most frequently. Typically, the number, 5^2, is read "5 squared." Hence to "square a number" means to multiply it by itself.

CALCULATOR HINT. This can be done quite easily on most calculators. To find 5^2, you can punch the keys

[5] [X] [=]

This is handy if you do not have an [X^2] key on your calculator.

Using the Terms

To illustrate how you can use your knowledge of mathematical terms to assist you in understanding statistics, consider the notion of the "sum of squared deviations from the mean" for a list of numbers. Let's start with the deviations. Each deviation can be found by *subtracting* the mean from each number. Now you have a list of deviations. Next, each of the deviations needs to be multiplied by itself, resulting in a list of *squared* deviations. Finally, we can add to find the "sum of the squared deviations." So this rather awesome-sounding phrase and its equally intimidating formula can be understood, remembered, and reduced to subtraction, squaring, and adding.

Summary

Operation	**Notation**	**Literal Meaning**
Addition	$a + b$	The sum of *a* and *b*
Subtraction	$a - b$	The difference of *a* and *b*
Multiplication	$a \times b$ $a \cdot b$ $a(b)$	The product of *a* and *b*; *a* and *b* are factors

Division	$a \div b$ $\frac{a}{b}$ a/b	The quotient of a and b; the ratio of a to b; a divided by b
Exponentiation	a^n	The nth power of a; the product of n factors of a; a is the base n is the exponent

Decimals

You'll find decimals in nearly every problem in this book, so it's worthwhile to learn how to work with them. Even with the availability of calculators, you'll find a good working knowledge of decimals can be a real time-saver and can improve your accuracy. A sampling of problems taken from the text narrative follows.

Example 1 from Chapter 3, $Med = 93.5 + .83$

Example 2 from Chapter 8, $r = 1 - .69$

Example 3 from Chapter 6, $z\sigma = 2.58(2.5)$

Example 4 from Chapter 11, $F = \frac{30.365}{1.82}$

You'll Need to Know

- How to add and subtract decimal numbers.
- How to multiply and divide decimal numbers.
- How to round off decimal numbers.

Here's How

Before you start any of the problems, get in the habit of making a rough guess for each answer. This is particularly important if you'll be using a calculator.

ESTIMATES. Example 1 $\cong 93 + 1 = 94$. Example 2 $\cong 1 - .7 = .3$. Example 3 will be between $2 \times 2 = 4$ and $3 \times 3 = 9$. Example 4 $\cong 30 \div 2 = 15$.

TO ADD OR SUBTRACT DECIMAL NUMBERS. Line up the decimal points and proceed as you would with whole numbers.

line up
↓

Example 1	93.50 .83 94.33	There are an infinite number of trailing zeros. Add as many as you need to perform the operation.

implied
↓

Example 2	1.00 − .69 .31	If no decimal appears, it is assumed to be behind the last digit. For example, 19 is the same as 19. or 19.0 or 19.00.

TO MULTIPLY DECIMAL NUMBERS. Carry out the calculations ignoring the decimals. To place the decimal point in your answer, count the total number of digits behind the decimal point for each of the numbers involved in the calculation. That will be the total number of digits behind the decimal point in your answer.

Example 3	2.58	2 digits behind decimal point
	× 2.5	1 digit behind decimal point
	1290	
	516	
	6.450	3 digits behind decimal point

TO DIVIDE DECIMAL NUMBERS. Terms:

$$\begin{array}{r}\text{quotient}\\ \text{divisor}\overline{)\text{dividend}}\end{array}$$

Move the decimal point in the divisor to the right so it's behind the last digit. Move the

1.82.)30.36.5 — Decimal point in the dividend the same number of places. This is where the decimal will appear in the answer.

$$\begin{array}{r}16.684\\ 182\overline{)3036.5}\\ \underline{182}\\ 1216\\ \underline{1092}\\ 1245\\ \underline{1092}\\ 1530\\ \underline{1456}\\ 740\\ \underline{728}\end{array}$$

Divide

The question of when to terminate this seemingly endless task leads to our next topic.

TO ROUND OFF DECIMAL NUMBERS. How many places to leave after the decimal may vary depending on the application. For instance, in the last example, we were calculating a value for an F-statistic; such values are commonly reported with 2 decimal places. For more information on how many places to include, consult Note 3.2 in Chapter 3. Regardless of how many places you may choose to have beyond the decimal, the technique for rounding is virtually the same.

- Identify the number of places needed.
- Note the digit to the right of the last place you need. If it is 5 or greater, increase the number to its left by 1. If it's less than 5, leave the number to its left as it is.[†]

Example 4 Suppose that we want to round our answer to two places.

16.68<u>4</u> ← note that this digit is less than 5

answer rounded to two places 16.68

Suppose that we had arrived at

16.68<u>7</u> ← this digit is greater than 5

so our answer rounded to two places would have been 16.69

EXERCISES

1. Find the sum, difference, product and quotient for each of the following pairs of numbers:
   (a) 3.5, 1.39
   (b) 1.96, 0.85
   (c) 5, 3.5
2. $1.39(3.5) + 7.12 =$ [Chapter 5]

[†]There are other methods championed by some, but this method is adequate for most purposes.

3. $.4861 - .3944 =$ [Chapter 6]

4. $\frac{37.5}{15} =$ [Chapter 4]

5. $2.33(2.5) =$ [Chapter 6]

6. $102 + 1.96(0.85) =$ [Chapter 7]

7. $1.11(10) + 6.57 =$ [Chapter 9]

8. $93.5 + \left(\frac{8.5 - 6}{3}\right)$ [Chapter 3]

Fractions

Example 1 [Chapter 3]

$$Med = 1.5 + \tfrac{3}{13}$$

Solving this problem in Chapter 3 does not require any extensive knowledge of the arithmetic of fractions. This also generally holds true for most of the other problems involving fractions in this book. As is the case in the majority of real-life applications, the fraction in our example would be converted to a decimal number with the much appreciated aid of a calculator. Converting fractions to decimals will also be used in your work with probabilities in Chapter 6. Once fractions have been converted to decimals, you need only know how to perform operations with decimals to finish up your calculations. Even so, some of your work in this book can be simplified if you know how to reduce, add, and multiply with fractions.

More examples involving fractions from the text follow.

Example 2 [Chapter 8]

$$r_s = 1 - \frac{2334}{3360}$$

Example 3

$$\sqrt{\frac{5.9 + 28.9}{18}\left(\frac{1}{10} + \frac{1}{10}\right)}$$

Example 4 [Chapter 10]

$$\sqrt{\frac{12(3.75) + 10(1.21)}{12 + 10 - 2}\left(\frac{1}{12} + \frac{1}{10}\right)}$$

Example 5 [Chapter 5]

"To check your calculations, take $\frac{1}{4}$ of 14"

You'll Need to Know

- How to convert a fraction to a decimal.

Here's How

TO CONVERT A FRACTION TO A DECIMAL. Divide the numbers *below* the quotient bar *into* the number *above* the quotient bar.

Example 1 $\frac{3}{13}$ is the same as $3 \div 13$.

$$13\overline{)3.}\quad .2307\ldots$$

You can use the same method discussed in the decimals section to round this nonterminating decimal. So

$$Med = 1.5 + .23 = 1.73$$

Example 2 $\frac{2334}{3360}$ is the same as $2334 \div 3360$.

$$3360\overline{)2334.}\quad .694\ldots$$

So, $r_s = 1 - .69 = .31$

It's Nice to Know

- How to add fractions.
- How to multiply fractions.

Here's How

TO ADD OR SUBTRACT FRACTIONS. You need a common denominator.

Then, you can add (or subtract) the numerators. With this information in hand, you can save yourself a little work on Example 3 and avoid a common error on Example 4.

Example 3 $(\frac{1}{10} + \frac{1}{10})$ Since you have a common denominator, you can add the numerators and write

$(\frac{2}{10})$. Then convert to a decimal and finish the calculations.

Example 4 $(\frac{1}{12} + \frac{1}{10})$ There is not a common denominator here, so you cannot add yet.

A convenient approach here, if you're working with a calculator, is to convert each fraction to a decimal.

$$(0.83 + .100)$$

Examples 3 and 4 will be worked in their entirety in the Order of Operations section.

TO MULTIPLY FRACTIONS

1. Multiply the numerators.
2. Multiply the denominators.
3. Divide.

Example 5 $\frac{1}{4}$ of 14 [Chapter 5]

"of" can be translated as multiplication

$\frac{1}{4} \cdot \frac{14}{1}$

any whole number can be written in fractional form in this way

$\frac{1 \cdot 14}{4 \cdot 1} = \frac{14}{4}$ is the same as $14 \div 4$ or 3.5

EXERCISES

Write your answers using decimal numbers. Round to two decimal places, when necessary.

1. $\frac{1}{6}$ = [Chapter 7]
2. $\frac{12}{14}$ = [Chapter 5]
3. $\frac{30}{8}$ = [Chapter 5]
4. Med = 59.5 + $(\frac{5}{14})$ [Chapter 3]

5. $\frac{1}{4}$ of 64 [Chapter 5]

6. $\frac{1}{5}$ of 64 [Chapter 5]

7. $\frac{1}{6} + \frac{1}{6}$ [Chapter 11]

8. $\frac{1}{6} + \frac{1}{3}$ [Chapter 11]

9. $\frac{1}{3} + \frac{1}{5}$ [Chapter 11]

Percents

You'll find that applications of the notion of percent keep appearing throughout your study of probability and statistics in a variety of forms. The word "per cent" is derived from the Latin "per centum" meaning "for a hundred." Technically, everywhere you read "percent" you could translate it to "hundredths." So, 57% could also be written as $\frac{57}{100}$. Therefore, if you take a proportion in decimal form and multiply it by 100, you'll obtain a percent. Similarly, if you want to convert a percent back to a proportion, you need only divide by 100. Advantages of using percents are that the concept is widely understood and it allows for numerous, diverse comparisons to be made. Percents are used in this book in some of the following ways:

Example 1 In Chapter 6, the probability of rolling a "4" in one toss of the die is said to be $\frac{1}{6}$. What percentage of the time can we expect to observe a "4" in the long run? That is, convert the proportion $\frac{1}{6}$ to a percent.

Example 2 From Chapter 6, if 3.01% of the people in a population have an IQ less than 70, approximately how many with this low IQ would you expect to find in a sample of 15,000? That is, find 3.01% of 15,000.

Example 3 In Chapter 4, while calculating cumulative percentages, you need to determine what percent of sixty 52 represents.

Example 4 If 15 persons constitute the top 2% of all graduates at a college, what is the size of the graduating class?

All of the percent problems you deal with in this text can be transformed into the following:

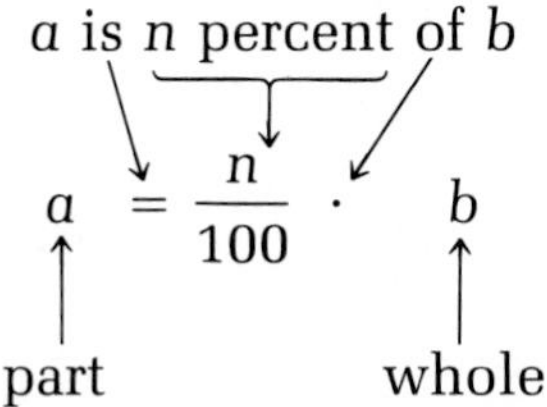

You'll Need to Know

- How to convert a proportion to a percent.
- How to convert a percent to a proportion.
- How to find the "part" given the percent amount and "whole" amount.
- How to find the percent amount given the "part" and "whole" amounts.
- How to find the "whole" given the "part" and the percent amounts.

Here's How

TO CONVERT A PROPORTION TO A PERCENT

- Convert the proportion to a decimal.
- Multiply by 100.

Example 1 $\frac{1}{6}$ can be converted and rounded off to .17
$.17 \times 100 = 17\%$

Proportion (decimal form)	$\times$ 100 → / ← $\div$ 100	Percent

Estimation can also be very helpful in working percent problems. *Before* you do any calculating, guess a "ball park" figure for your answer.

TO FIND THE "PART" GIVEN THE PERCENT AND "WHOLE"

- Identify the percent amount, n, and the "whole" amount, b, and plug in the equation.
- Solve for a.

Example 2

$$\begin{aligned} a &= \frac{3.01}{100} \cdot 15{,}000 \\ &= .0301 \cdot 15{,}000 \\ &= 451.5 \end{aligned}$$

TO FIND THE PERCENT AMOUNT GIVEN THE "PART" AND "WHOLE" AMOUNTS

- Identify the "part," a, and "whole" amounts and plug in the equation.
- Solve for n.

Example 3

$$52 = \frac{n}{100} \cdot 60$$

$$52 = n \cdot \frac{60}{100}$$

$$52 = n \cdot .6$$

$$\frac{52}{.6} = n \qquad \text{Divide both sides by .6.}$$

$$86.7\% = n$$

TO FIND THE "WHOLE" AMOUNT GIVEN THE "PART" AND PERCENT AMOUNTS

- Identify the "part" and percent amounts and plug them into the equation.
- Solve for b.

Example 4

$$15 = \frac{2}{100} \cdot b$$

$$15 = .02 \cdot b$$

$$\frac{15}{.02} = \frac{.02 \cdot b}{.02} \qquad \text{Divide both sides by .02.}$$

$$750 = b$$

Another method for solving percent problems that does not lend itself quite so readily to the word problems is the ratio-proportion method.

$$\frac{a}{b} = \frac{n}{100}$$

Cross multiplying, $100 \cdot a = b \cdot n$, where a, b, and n have the same interpretations as previously discussed.

If you are more comfortable with this method, by all means continue to use it. The important points are

(i) To make a rough guess beforehand and later check your answer to see if it is consistent with your guess.
(ii) To become adept at setting up and solving percent problems.

EXERCISES

Make an estimate before calculating each answer.

1. Express $\frac{1}{9}$ as a percent.

2. Express $\frac{5}{7}$ as a percent.
3. Express .0062 as a percent.
4. Express .0301 as a percent
5. Express 37% as a proportion
6. Express 125% as a proportion.
7. Express 0.5% as a proportion.
8. Express 12.14% as a proportion.
9. Determine 65% of 50. [Chapter 4]
10. Determine 25% of 120.
11. Determine 80% of 410.
12. Determine 0.3% of 200.
13. What percent of 50 is 13? [Chapter 4]
14. What percent of 480 is 200?
15. What percent of 10,000 is 2?
16. What percent of 15,000 is 9?
17. If 31 represents 90 percent of the students in the class, what is the total number of students in the class?
18. If 146 represents 15 percent of the gifted persons in a population, how many persons are there in the population?

Signed Numbers and Absolute Value

You have, no doubt, been introduced, at one time, to the idea that every number has an opposite. For example, the opposite of 5 is −5. The opposite of −1.2 is 1.2, and so on. In statistics, it will be important to recall how these signed numbers appear on a number line as well as how to perform the basic operations with them. Absolute values will assist us in this discussion, but will also be of interest in their own right in a few statistical applications.

Example 1

X	$X - \overline{X}$	$\lvert X - \overline{X}\rvert$
14	$14 - 12 = 2$	2
12	$12 - 12 = 0$	0
9	$9 - 12 = -3$	3
17	$17 - 12 = 5$	5
8	$8 - 12 = -4$	4

[Chapter 5]

Example 2

[Chapter 6]

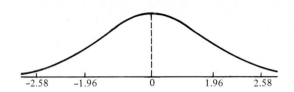

Example 3 $z = \dfrac{2 - 6}{2.68} =$ [Chapter 5]

Example 4 $\chi^2 = \dfrac{(|2 - 10| - 1)^2}{2 + 10} =$ [Chapter 13]

You'll Need to Know

- How to find the absolute value of a number.
- How to locate signed numbers along a number line.
- How to add, subtract, multiply, and divide signed numbers.

Here's How

TO FIND THE ABSOLUTE VALUE. Determine the magnitude of the number without regard to its sign. *Hint:* An absolute value will always be non-negative.

Example 1

$$|2| = 2$$
$$|0| = 0$$
$$|-3| = 3$$

and so on

TO LOCATE SIGNED NUMBERS ALONG A NUMBER LINE

- Recall that all negative numbers are to the left of zero and all positive are to the right.

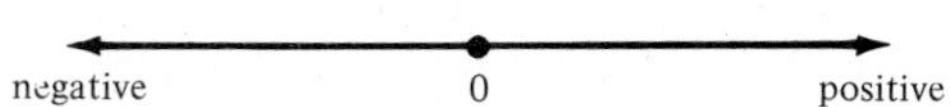

- The absolute value of a number indicates the number of units the point is from zero.

Example 2

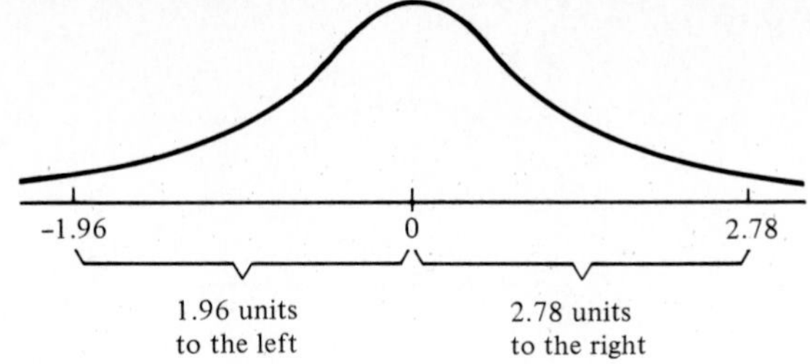

TO SUBTRACT SIGNED NUMBERS

- Write as addition problems and follow the rules for adding the following signed numbers.

Example 1 [continued]

$$14 - 12 = 14 + -12$$
$$12 - 12 = 12 + -12$$
$$9 - 12 = 9 + -12$$
$$17 - 12 = 17 + -12$$
$$8 - 12 = 8 + -12$$

TO ADD SIGNED NUMBERS

- If the signs are the same, add as always and attach the common sign. For example,

$$-3 + -4 = -7$$
$$5 + 6 = 11$$

- If the signs are different, subtract and attach the sign of the number with the larger absolute value.

Example 1 [continued]

$14 + -12 = 2$	$\|14\|$ is the larger
$12 + -12 = 0$	
$9 + -12 = -3$	$\|-12\|$ is the larger
$17 + -12 = 5$	
$8 + -12 = -4$	

TO MULTIPLY OR DIVIDE SIGNED NUMBERS

- Proceed as always working with the absolute values.
- To determine the sign of the answer: If both signs are the *same*, the answer is *positive*.

$$\frac{-10}{-2} = 5 \quad \text{or} \quad \tfrac{4}{5} = 0.8$$

If the signs differ, the answer is *negative*.

$$\frac{-10}{2} = -5 \quad \text{or} \quad \frac{4}{-5} = -0.8$$

Example 3

$$Z = \frac{2 - 6}{2.68} = \frac{2 + -6}{2.68} = \frac{-4}{2.68} = -1.49$$

Example 4

$$\begin{aligned} \chi^2 &= \frac{(|2 + -10| - 1)^2}{12} \\ &= \frac{(|-8| - 1)^2}{12} \\ &= \frac{(8 - 1)^2}{12} \\ &= \frac{(7)^2}{12} \\ &= \frac{49}{12} = 4.08 \end{aligned}$$

EXERCISES

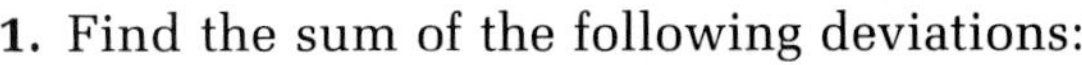

1. Find the sum of the following deviations:

2, 0, −3, 5, −4 [Chapter 5]

2. Find the difference of the chemistry and biology scores for each of the students. [Chapter 10]

	Chemistry Score	Biology Score	Difference
Student 1	73.5	73.9	
2	74.6	72.1	
3	72.8	71.3	
4	73.4	74.6	
5	72.9	72.9	

3. Multiply $-1.96(0.85) =$ [Chapter 7]

4. Multiply $-2.58(0.36) =$ [Chapter 7]

5. Multiply $-3(-3) =$ [Chapter 5]

6. Multiply $-4(-4) =$

7. $\dfrac{62 - 72}{8} =$ [Chapter 5]

8. $\frac{437 - 500}{100} =$ [Chapter 5]

9. [Chapter 5]

X	$X - 12$	$\|X - 12\|$
14		
12		
9		
17		
8		

Exponents and Square Roots

The widespread use of calculators has dramatically improved the way exponents and square roots are handled. Students of statistics will appreciate these improvements, since statisticians make frequent use of exponents and square roots.

In addition to learning the definitions, one of your primary tasks here will be to master the proper operation of your calculator. As with all your calculator work, make an estimate before you start. Then check to see if your calculated answer is consistent with your estimate.

Example 1 (Table 5.2) [Chapter 5]

X	$(X - \overline{X})$	$(X - \overline{X})^2$
6	1	1
8	3	9
2	−3	9

Example 2 $S = 3.38$ so $S^2 = (3.38)^2 =$ [Chapter 5]

Example 3 $S = \frac{30}{8} =$ [Chapter 5]

You'll Need to Know

- How to "square" a number.
- How to find the square root of a number.

Here's How

TO SQUARE A NUMBER

- Multiply it by itself

7^2 is the same as $7 \cdot 7$ or 49

$(-4)^2$ is the same as $-4 \cdot -4$ or 16

Hint: Squared numbers are always positive (except for 0^2).

Example 1

$$1^2 = 1 \cdot 1 = 1$$
$$3^2 = 3 \cdot 3 = 9$$
$$(-3)^2 = -3 \cdot -3 = 9$$

Example 2 $(3.38)^2$ can be rounded to 11.42.

Terminology

Radical sign $\sqrt{25}$ Radicand

TO FIND THE SQUARE ROOT OF A NUMBER

- Find the number that multiplied by itself yields the radicand

$$\sqrt{25} \text{ is the same as } \pm 5$$

since $5 \cdot 5$ is 25 or $(-5) \cdot (-5)$ is 25[†]
- Or use the square root table in Appendix A.
- Or use a calculator with a $\boxed{\sqrt{x}}$ key.

Example 3 $\sqrt{\dfrac{30}{8}}$ is the same as $\sqrt{3.75}$. Using a calculator $\sqrt{3.75}$ can be rounded to 1.94.

EXERCISES

1. $S = 1.94$. Find S^2. [Chapter 5]

2. $S = 6.34$. Find S^2. [Chapter 5]

3. Evaluate $\sqrt{720}$. [Chapter 5]

4. Evaluate $\sqrt{\frac{16}{40}}$. [Chapter 7]

Order of Operations

For a sample problem, let us evaluate the expression

[†]In this book, we will be concerned only with positive square roots.

$$69.5 + \frac{37.5}{15} =$$

Use your calculator to evaluate (find the number for) this expression. If you first divided 37.5 by 15 to obtain 2.5, and then added this result to 69.5, you would have correctly found the expression equal to 72.

If this isn't the answer you calculated, read on. Many a person will innocently (and incorrectly) add 69.5 to 37.5 and then divide the result (107) by 15. Their (wrong) answer of 7.13 suggests some ambiguity inherent in mathematics notation.

Both parties could argue that their calculations were done without error. How does one determine, then, which answer is the correct one? Mathematicians solved this dilemma by agreeing to perform the operations in a specific order.

You'll Need to Know

- Which order to perform the operations of exponentiation, multiplication, division, addition, and subtraction.

Here's How

To perform the operations in the correct order, work from left to right. First, do all the work inside parentheses, then evaluate all the exponents and square roots, and then carry out all the multiplication and division. Finally, you can perform all addition and subtraction.

BACK TO OUR SAMPLE PROBLEM. We need to add and we need to divide.

$$69.5 + \frac{37.5}{15}.$$

Summary

Parentheses	First
Exponents and Square Roots	
Multiplication and Division	
Addition and Subtraction	Last

The established order of operations instructs us to divide first and then add. So, 72 is the correct value for the expression.

1. $1.11(10) + 6.57 =$

Multiply, to obtain

$$11.1 + 6.57 =$$

Then, add

$$11.1 + 6.57 = 17.67$$

2. $\dfrac{\dfrac{973}{15} - (7.73)(7.47)}{(4.03)(2.24)} =$

Parentheses are implied above and below the quotient bar, so evaluate what you have above and below the bar first, then divide.
Above: Carry out the division and multiplication first

$$64.87 - 57.74$$

Then subtract to obtain 7.13
Below:

$$(4.03)(2.24) = 9.03$$

So, we have:

$$\frac{7.13}{9.03} = 0.74$$

3. $\dfrac{3}{\sqrt{100 - 1}} =$

Be sure and carry out the subtraction under the radical *before* taking the square root.

$$\frac{3}{\sqrt{99}} = \frac{3}{9.95} = 0.30$$

4. $\dfrac{1}{10}\sqrt{10(432) - (60)^2}$

Again make sure the work under the radical is done *before* finding the square root. Under the radical use the same order discussed.
Under the radical:
Exponentiation first

$$\tfrac{1}{10}\sqrt{10(432) - 3600}$$

Then multiplication

$$\tfrac{1}{10}\sqrt{4320 - 3600}$$

Finally, subtraction

$$\tfrac{1}{10}\sqrt{720}$$

Now take the square root

$$\tfrac{1}{10}\,(26.83)$$

Divide

$$2.68$$

FOR MORE PRACTICE

1. $72 + 5\left(\frac{-30}{50}\right)$ [Appendix]

2. $59.5 + 5\left(\frac{1}{14}\right)$ [Appendix]

3. $1.11(16) + 6.57$ [Chapter 9]

4. $93.5 + \left(\frac{8.5 - 6}{3}\right)$ [Chapter 3]

5. $\frac{9 - 6}{2.68}$ [Chapter 5]

6. $\frac{437 - 500}{100}$ [Chapter 5]

7. $\frac{10(17) + 30(20) + 40(23)}{10 + 30 + 40}$ [Chapter 3]

8. $\sqrt{\frac{432}{10} - (6)^2}$ [Chapter 10]

9. $15(6.0)^2 + 15(3.33)^2 - 30(4.67)^2$ [Chapter 11]

10. $\frac{15(973) - (116)(112)}{\sqrt{15(1140) - 116^2}\sqrt{15(912) - (112)^2}}$ [Chapter 8]

11. Find S given $S = \frac{1}{10}\sqrt{10(10{,}677) - (325)^2}$ [Chapter 5]

12. $S = \sqrt{\frac{432}{10} - (6)^2}$ [Chapter 5]

13. $q = \frac{9 - 6.4}{\sqrt{\frac{8.7}{2}\left(\frac{1}{6} + \frac{1}{5}\right)}}$ [Chapter 11]

14. $r = \frac{10(2799) - (130)(210)}{\sqrt{10(1752) - (130)^2}\sqrt{10(4498) - (210)^2}}$ [Chapter 8]

15. $S_E = 0.67\sqrt{1 - (.53)^2}$ [Chapter 9]

EXERCISE FOR FUN

$$\frac{12}{28(28 + 1)}\left[\frac{(72.5)^2}{9} + \frac{(71.5)^2}{7} + \frac{(114.5)^2}{6} + \frac{(147.5)^2}{6}\right] - 3(28 + 1)$$

[Chapter 13]

You should be pleased to find out that you actually will have a chance to put your high school training in algebra to work in statistics!

While studying algebra, you gained some skills in working with symbols. No matter how accomplished you may have become with symbols, you'll find statistics easier and more interesting if you recognize that symbols are used as an efficient way to communicate an idea. When working with symbols, you may find it helpful to translate the ideas into words. Here's an example:

$$\overline{X} = \frac{\Sigma X}{N}$$

Translation:

$$\begin{array}{ccccccccc} \overline{X} & = & \Sigma & & X & & \div & & N \\ \downarrow & & \downarrow & & \downarrow & & \downarrow & & \downarrow \end{array}$$

"The mean is the sum of the observations divided by the number of observations."

Not only will the "translations" make the formulas easier to remember and use, but the concepts and similarities among the formulas will become apparent.

Chapter 1 begins with a discussion of one of the most fundamental concepts in statistics: variables. Statisticians use the concept of a variable in very specific ways, but the general notion of variables you are familiar with will be helpful in statistics.

Terms

- A *variable* is a letter used to represent a set of numbers. In statistics, variables are used to denote quantities that may vary, such as test scores, IQ, or weight.
- A *formula* is an equation relating two or more variables. You'll find an entire section of the Appendix devoted entirely to formulas used in statistics. The next examples from the text illustrate some uses of formulas in statistics.
- *Inequalities* are statements involving the *greater than*, $>$, or *less than*, $<$, symbols. The solution set for an inequality typically includes many values, although it may have no values (the empty set). In elementary statistics, inequalities are used to describe the size of some

values, such as $p < .01$, and to define an interval, such as the confidence intervals in Chapter 7.

Example 1 The formula for the combined mean is

$$\overline{X} = \frac{N_1\overline{X}_1 + N_2\overline{X}_2 + N_3\overline{X}_3 + \cdots}{N_1 + N_2 + N_3 + \cdots}$$

Find the combined mean, $\overline{X}$, given

$$N_1 = 10 \qquad N_2 = 30 \qquad N_3 = 40$$
$$\overline{X}_1 = 17 \qquad \overline{X}_2 = 20 \qquad \overline{X}_3 = 23$$

[Chapter 3]

Example 2 Given the formula

$$Z = \frac{X - \overline{X}}{S}$$

Find X in terms of Z, $\overline{X}$, and S. [Chapter 6]

Example 3 Suppose that $Y' = 1.11X + 6.57$.
What does the graph look like? [Chapter 9]
What is Y' when X is 10?

Example 4 Write the inequality

$$-1.96 \leq \frac{\overline{X} - \overline{X}_{\overline{X}}}{\sigma_{\overline{X}}} \leq 1.96$$

in the form

$$\square \leq \overline{X}_{\overline{X}} \leq \square$$

You'll Need to Know

- How to perform a substitution, that is, find the value of one variable given specific numerical values for the other variables in a formula.
- How to solve a formula for one of the variables in terms of the others.
- How to recognize a linear equation, identify the slope of its graph, and use the equation to find other points on the line.
- How to solve an inequality.

TO PERFORM A SUBSTITUTION

- Plug the given numerical values in for each corresponding letter and carry out the calculations.

Example 1

$$\begin{aligned}\overline{X} &= \frac{10(17) + 30(20) + 40(23)}{10 + 30 + 40}\\ &= \frac{170 + 600 + 920}{80}\\ &= \frac{1690}{80}\\ &= 21.125\end{aligned}$$

TO SOLVE A FORMULA FOR ONE VARIABLE
IN TERMS OF THE OTHER VARIABLES

- Isolate the variable of interest on one side of the equation by using one or more of the following operations:
  (a) Multiply (or divide) both sides by the same nonzero quantity.
  (b) Add (or subtract) the same quantity from each side of the equation.

Example 2

$$\begin{aligned}Z &= \frac{X - \overline{X}}{S} && \text{solve for } X\\ Z(S) &= X - \overline{X} && \text{multiply both sides by } S\\ Z(S) + \overline{X} &= X - \overline{X} + \overline{X} && \text{add } \overline{X} \text{ to each side}\\ Z(S) + \overline{X} &= X\end{aligned}$$

TO WORK WITH A LINEAR EQUATION

- The graph of all linear equations is a line.
- Equations that can be written in the form $y = mX + b$, where m is the slope and b is the y-intercept (the place where the line crosses the y-axis), are referred to as linear equations. In our example, $Y' = 1.11X + 6.57$. The graph of all ordered pairs (X, Y'), which satisfies this equation is a straight line with a slope of 1.11 crossing the y-axis at 6.57.
- To find Y' when $X = 10$, just plug 10 in for X in the equation and calculate.

Example 3

$$\begin{aligned}Y' &= 1.11(10) + 6.57\\ Y' &= 11.1 + 6.57\\ Y' &= 17.67\end{aligned}$$

So the ordered pair (10, 17.67) represents another point on the line.

TO SOLVE AN INEQUALITY

- Isolate the variable of interest on one side by using one or more of the following properties of inequality:
  (a) Add the same quantity to each side of the inequality.
  (b) Multiply each side of the inequality by the same nonnegative quantity.

The example worked out below is one of the rare instances where you need to know how to solve an inequality.

Example 4

$$-1.96 \leq \frac{\overline{X} - X_{\overline{X}}}{\sigma_{\overline{X}}} \leq 1.96$$

Left Side	Right Side	
$-1.96\sigma_{\overline{X}} \leq \overline{X} - X_{\overline{X}}$	$\overline{X} - X_{\overline{X}} \leq 1.96\sigma_{\overline{X}}$	Multiply each side by $\sigma_{\overline{X}}$
$X_{\overline{X}} - 1.96\sigma_{\overline{X}} \leq \overline{X}$	$\overline{X} \leq X_{\overline{X}} + 1.96\sigma_{\overline{X}}$	Add $X_{\overline{X}}$ to each side
$X_{\overline{X}} \leq \overline{X} + 1.96\sigma_{\overline{X}}$	$\overline{X} - 1.96\sigma_{\overline{X}} \leq X_{\overline{X}}$	(Left) Add $+1.96\sigma_{\overline{X}}$ to each side (Right) Add $-1.96\sigma_{\overline{X}}$ to each side

$$\overline{X} - 1.96\sigma_{\overline{X}} \leq \underbrace{X_{\overline{X}}}_{\text{variable of interest}} \leq \overline{X} + 1.96\sigma_{\overline{X}} \qquad \text{Combining}$$

EXERCISES

1. Use the combined mean formula given in Example 1 to determine the value of the combined mean when

   $N_1 = 42$ $\quad N_2 = 15$ $\quad N_3 = 9$ $\quad N_4 = 14$

   $\overline{X}_1 = 26{,}800$ $\quad \overline{X}_2 = 20{,}480$ $\quad \overline{X}_3 = 36{,}600$ $\quad \overline{X}_4 = 18{,}900$ [Chapter 3]

2. If the upper and lower limits for an interval are defined by

   $$Y' \pm 1.96S_E$$

   find the limits if

   $Y' = 23.2$ and $S_E = 1.10$ [Chapter 9]

3. If the upper and lower limits are defined by

$$Y' \pm 2.58S_E$$

find the limits if

$$Y' = 23.2 \text{ and } S_E = 1.10$$ [Chapter 9]

4. If $r = \dfrac{\dfrac{\Sigma XY}{N} - \overline{X}\overline{Y}}{S_x S_y}$

and

$\Sigma XY = 27{,}884.06$
$N = 484$
$\overline{X} = 22.21$
$\overline{Y} = 2.53$
$S_X = 4$
$S_Y = 0.67$, find r.

[Chapter 9]

5. $S_{D\overline{X}} = \sqrt{\dfrac{N_1S_1^2 + N_2S_2^2}{N_1 + N_2 - 2}\left(\dfrac{1}{N_1} + \dfrac{1}{N_2}\right)}$ [Chapter 10]

where

$\overline{X}_1 = 108.6$
$S_1 = 14.8$
$N_1 = 50$
$\overline{X}_2 = 110.2$
$S_2 = 13.9$
$N_2 = 40$

6. If $S_E = S_Y\sqrt{1 - r^2}$ and $S_Y = 0.67$, $r = .53$, calculate the value for S_E.
[Chapter 9]

7. If $z = \dfrac{X - \mu}{\sigma}$, find X in terms of z, σ, and μ. [Chapter 6]

8. Show that $N_T - 1 = (k - 1) + (N_T - k)$ [Chapter 11]

9. Given the formula

$$\chi^2 = \frac{N(AD - BC)^2}{(A + B)(C + D)(A + C)(B + D)}$$

calculate the value of χ^2 if

$A = 12$

$B = 19$

$C = 28$

$D = 6$

$N = 65$

10. $p < .01$. The actual value of p could be
(a) .15.
(b) .07.
(c) .021.
(d) .003.

11. $p > .10$. The actual value of p could be
(a) .15.
(b) .07.
(c) .021.
(d) .003.

EXERCISE SOLUTIONS

Decimals

1.

	Sum	Difference	Product	Quotient
(a)	4.89	2.11	4.865	2.52
(b)	2.81	1.11	1.666	2.31
(c)	8.5	1.5	17.5	1.43

2. 11.985

3. 0.0917

4. 2.5

5. 5.825

6. $102 + 1.666 = 103.666$

7. $11.1 + 6.57 = 17.67$

8. $93.5 + \dfrac{2.5}{3} = 93.5 + 0.83 = 94.33$

Fractions

1. 0.17
2. 0.86
3. 3.75
4. $59.5 + \frac{5}{14} = 59.5 + 0.36 = 59.86$
5. $\frac{1}{4} \cdot 64 = \frac{64}{4} = 16$
6. $\frac{1}{5} \cdot 64 = \frac{64}{5} = 12.8$
7. $\frac{2}{6} = 0.33$
8. $0.17 + 0.33 = 0.5$
9. $0.33 + 0.2 = 0.53$

Percents

1. $0.1111 \times 100 = 11.11\%$
2. $0.7143 \times 100 = 71.43\%$
3. $0.0062 \times 100 = 0.62\%$
4. $0.0301 \times 100 = 3.01\%$
5. $\frac{37}{100}$ or 0.37
6. $\frac{125}{100}$ or 1.25
7. $\dfrac{.5}{100}$ or 0.005
8. $\dfrac{12.14}{100}$ or 0.1214
9. $\frac{50}{100} \times 65 = .5 \times 65 = 32.5$
10. $\frac{25}{100} \times 120 = .25 \times 120 = 30$
11. $\frac{80}{100} \times 410 = .80 \times 410 = 328$
12. $\dfrac{.3}{100} \times 200 = .003 \times 200 = 0.6$
13. $13 = \dfrac{n}{100} \times 50 \qquad n = \dfrac{13}{.5} = 26\%$
14. $200 = \dfrac{n}{100} \times 480 \qquad 200 = 4.8n \qquad n = \dfrac{200}{4.8} = 41.67\%$
15. $2 = \dfrac{n}{100} \times 10{,}000 \qquad 2 = 100n \qquad n = \dfrac{2}{100} = 0.02\%$

16. $9 = \frac{n}{100} \times 15{,}000 \qquad 9 = 150n \qquad n = \frac{9}{150} = 0.06\%$

17. $31 = \frac{90}{100} \times b \qquad 31 = .9b \qquad b = \frac{31}{.9} = 34.44$ students

18. $146 = \frac{15}{100} \times b \qquad 146 = .15b \qquad b = \frac{146}{.15} = 973.33$ persons

Signed Numbers and Absolute Values

1. $2 + 0 + (-3) + 5 + (-4) = 0$

2. $73.5 - 73.9 = -0.4$
$74.6 - 72.1 = 2.5$
$72.8 - 71.3 = 1.5$
$73.4 - 74.6 = -1.2$
$72.9 - 72.9 = 0$

3. -1.666

4. -0.9288

5. 9

6. 16

7. $\frac{62 + -72}{8} = \frac{-10}{8} = -1.25$

8. $\frac{437 + -500}{100} = \frac{-63}{100} = -0.63$

9.

X	$X - 12$	$\lvert X - 12 \rvert$
14	2	2
12	0	0
9	−3	3
17	5	5
8	−4	4

Exponents and Square Roots

1. $S^2 = (1.94)^2 = 3.7636$

2. $S^2 = (6.34)^2 = 40.1956$

3. 26.83

4. $\sqrt{\frac{16}{40}} = \sqrt{0.4} = 0.63$

1. $72 + \frac{-150}{50} = 72 + (-3) = 69$

2. $59.5 + \frac{5}{14} = 59.5 + 0.36 = 59.86$

3. $17.76 + 6.57 = 24.33$

4. $93.5 + \frac{2.5}{3} = 93.5 + 0.83 = 94.33$

5. $\frac{3}{2.68} = 1.12$

6. $\frac{-63}{100} = -0.63$

7. $\frac{170 + 600 + 920}{80} = \frac{1690}{80} = 21.125$

8. $\sqrt{43.2 - 36} = \sqrt{7.2} = 2.68$

9. $15(36) + 15(11.0889) + -30(21.8089) = 540 + 166.3335 + (-654.267)$
$= 52.0665$

10. $\frac{14{,}595 - 12{,}992}{\sqrt{17{,}100 - 13{,}456}\sqrt{13{,}680 - 12{,}544}} = \frac{1{,}603}{\sqrt{3{,}644}\sqrt{1{,}136}}$

$\frac{1{,}603}{(60.37)(33.70)} = \frac{1{,}603}{2{,}034.47} = 0.79$

11. $S = \frac{1}{10}\sqrt{106{,}770 - 105{,}625} = \frac{1}{10}\sqrt{1{,}145} = \frac{33.84}{10} = 3.38$

12. $S = \sqrt{43.2 - 36} = \sqrt{7.2} = 2.68$

13. $q = \frac{2.6}{\sqrt{4.35(0.17 + 0.20)}} = \frac{2.6}{\sqrt{4.35(0.37)}}$

$= \frac{2.6}{\sqrt{1.61}} = \frac{2.6}{1.27} = 2.05$

14. $r = \frac{27{,}990 - 27{,}300}{\sqrt{17{,}520 - 16{,}900}\sqrt{44{,}980 - 44{,}100}}$

$= \frac{690}{\sqrt{620}\sqrt{880}} = \frac{690}{(24.90)(29.66)} = \frac{690}{738.53} = 0.93$

15. $S_E = 0.67\sqrt{1 - 0.28}$

$= 0.67\sqrt{0.72}$

$= 0.67(0.85)$

$= 0.57$

$$\tfrac{12}{812}(584.03 + 730.32 + 2185.04 + 3626.04) - 87$$
$$= 0.0148(7125.43) - 87$$
$$= 105.46 - 87$$
$$= 18.46$$

Algebra

1. $$\overline{X} = \frac{42(26{,}800) + 15(20{,}480) + 9(36{,}600) + 14(18{,}900)}{42 + 15 + 9 + 14}$$
$$= \frac{1{,}125{,}600 + 307{,}200 + 329{,}400 + 264{,}600}{80}$$
$$= \frac{2{,}026{,}800}{80} = 25{,}335$$

2. $23.2 \pm 1.96(1.10)$
Lower limit: $23.2 - 1.96(1.10)$
$23.2 - 2.16$ 21.04
Upper limit: $23.2 + 1.96(1.10)$
$23.2 + 2.16$ 25.36

3. $23.2 \pm 2.58(1.10)$
Lower limit: $23.2 - 2.84 = 20.36$
Upper limit: $23.2 + 2.84 = 26.04$

4. $$r = \frac{\dfrac{27{,}884.06}{484} - (22.21)(2.53)}{4(0.67)}$$
$$= \frac{57.6117 - 56.1913}{2.68} = \frac{1.42}{2.68} = .53$$

5. $$S_{D\overline{X}} = \sqrt{\frac{50(14.8)^2 + 40(13.9)^2}{50 + 40 - 2}\left(\frac{1}{50} + \frac{1}{40}\right)}$$
$$= \sqrt{\frac{10{,}952 + 7728.4}{88}(0.02 + 0.025)}$$
$$= \sqrt{212.28(0.045)}$$
$$= \sqrt{9.5526} = 3.09$$

6. $S_E = 0.67\sqrt{1 - (.53)^2} = 0.67(0.85) = 0.57$

7. $z\sigma = X - \mu$ multiplying both sides by σ
$\mu + z\sigma = X$ adding μ to each side

8. $N_T - 1 = N_T - 1 + (k - k)$ rearranging terms on the right-hand side

9. $$\chi^2 = \frac{65(12 \cdot 6 - 19 \cdot 28)^2}{(12 + 19)(28 + 6)(12 + 28)(19 + 6)}$$

$$= \frac{65(-460)^2}{31 \cdot 34 \cdot 40 \cdot 25}$$

$$= \frac{13{,}754{,}000}{(1{,}054)(1{,}000)}$$

$$= 13.05$$

10. (d)

11. (a)

APPENDIX 3

Calculating the Mean, Median, Mode, and Standard Deviation from Grouped Frequency Distributions

There are times when it is convenient to be able to calculate statistics for data that are in the form of a grouped frequency distribution. The computational techniques for grouped data are valuable as a shortcut method if a calculator is not available, and they may be an absolute necessity for determining measures of central tendency and variability when you have only the grouped frequency distribution from a newspaper article or some other summary report. You will quite often encounter grouped frequency distributions in news magazines and annual reports of corporations.

The computational formulas use terms that were introduced in the discussion of grouped frequency distributions in Chapter 2, and it might be a good idea to review the meaning of such terms as *MP*, *L*, *i*, *f*, and *N*. The calculation of the three measures of central tendency and the standard deviation is illustrated below.

Mean

The calculation of the mean from a grouped frequency distribution of 50 reading readiness scores is shown in Table A.1. Note that this is an ordinary frequency distribution, with the addition of two more columns, *d* and *fd*.

To obtain the entries in the *d* column, it is necessary to choose one of the intervals as a starting point, so a 0 is placed in the *d* column opposite the interval chosen. *Any* interval can be chosen, but it is usually preferred to choose one in the center of the distribution, since this

TABLE A.1. **Calculation of the Mean from Grouped Reading Readiness Scores**

Scores	f	d	fd	
85–89	1	3	3	
80–84	3	2	6	+15
75–79	6	1	6	
70–74	15	0		
65–69	12	−1	−12	
60–64	8	−2	−16	−45
55–59	3	−3	−9	
50–54	2	−4	−8	
	$N = 50$		$\Sigma fd = -30$	

$$\overline{X} = MP + i\left(\frac{\Sigma fd}{N}\right)$$

$$= 72 + 5\left(\frac{-30}{50}\right)$$

$$= 72 + (-3)$$

$$\overline{X} = 69$$

results in smaller numbers for your calculations. In Table A.1 we have picked the 70–74 interval and have placed a 0 in the d column.

The next step is to count up by units from this 0 (+1, +2, etc.) until you reach the top of the distribution. Then count down from the 0 (−1, −2, etc.) until you reach the bottom of the distribution. These numbers simply tell us how many intervals a given interval is from the starting point.

To obtain the entries in the fd column, simply multiply each f by each d for every interval. Thus, in Table A.1, $1 \times 3 = 3$, $3 \times 2 = 6$, and so on. Do the same for all the intervals, remembering that some of the fd values will be negative, since some of the d values are negative.

The next step is to obtain the algebraic sum of the fd column. Add the positive values (15) and the negative values (−45) and obtain the sum. This quantity is known as Σfd, and in Table A.1 $\Sigma fd = -30$.

The formula for finding the mean from grouped data (as in Table A.1) is

$$\overline{X} = MP + i\left(\frac{\Sigma fd}{N}\right)$$

where

MP is the midpoint of the interval at the starting point
i is the size of the interval

N is the number of scores
Σfd is the sum of the fd column

Substituting the data from Table A.1 into the formula, we get

$$\overline{X} = 72 + 5\left(\frac{-30}{50}\right)$$
$$\overline{X} = 72 + (-3)$$
$$\overline{X} = 69$$

The starting point for the 0 in the d column of Table A.1 was the 70–74 interval, but any interval would have resulted in a mean of 69. As a check on the accuracy of calculations, you might choose a different interval as the starting point and see if you obtain a mean of 69. For example, if you started at the 55–59 interval, then MP would be 57, i would be 5, N would be 50, and Σfd would be 120. So, no matter which interval is used for the starting point, the mean will always be the same.

It was stated in Chapter 2 that in using a grouped frequency distribution we lost a little accuracy because we did not know the identity of the individual scores. What effect does this have on the accuracy of a mean calculated from grouped data? The formula for the mean from grouped data assumes that the average of all the scores in any interval falls at the midpoint of that interval, so we expect a small, but usually not serious, discrepancy between a mean calculated from the raw data and one calculated after the data have been grouped into a frequency distribution. The mean for the raw data on which Table A.1 was based is actually 68.94, compared with 69 using the grouped-data formula. So, in this case the discrepancy is negligible.

Median

Since the median is the 50th percentile, we want to find the point in a frequency distribution that separates the top half of the scores from the bottom half.

MEDIAN FOR THE GROUPED FREQUENCY DISTRIBUTION. Table A.2 shows the calculation of the median, or the 50th percentile (P_{50}), for a group of 54 driver-training test scores. Note the *cum f* (cumulative frequency) column, which is again the addition of the frequencies in each interval as you count up from the bottom. For example, there are 2 scores in the 40–44 interval and 3 scores in the 45–49 interval, so the cumulative frequency for the 45–49 interval is 5. There are 6 scores in the 50–54 interval, so the cumulative frequency for the 50–54 interval is 11, since in *that interval and below* there are 11 scores. Technically speaking, there are 11 scores below 54.5, the exact upper limit of that interval.

TABLE A.2. **Calculation of the Median from Grouped Driver-Training Scores**

Scores	f	*cum f*	Calculations
75–79	2	54	(A) 50% of 54 = 27
70–74	4	52	
65–69	8	48	(B) 2 + 3 + 6 + 15 = 26
60–64	14	40	
55–59	15	26	(C) $0.5N - cum\ f = 27 - 26 = 1$
50–54	6	11	
45–49	3	5	(D) $f_{MED} = 14$
40–44	2	2	
	N = 54		(E) L = 59.5

$$Med = L + i\left(\frac{0.5N - cum\ f}{f_{MED}}\right)$$

$$= 59.5 + 5\left(\frac{5}{14}\right)$$

$$= 59.5 + .4$$

$$Med = 59.9$$

The first step in the calculation of the median is to take 50% of N. In step A, $0.50 \times 54 = 27$. The next step is to count up from the bottom in the *cum f* column by adding each frequency until we get as close to 27 as we can, without exceeding 27. In step B we see that 2 + 3 + 6 + 15 = 26. If we went up one more *f* value, we would go beyond 27, so we must stop here. This tells us that the median, or P_{50}, is in the *next interval*—somewhere between 59.5 and 64.5.

The next step is to subtract the frequencies that we have just totaled from the value that we actually needed. We needed 50% of N, or 27, but we could total only 26. So, in step C, 27 − 26 = 1.

Next we inspect the interval that contains the median and find out the frequency for this interval. Looking at Table A.2 we note that there are 14 scores in the 60–64 interval. We call this value f_{MED}, the frequency of the interval containing the median. Thus, in step D, $f_{MED} = 14$.

Finally, we determine the exact lower limit, *L*, for the interval containing the median. Since this interval is 60–64, *L* = 59.5.

The formula for the median is

$$Med = L + i\left(\frac{0.5N - cum\ f}{f_{MED}}\right)$$

where

L is the exact lower limit of class containing the median
i is the size of the interval
$0.5N$ is 50% of the size of the group
$cum\ f$ is the number of scores below the interval containing the median
f_{MED} is the frequency of the interval containing the median

From the example in Table A.2

$$\begin{aligned} Med &= 59.5 + 5\left(\frac{27 - 26}{14}\right) = 59.5 + 5\left(\frac{1}{14}\right) \\ &= 59.5 + \frac{5}{14} \\ &= 59.5 + 0.4 \\ Med &= 59.9 \end{aligned}$$

So, the median, or P_{50}, is 59.9, and we say that 50% of the distribution falls below 59.9.

The preceding section on the calculation of the median will be much clearer if we look at a graphical interpretation as shown in the histogram of the driver-training scores in Figure A.1.

Note that 50% of N, or 27, includes the four lowest intervals, with frequencies of 2, 3, 6, and 15, whose sum is 26. Since this leaves us 1 short of the required 27, we must go part of the way into the next interval to obtain the value of the median. There are 14 scores in this interval, so we go $\frac{1}{14}$ of the way into the 60–64 interval. (This leaves a single observation below the dotted line of Figure A.1 and 13 above.)

Since this interval is 5 units wide ($i = 5$), the median is the lower limit of $59.5 + \frac{1}{14}$ of the interval of 5, which would be $59.5 + (\frac{1}{14})(5) = 59.9$. The dotted vertical line shows the location of the median, and, as you can see, this point divides the distribution into two equal halves, with 27 observations on each side and a score of 59.9 as the dividing point.

An important assumption that is necessary when the median is calculated from grouped data is that in a given interval the scores are spread evenly over the interval. For example, in the 55–59 interval in

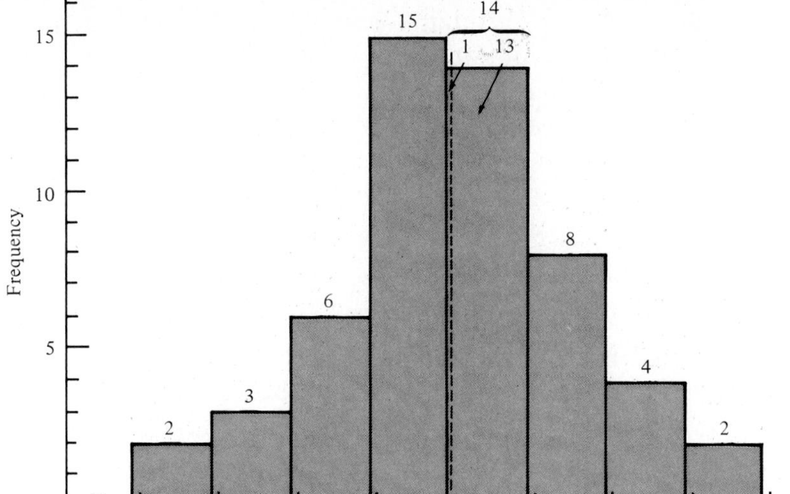

$$Med = 59.5 + \frac{1}{14} \text{ of the interval of } 5$$

$$Med = 59.5 + \frac{5}{14} = 59.9$$

FIGURE A.1. **Histogram of driver-training scores showing calculation of the median.**

Table A.2, it is assumed that of the 15 scores in this interval, there would be three 55's, three 56's, and so on. Since the formula for the median is one that involves *linear interpolation,* this assumption is necessary. Though this condition is not always present in every interval, it appears that slight deviations from this assumption will not greatly affect the accuracy of the median calculated in this fashion.

Mode

As we noted earlier, the mode is the score that appears most frequently in a distribution of scores. Since the actual identity of each individual score is lost when the scores are grouped in a frequency distribution, we can only make an assumption concerning the value of the mode when we use grouped data.

This value is called the *crude mode,* and it is simply the midpoint of the interval with the greatest frequency. In other words we find the interval that contains the greatest number of scores and assume that the midpoint of the interval was the score made most often. You can see why this value is called the *crude* mode. Since the mode itself is crude, the term *crude mode* is redundant—something like *dirty dirt.*

In Table A.1, the interval containing the greatest number of scores is 70–74, so the crude mode would be 72. In Table A.2, the crude mode would be the midpoint of the 55–59 interval, or 57.

Standard Deviation

The calculation of S from a frequency distribution involves only one step beyond those required for calculating the mean. Table A.3 shows the same data that were used to illustrate the calculation of the mean in Table A.1, but a new column, fd^2, has been added.

TABLE A.3. **Calculation of the Standard Deviation for Grouped Reading Readiness Scores**

Scores	f	d	fd		fd^2
85–89	1	3	3		9
80–84	3	2	6	+15	12
75–79	6	1	6		6
70–74	15	0			
65–69	12	−1	−12		12
60–64	8	−2	−16	−45	32
55–59	3	−3	−9		27
50–54	2	−4	−8		32
	N = 50		Σfd = −30		Σfd^2 = 130

After you have completed the steps described earlier in the calculation of the mean, it is only necessary to compute the entries in the fd^2 column. To obtain each fd^2 entry, multiply each fd by the corresponding d value. In Table A.3, $3 \times 3 = 9$, $2 \times 6 = 12$, and so on for the rest of the distribution. Remember that fd^2 means fd times d, not fd times fd. For example, the entry in the 80–84 interval is $2 \times 6 = 12$, not $6 \times 6 = 36$. Also note that all the entries in the fd^2 column are positive. There are no negative numbers, because multiplying a negative d by a negative fd results in a positive answer. For example, in the 55–59 interval, $-3 \times -9 = 27$. The next step is to add the fd^2 column to obtain $\Sigma fd^2 = 130$.

The formula for the standard deviation from grouped data is

$$S = i\sqrt{\frac{N\Sigma fd^2 - (\Sigma fd)^2}{N^2}}$$

where

S is the standard deviation
i is the size of the interval
N is the number of scores
Σfd^2 is the sum of the fd^2 column
Σfd is the sum of the fd column

For the frequency distribution of Table A.3

$$S = i\sqrt{\frac{N\Sigma fd^2 - (\Sigma fd)^2}{N^2}}$$

$$= 5\sqrt{\frac{50(130) - (-30)^2}{(50)^2}}$$

$$= 5\sqrt{\frac{6500 - 900}{2500}}$$

$$= 5\sqrt{\frac{5600}{2500}}$$

$$= 5\sqrt{2.24}$$

$$= 5(1.5)$$

$$S = 7.50$$

APPENDIX 4

Statistical Tables

A. Squares and Square Roots of Numbers from 1 to 1000

B. Areas Under the Normal Curve Between the Mean and z

C. Values of r for the .05 and .01 Levels of Significance

D. Values of Spearman r_s for the .05 and .01 Levels of Significance

E. Values of t at the .05, .01, and .001 Levels of Significance

F. Values of F at the 5% and 1% Levels of Significance

G. Values of Chi Square at the .05, .01, and .001 Levels of Significance

H. Probabilities Associated with Values as Small as Observed Values of U in the Mann–Whitney U Test

I. Critical Values of U in the Mann–Whitney U Test

J. Probabilities Associated with Values as Small as Observed Values of x in the Binomial Test

K. Critical Values of T in the Wilcoxon Matched-Pairs Signed-Ranks Test

L. Random Numbers

M. Critical Values of the Studentized Range Statistic, q

N	N^2	$\sqrt{N}$	$\sqrt{10N}$	N	N^2	$\sqrt{N}$	$\sqrt{10N}$
1	1.	1.000	3.162	45	2025.	6.708	21.213
2	4.	1.414	4.472	46	2116.	6.782	21.448
3	9.	1.732	5.477	47	2209.	6.856	21.679
4	16.	2.000	6.325	48	2304.	6.928	21.909
5	25.	2.236	7.071	49	2401.	7.000	22.136
6	36.	2.449	7.746	50	2500.	7.071	22.361
7	49.	2.646	8.367	51	2601.	7.141	22.583
8	64.	2.828	8.944	52	2704.	7.211	22.804
9	81.	3.000	9.487	53	2809.	7.280	23.022
10	100.	3.162	10.000	54	2916.	7.348	23.238
11	121.	3.317	10.488	55	3025.	7.416	23.452
12	144.	3.464	10.954	56	3136.	7.483	23.664
13	169.	3.606	11.402	57	3249.	7.550	23.875
14	196.	3.742	11.832	58	3364.	7.616	24.083
15	225.	3.873	12.247	59	3481.	7.681	24.290
16	256.	4.000	12.649	60	3600.	7.746	24.495
17	289.	4.123	13.038	61	3721.	7.810	24.698
18	324.	4.243	13.416	62	3844.	7.874	24.900
19	361.	4.359	13.784	63	3969.	7.937	25.100
20	400.	4.472	14.142	64	4096.	8.000	25.298
21	441.	4.583	14.491	65	4225.	8.062	25.495
22	484.	4.690	14.832	66	4356.	8.124	25.690
23	529.	4.796	15.166	67	4489.	8.185	25.884
24	576.	4.899	15.492	68	4624.	8.246	26.077
25	625.	5.000	15.811	69	4761.	8.307	26.268
26	676.	5.099	16.125	70	4900.	8.367	26.458
27	729.	5.196	16.432	71	5041.	8.426	26.646
28	784.	5.292	16.733	72	5184.	8.485	26.833
29	841.	5.385	17.029	73	5329.	8.544	27.019
30	900.	5.477	17.321	74	5476.	8.602	27.203
31	961.	5.568	17.607	75	5625.	8.660	27.386
32	1024.	5.657	17.889	76	5776.	8.718	27.568
33	1089.	5.745	18.166	77	5929.	8.775	27.749
34	1156.	5.831	18.439	78	6084.	8.832	27.928
35	1225.	5.916	18.708	79	6241.	8.888	28.107
36	1296.	6.000	18.974	80	6400.	8.944	28.284
37	1369.	6.083	19.235	81	6561.	9.000	28.460
38	1444.	6.164	19.494	82	6724.	9.055	28.636
39	1521.	6.245	19.748	83	6889.	9.110	28.810
40	1600.	6.325	20.000	84	7056.	9.165	28.983
41	1681.	6.403	20.248	85	7225.	9.220	29.155
42	1764.	6.481	20.494	86	7396.	9.274	29.326
43	1849.	6.557	20.736	87	7569.	9.327	29.496
44	1936.	6.633	20.976	88	7744.	9.381	29.665

(*continued*)

TABLE A (continued)

N	N^2	$\sqrt{N}$	$\sqrt{10N}$	N	N^2	$\sqrt{N}$	$\sqrt{10N}$
89	7921.	9.434	29.833	134	17956.	11.576	36.606
90	8100.	9.487	30.000	135	18225.	11.619	36.742
91	8281.	9.539	30.166	136	18496.	11.662	36.878
92	8464.	9.592	30.332	137	18769.	11.705	37.014
93	8649.	9.644	30.496	138	19044.	11.747	37.148
94	8836.	9.695	30.659	139	19321.	11.790	37.283
95	9025.	9.747	30.822	140	19600.	11.832	37.417
96	9216.	9.798	30.984	141	19881.	11.874	37.550
97	9409.	9.849	31.145	142	20164.	11.916	37.683
98	9604.	9.899	31.305	143	20449.	11.958	37.815
99	9801.	9.950	31.464	144	20736.	12.000	37.947
100	10000.	10.000	31.623	145	21025.	12.042	38.079
101	10201.	10.050	31.780	146	21316.	12.083	38.210
102	10404.	10.100	31.937	147	21609.	12.124	38.341
103	10609.	10.149	32.094	148	21904.	12.166	38.471
104	10816.	10.198	32.249	149	22201.	12.207	38.601
105	11025.	10.247	32.404	150	22500.	12.247	38.730
106	11236.	10.296	32.558	151	22801.	12.288	38.859
107	11449.	10.344	32.711	152	23104.	12.329	38.987
108	11664.	10.392	32.863	153	23409.	12.369	39.115
109	11881.	10.440	33.015	154	23716.	12.410	39.243
110	12100.	10.488	33.166	155	24025.	12.450	39.370
111	12321.	10.536	33.317	156	24336.	12.490	39.497
112	12544.	10.583	33.466	157	24649.	12.530	39.623
113	12769.	10.630	33.615	158	24964.	12.570	39.749
114	12996.	10.677	33.764	159	25281.	12.610	39.875
115	13225.	10.724	33.912	160	25600.	12.649	40.000
116	13456.	10.770	34.059	161	25921.	12.689	40.125
117	13689.	10.817	34.205	162	26244.	12.728	40.249
118	13924.	10.863	34.351	163	26569.	12.767	40.373
119	14161.	10.909	34.496	164	26896.	12.806	40.497
120	14400.	10.954	34.641	165	27225.	12.845	40.620
121	14641.	11.000	34.785	166	27556.	12.884	40.743
122	14884.	11.045	34.928	167	27889.	12.923	40.866
123	15129.	11.091	35.071	168	28224.	12.961	40.988
124	15376.	11.136	35.214	169	28561.	13.000	41.110
125	15625.	11.180	35.355	170	28900.	13.038	41.231
126	15876.	11.225	35.496	171	29241.	13.077	41.352
127	16129.	11.269	35.637	172	29584.	13.115	41.473
128	16384.	11.314	35.777	173	29929.	13.153	41.593
129	16641.	11.358	35.917	174	30276.	13.191	41.713
130	16900.	11.402	36.056	175	30625.	13.229	41.833
131	17161.	11.446	36.194	176	30976.	13.266	41.952
132	17424.	11.489	36.332	177	31329.	13.304	42.071
133	17689.	11.533	36.469	178	31684.	13.342	42.190

TABLE A (continued)

N	N^2	$\sqrt{N}$	$\sqrt{10N}$	N	N^2	$\sqrt{N}$	$\sqrt{10N}$
179	32041.	13.379	42.308	224	50176.	14.967	47.329
180	32400.	13.416	42.426	225	50625.	15.000	47.434
181	32761.	13.454	42.544	226	51076.	15.033	47.539
182	33124.	13.491	42.661	227	51529.	15.067	47.645
183	33489.	13.528	42.778	228	51984.	15.100	47.749
184	33856.	13.565	42.895	229	52441.	15.133	47.854
185	34225.	13.601	43.012	230	52900.	15.166	47.958
186	34596.	13.638	43.128	231	53361.	15.199	48.062
187	34969.	13.675	43.243	232	53824.	15.232	48.166
188	35344.	13.711	43.359	233	54289.	15.264	48.270
189	35721.	13.748	43.474	234	54756.	15.297	48.374
190	36100.	13.784	43.589	235	55225.	15.330	48.477
191	36481.	13.820	43.704	236	55696.	15.362	48.580
192	36864.	13.856	43.818	237	56169.	15.395	48.683
193	37249.	13.892	43.932	238	56644.	15.427	48.785
194	37636.	13.928	44.045	239	57121.	15.460	48.888
195	38025.	13.964	44.159	240	57600.	15.492	48.990
196	38416.	14.000	44.272	241	58081.	15.524	49.092
197	38809.	14.036	44.385	242	58564.	15.556	49.193
198	39204.	14.071	44.497	243	59049.	15.588	49.295
199	39601.	14.107	44.609	244	59536.	15.620	49.396
200	40000.	14.142	44.721	245	60025.	15.652	49.497
201	40401.	14.177	44.833	246	60516.	15.684	49.598
202	40804.	14.213	44.944	247	61009.	15.716	49.699
203	41209.	14.248	45.056	248	61504.	15.748	49.800
204	41616.	14.283	45.166	249	62001.	15.780	49.900
205	42025.	14.318	45.277	250	62500.	15.811	50.000
206	42436.	14.353	45.387	251	63001.	15.843	50.100
207	42849.	13.387	45.497	252	63504.	15.875	50.200
208	43264.	14.422	45.607	253	64009.	15.906	50.299
209	43681.	14.457	45.717	254	64516.	15.937	50.398
210	44100.	14.491	45.826	255	65025.	15.969	50.498
211	44521.	14.526	45.935	256	65536.	16.000	50.596
212	44944.	14.560	46.043	257	66049.	16.031	50.695
213	45369.	14.595	46.152	258	66564.	16.062	50.794
214	45796.	14.629	46.260	259	67081.	16.093	50.892
215	46225.	14.663	46.368	260	67600.	16.125	50.990
216	46656.	14.697	46.476	261	68121.	16.155	51.088
217	47089.	14.731	46.583	262	68644.	16.186	51.186
218	47524.	14.765	46.690	263	69169.	16.217	51.284
219	47961.	14.799	46.797	264	69696.	16.248	51.381
220	48400.	14.832	46.904	265	70225.	16.279	51.478
221	48841.	14.866	47.011	266	70756.	16.310	51.575
222	49284.	14.900	47.117	267	71289.	16.340	51.672
223	49729.	14.933	47.223	268	71824.	16.371	51.769

(continued)

TABLE A (continued)

N	N^2	$\sqrt{N}$	$\sqrt{10N}$	N	N^2	$\sqrt{N}$	$\sqrt{10N}$
269	72361.	16.401	51.865	314	98596.	17.720	56.036
270	72900.	16.432	51.962	315	99225.	17.748	56.125
271	73441.	16.462	52.058	316	99856.	17.776	56.214
272	73984.	16.492	52.154	317	100489.	17.804	56.303
273	74529.	16.523	52.249	318	101124.	17.833	56.391
274	75076.	16.553	52.345	319	101761.	17.861	56.480
275	75625.	16.583	52.440	320	102400.	17.889	56.569
276	76176.	16.613	52.536	321	103041.	17.916	56.657
277	76729.	16.643	52.631	322	103684.	17.944	56.745
278	77284.	16.673	52.726	323	104329.	17.972	56.833
279	77841.	16.703	52.820	324	104976.	18.000	56.921
280	78400.	16.733	52.915	325	105625.	18.028	57.009
281	78961.	16.763	53.009	326	106276.	18.055	57.096
282	79524.	16.793	53.104	327	106929.	18.083	57.184
283	80089.	16.823	53.198	328	107584.	18.111	57.271
284	80656.	16.852	53.292	329	108241.	18.138	57.359
285	81225.	16.882	53.385	330	108900.	18.166	57.446
286	81796.	16.912	53.479	331	109561.	18.193	57.533
287	82369.	16.941	53.572	332	110224.	18.221	57.619
288	82944.	16.971	53.666	333	110889.	18.248	57.706
289	83521.	17.000	53.759	334	111556.	18.276	57.793
290	84100.	17.029	53.852	335	112225.	18.303	57.879
291	84681.	17.059	53.944	336	112896.	18.330	57.966
292	85264.	17.088	54.037	337	113569.	18.358	58.052
293	85849.	17.117	54.129	338	114244.	18.385	58.138
294	86436.	17.146	54.222	339	114921.	18.412	58.224
295	87025.	17.176	54.314	340	115600.	18.439	58.310
296	87616.	17.205	54.406	341	116281.	18.466	58.395
297	88209.	17.234	54.498	342	116964.	18.493	58.481
298	88804.	17.263	54.589	343	117649.	18.520	58.566
299	89401.	17.292	54.681	344	118336.	18.547	58.652
300	90000.	17.321	54.772	345	119025.	18.574	58.737
301	90601.	17.349	54.863	346	119716.	18.601	58.822
302	91204.	17.378	54.955	347	120409.	18.628	58.907
303	91809.	17.407	55.045	348	121104.	18.655	58.992
304	92416.	17.436	55.136	349	121801.	18.682	59.076
305	93025.	17.464	55.227	350	122500.	18.708	59.161
306	93636.	17.493	55.317	351	123201.	18.735	59.245
307	94249.	17.521	55.408	352	123904.	18.762	59.330
308	94864.	17.550	55.498	353	124609.	18.788	59.414
309	95481.	17.578	55.588	354	125316.	18.815	59.498
310	96100.	17.607	55.678	355	126025.	18.841	59.582
311	96721.	17.635	55.767	356	126736.	18.868	59.666
312	97344.	17.664	55.857	357	127449.	18.894	59.749
313	97969.	17.692	55.946	358	128164.	18.921	59.833

N	N^2	$\sqrt{N}$	$\sqrt{10N}$	N	N^2	$\sqrt{N}$	$\sqrt{10N}$
359	128881.	18.947	59.917	404	163216.	20.100	63.561
360	129600.	18.974	60.000	405	164025.	20.125	63.640
361	130321.	19.000	60.083	406	164836.	20.149	63.718
362	131044.	19.026	60.166	407	165649.	20.174	63.797
363	131769.	19.053	60.249	408	166464.	20.199	63.875
364	132496.	19.079	60.332	409	167281.	20.224	63.953
365	133225.	19.105	60.415	410	168100.	20.248	64.031
366	133956.	19.131	60.498	411	168921.	20.273	64.109
367	134689.	19.157	60.581	412	169744.	20.298	64.187
368	135424.	19.183	60.663	413	170569.	20.322	64.265
369	136161.	19.209	60.745	414	171396.	20.347	64.343
370	136900.	19.235	60.828	415	172225.	20.372	64.420
371	137641.	19.261	60.910	416	173056.	20.396	64.498
372	138384.	19.287	60.992	417	173889.	20.421	64.576
373	139129.	19.313	61.074	418	174724.	20.445	64.653
374	139876.	19.339	61.156	419	175561.	20.469	64.730
375	140625.	19.365	61.237	420	176400.	20.494	64.807
376	141376.	19.391	61.319	421	177241.	20.518	64.885
377	142129.	19.416	61.400	422	178084.	20.543	64.962
378	142884.	19.442	61.482	423	178929.	20.567	65.038
379	143641.	19.468	61.563	424	179776.	20.591	65.115
380	144400.	19.494	61.644	425	180625.	20.616	65.192
381	145161.	19.519	61.725	426	181476.	20.640	65.269
382	145924.	19.545	61.806	427	182329.	20.664	65.345
383	146689.	19.570	61.887	428	183184.	20.688	65.422
384	147456.	19.596	61.968	429	184041.	20.712	65.498
385	148225.	19.621	62.048	430	184900.	20.736	65.574
386	148996.	19.647	62.129	431	185761.	20.761	65.651
387	149769.	19.672	62.209	432	186624.	20.785	65.727
388	150544.	19.698	62.290	433	187489.	20.809	65.803
389	151321.	19.723	62.370	434	188356.	20.833	65.879
390	152100.	19.748	62.450	435	189225.	20.857	65.955
391	152881.	19.774	62.530	436	190096.	20.881	66.030
392	153664.	19.799	62.610	437	190969.	20.905	66.106
393	154449.	19.824	62.690	438	191844.	20.928	66.182
394	155236.	19.849	62.769	439	192721.	20.952	66.257
395	156025.	19.875	62.849	440	193600.	20.976	66.332
396	156816.	19.900	62.929	441	194481.	21.000	66.408
397	157609.	19.925	63.008	442	195364.	21.024	66.483
398	158404.	19.950	63.087	443	196249.	21.048	66.558
399	159201.	19.975	63.166	444	197136.	21.071	66.633
400	160000.	20.000	63.246	445	198025.	21.095	66.708
401	160801.	20.025	63.325	446	198916.	21.119	66.783
402	161604.	20.050	63.403	447	199809.	21.142	66.858
403	162409.	20.075	63.482	448	200704.	21.166	66.933

(continued)

TABLE A (continued)

N	N^2	$\sqrt{N}$	$\sqrt{10N}$	N	N^2	$\sqrt{N}$	$\sqrt{10N}$
449	201601.	21.190	67.007	494	244036.	22.226	70.285
450	202500.	21.213	67.082	495	245025.	22.249	70.356
451	203401.	21.237	67.157	496	246016.	22.271	70.427
452	204304.	21.260	67.231	497	247009.	22.293	70.498
453	205209.	21.284	67.305	498	248004.	22.316	70.569
454	206116.	21.307	67.380	499	249001.	22.338	70.640
455	207025.	21.331	67.454	500	250000.	22.361	70.711
456	207936.	21.354	67.528	501	251001.	22.383	70.781
457	208849.	21.378	67.602	502	252004.	22.405	70.852
458	209764.	21.401	67.676	503	253009.	22.428	70.922
459	210681.	21.424	67.750	504	254016.	22.450	70.993
460	211600.	21.448	67.823	505	255025.	22.472	71.063
461	212521.	21.471	67.897	506	256036.	22.494	71.134
462	213444.	21.494	67.971	507	257049.	22.517	71.204
463	214369.	21.517	68.044	508	258064.	22.539	71.274
464	215296.	21.541	68.118	509	259081.	22.561	71.344
465	216225.	21.564	68.191	510	260100.	22.583	71.414
466	217156.	21.587	68.264	511	261121.	22.605	71.484
467	218089.	21.610	68.337	512	262144.	22.627	71.554
468	219024.	21.633	68.411	513	263169.	22.650	71.624
469	219961.	21.656	68.484	514	264196.	22.672	71.694
470	220900.	21.679	68.557	515	265225.	22.694	71.764
471	221841.	21.703	68.629	516	266256.	22.716	71.833
472	222784.	21.726	68.702	517	267289.	22.738	71.903
473	223729.	21.749	68.775	518	268324.	22.760	71.972
474	224676.	21.772	68.848	519	269361.	22.782	72.042
475	225625.	21.794	68.920	520	270400.	22.804	72.111
476	226576.	21.817	68.993	521	271441.	22.825	72.180
477	227529.	21.840	69.065	522	272484.	22.847	72.250
478	228484.	21.863	69.138	523	273529.	22.869	72.319
479	229441.	21.886	69.210	524	274576.	22.891	72.388
480	230400.	21.909	69.282	525	275625.	22.913	72.457
481	231361.	21.932	69.354	526	276676.	22.935	72.526
482	232324.	21.954	69.426	527	277729.	22.956	72.595
483	233289.	21.977	69.498	528	278784.	22.978	72.664
484	234256.	22.000	69.570	529	279841.	23.000	72.732
485	235225.	22.023	69.642	530	280900.	23.022	72.801
486	236196.	22.045	69.714	531	281961.	23.043	72.870
487	237169.	22.068	69.785	532	283024.	23.065	72.938
488	238144.	22.091	69.857	533	284089.	23.087	73.007
489	239121.	22.113	69.929	534	285156.	23.108	73.075
490	240100.	22.136	70.000	535	286225.	23.130	73.144
491	241081.	22.159	70.071	536	287296.	23.152	73.212
492	242064.	22.181	70.143	537	288369.	23.173	73.280
493	243049.	22.204	70.214	538	289444.	23.195	73.348

N	N^2	$\sqrt{N}$	$\sqrt{10N}$	N	N^2	$\sqrt{N}$	$\sqrt{10N}$
539	290521.	23.216	73.417	584	341056.	24.166	76.420
540	291600.	23.238	73.485	585	342225.	24.187	76.485
541	292681.	23.259	73.553	586	343396.	24.207	76.551
542	293764.	23.281	73.621	587	344569.	24.228	76.616
543	294849.	23.302	73.689	588	345744.	24.249	76.681
544	295936.	23.324	73.756	589	346921.	24.269	76.746
545	297025.	23.345	73.824	590	348100.	24.290	76.811
546	298116.	23.367	73.892	591	349281.	24.310	76.877
547	299209.	23.388	73.959	592	350464.	24.331	76.942
548	300304.	23.409	74.027	593	351649.	24.352	77.006
549	301401.	23.431	74.095	594	352836.	24.372	77.071
550	302500.	23.452	74.162	595	354025.	24.393	77.136
551	303601.	23.473	74.229	596	355216.	24.413	77.201
552	304704.	23.495	74.297	597	356409.	24.434	77.266
553	305809.	23.516	74.364	598	357604.	24.454	77.330
554	306916.	23.537	74.431	599	358801.	24.474	77.395
555	308025.	23.558	74.498	600	360000.	24.495	77.460
556	309136.	23.580	74.565	601	361201.	24.515	77.524
557	310249.	23.601	74.632	602	362404.	24.536	77.589
558	311364.	23.622	74.699	603	363609.	24.556	77.653
559	312481.	23.643	74.766	604	364816.	24.576	77.717
560	313600.	23.664	74.833	605	366025.	24.597	77.782
561	314721.	23.685	74.900	606	367236.	24.617	77.846
562	315844.	23.707	74.967	607	368449.	24.637	77.910
563	316969.	23.728	75.033	608	369664.	24.658	77.974
564	318096.	23.749	75.100	609	370881.	24.678	78.038
565	319225.	23.770	75.166	610	372100.	24.698	78.102
566	320356.	23.791	75.233	611	373321.	24.718	78.166
567	321489.	23.812	75.299	612	374544.	24.739	78.230
568	322624.	23.833	75.366	613	375769.	24.759	78.294
569	323761.	23.854	75.432	614	376996.	24.779	78.358
570	324900.	23.875	75.498	615	378225.	24.799	78.422
571	326041.	23.896	75.565	616	379456.	24.819	78.486
572	327184.	23.917	75.631	617	380689.	24.839	78.549
573	328329.	23.937	75.697	618	381924.	24.860	78.613
574	329476.	23.958	75.763	619	383161.	24.880	78.677
575	330625.	23.979	75.829	620	384400.	24.900	78.740
576	331776.	24.000	75.895	621	385641.	24.920	78.804
577	332929.	24.021	75.961	622	386884.	24.940	78.867
578	334084.	24.042	76.026	623	388129.	24.960	78.930
579	335241.	24.062	76.092	624	389376.	24.980	78.994
580	336400.	24.083	76.158	625	390625.	25.000	79.057
581	337561.	24.104	76.223	626	391876.	25.020	79.120
582	338724.	24.125	76.289	627	393129.	25.040	79.183
583	339889.	24.145	76.354	628	394384.	25.060	79.246

(continued)

TABLE A (continued)

N	N^2	$\sqrt{N}$	$\sqrt{10N}$	N	N^2	$\sqrt{N}$	$\sqrt{10N}$
629	395641.	25.080	79.310	674	454276.	25.962	82.098
630	396900.	25.100	79.373	675	455625.	25.981	82.158
631	398161.	25.120	79.436	676	456976.	26.000	82.219
632	399424.	25.140	79.498	677	458329.	26.019	82.280
633	400689.	25.159	79.561	678	459684.	26.038	82.341
634	401956.	25.179	79.624	679	461041.	26.058	82.401
635	403225.	25.199	79.687	680	462400.	26.077	82.462
636	404496.	25.219	79.750	681	463761.	26.096	82.523
637	405769.	25.239	79.812	682	465124.	26.115	82.583
638	407044.	25.259	79.875	683	466489.	26.134	82.644
639	408321.	25.278	79.937	684	467856.	26.153	82.704
640	409600.	25.298	80.000	685	469225.	26.173	82.765
641	410881.	25.318	80.062	686	470596.	26.192	82.825
642	412164.	25.338	80.125	687	471969.	26.211	82.885
643	413449.	25.357	80.187	688	473344.	26.230	82.946
644	414736.	25.377	80.250	689	474721.	26.249	83.006
645	416025.	25.397	80.312	690	476100.	26.268	83.066
646	417316.	25.417	80.374	691	477481.	26.287	83.126
647	418609.	25.436	80.436	692	478864.	26.306	83.187
648	419904.	25.456	80.498	693	480249.	26.325	83.247
649	421201.	25.475	80.561	694	481636.	26.344	83.307
650	422500.	25.495	80.623	695	483025.	26.363	83.367
651	423801.	25.515	80.685	696	484416.	26.382	83.427
652	425104.	25.534	80.747	697	485809.	26.401	83.487
653	426409.	25.554	80.808	698	487204.	26.420	83.546
654	427716.	25.573	80.870	699	488601.	26.439	83.606
655	429025.	25.593	80.932	700	490000.	26.458	83.666
656	430336.	25.612	80.994	701	491401.	26.476	83.726
657	431649.	25.632	81.056	702	492804.	26.495	83.785
658	432964.	25.652	81.117	703	494209.	26.514	83.845
659	434281.	25.671	81.179	704	495616.	26.533	83.905
660	435600.	25.690	81.240	705	497025.	26.552	83.964
661	436921.	25.710	81.302	706	498436.	26.571	84.024
662	438244.	25.729	81.363	707	499849.	26.589	84.083
663	439569.	25.749	81.425	708	501264.	26.608	84.143
664	440896.	25.768	81.486	709	502681.	26.627	84.202
665	442225.	25.788	81.548	710	504100.	26.646	84.261
666	443556.	25.807	81.609	711	505521.	26.665	84.321
667	444889.	25.826	81.670	712	506944.	26.683	84.380
668	446224.	25.846	81.731	713	508369.	26.702	84.439
669	447561.	25.865	81.792	714	509796.	26.721	84.499
670	448900.	25.884	81.854	715	511225.	26.739	84.558
671	450241.	25.904	81.915	716	512656.	26.758	84.617
672	451584.	25.923	81.976	717	514089.	26.777	84.676
673	452929.	25.942	82.037	718	515524.	26.796	84.735

N	N^2	$\sqrt{N}$	$\sqrt{10N}$	N	N^2	$\sqrt{N}$	$\sqrt{10N}$
719	516961.	26.814	84.794	764	583696.	27.641	87.407
720	518400.	26.833	84.853	765	585225.	27.659	87.464
721	519841.	26.851	84.912	766	586756.	27.677	87.521
722	521284.	26.870	84.971	767	588289.	27.695	87.579
723	522729.	26.889	85.029	768	589824.	27.713	87.636
724	524176.	26.907	85.088	769	591361.	27.731	87.693
725	525625.	26.926	85.147	770	592900.	27.749	87.750
726	527076.	26.944	85.206	771	594441.	27.767	87.807
727	528529.	26.963	85.264	772	595984.	27.785	87.864
728	529984.	26.981	85.323	773	597529.	27.803	87.920
729	531441.	27.000	85.381	774	599076.	27.821	87.977
730	532900.	27.019	85.440	775	600625.	27.839	88.034
731	534361.	27.037	85.499	776	602176.	27.857	88.091
732	535824.	27.055	85.557	777	603729.	27.875	88.148
733	537289.	27.074	85.615	778	605284.	27.893	88.204
734	538756.	27.092	85.674	779	606841.	27.911	88.261
735	540225.	27.111	85.732	780	608400.	27.928	88.318
736	541696.	27.129	85.790	781	609961.	27.946	88.374
737	543169.	27.148	85.849	782	611524.	27.964	88.431
738	544644.	27.166	85.907	783	613089.	27.982	88.487
739	546121.	27.185	85.965	784	614656.	28.000	88.544
740	547600.	27.203	86.023	785	616225.	28.018	88.600
741	549081.	27.221	86.081	786	617796.	28.036	88.657
742	550564.	27.240	86.139	787	619369.	28.054	88.713
743	552049.	27.258	86.197	788	620944.	28.071	88.769
744	553536.	27.276	86.255	789	622521.	28.089	88.826
745	555025.	27.295	86.313	790	624100.	28.107	88.882
746	556516.	27.313	86.371	791	625681.	28.125	88.938
747	558009.	27.331	86.429	792	627264.	28.142	88.994
748	559504.	27.350	86.487	793	628849.	28.160	89.051
749	561001.	27.368	86.545	794	630436.	28.178	89.107
750	562500.	27.386	86.603	795	632025.	28.196	89.163
751	564001.	27.404	86.660	796	633616.	28.213	89.219
752	565504.	27.423	86.718	797	635209.	28.231	89.275
753	567009.	27.441	86.776	798	636804.	28.249	89.331
754	568516.	27.459	86.833	799	638401.	28.267	89.387
755	570025.	27.477	86.891	800	640000.	28.284	89.443
756	571536.	27.495	86.948	801	641601.	28.302	89.499
757	573049.	27.514	87.006	802	643204.	28.320	89.554
758	574564.	27.532	87.063	803	644809.	28.337	89.610
759	576081.	27.550	87.121	804	646416.	28.355	89.666
760	577600.	27.568	87.178	805	648025.	28.373	89.722
761	579121.	27.586	87.235	806	649636.	28.390	89.778
762	580644.	27.604	87.293	807	651249.	28.408	89.833
763	582169.	27.622	87.350	808	652864.	28.425	89.889

(*continued*)

TABLE A (continued)

N	N^2	$\sqrt{N}$	$\sqrt{10N}$	N	N^2	$\sqrt{N}$	$\sqrt{10N}$
809	654481.	28.443	89.944	854	729316.	29.223	92.412
810	656100.	28.460	90.000	855	731025.	29.240	92.466
811	657721.	28.478	90.056	856	732736.	29.257	92.520
812	659344.	28.496	90.111	857	734449.	29.275	92.574
813	660969.	28.513	90.167	858	736164.	29.292	92.628
814	662596.	28.531	90.222	859	737881.	29.309	92.682
815	664225.	28.548	90.277	860	739600.	29.326	92.736
816	665856.	28.566	90.333	861	741321.	29.343	92.790
817	667489.	28.583	90.388	862	743044.	29.360	92.844
818	669124.	28.601	90.443	863	744769.	29.377	92.898
819	670761.	28.618	90.499	864	746496.	29.394	92.952
820	672400.	28.636	90.554	865	748225.	29.411	93.005
821	674041.	28.653	90.609	866	749956.	29.428	93.059
822	675684.	28.671	90.664	867	751689.	29.445	93.113
823	677329.	28.688	90.719	868	753424.	29.462	93.167
824	678976.	28.705	90.774	869	755161.	29.479	93.220
825	680625.	28.723	90.830	870	756900.	29.496	93.274
826	682276.	28.740	90.885	871	758641.	29.513	93.327
827	683929.	28.758	90.940	872	760384.	29.530	93.381
828	685584.	28.775	90.995	873	762129.	29.547	93.434
829	687241.	28.792	91.049	874	763876.	29.563	93.488
830	688900.	28.810	91.104	875	765625.	29.580	93.541
831	690561.	28.827	91.159	876	767376.	29.597	93.595
832	692224.	28.844	91.214	877	769129.	29.614	93.648
833	693889.	28.862	91.269	878	770884.	29.631	93.702
834	695556.	28.879	91.324	879	772641.	29.648	93.755
835	697225.	28.896	91.378	880	774400.	29.665	93.808
836	698896.	28.914	91.433	881	776161.	29.682	93.862
837	700569.	28.931	91.488	882	777924.	29.698	93.915
838	702244.	28.948	91.542	883	779689.	29.715	93.968
839	703921.	28.965	91.597	884	781456.	29.732	94.021
840	705600.	28.983	91.652	885	783225.	29.749	94.074
841	707281.	29.000	91.706	886	784996.	29.766	94.128
842	708964.	29.017	91.761	887	786769.	29.783	94.181
843	710649.	29.034	91.815	888	788544.	29.799	94.234
844	712336.	29.052	91.869	889	790321.	29.816	94.287
845	714025.	29.069	91.924	890	792100.	29.833	94.340
846	715716.	29.086	91.978	891	793881.	29.850	94.393
847	717409.	29.103	92.033	892	795664.	29.866	94.446
848	719104.	29.120	92.087	893	797449.	29.883	94.499
849	720801.	29.138	92.141	894	799236.	29.900	94.552
850	722500.	29.155	92.195	895	801025.	29.917	94.604
851	724201.	29.172	92.250	896	802816.	29.933	94.657
852	725904.	29.189	92.304	897	804609.	29.950	94.710
853	727609.	29.206	92.358	898	806404.	29.967	94.763

N	N^2	$\sqrt{N}$	$\sqrt{10N}$	N	N^2	$\sqrt{N}$	$\sqrt{10N}$
899	808201.	29.983	94.816	944	891136.	30.725	97.160
900	810000.	30.000	94.868	945	893025.	30.741	97.211
901	811801.	30.017	94.921	946	894916.	30.757	97.263
902	813604.	30.033	94.974	947	896809.	30.773	97.314
903	815409.	30.050	95.026	948	898704.	30.790	97.365
904	817216.	30.067	95.079	949	900601.	30.806	97.417
905	819025.	30.083	95.131	950	902500.	30.822	97.468
906	820836.	30.100	95.184	951	904401.	30.838	97.519
907	822649.	30.116	95.237	952	906304.	30.854	97.570
908	824464.	30.133	95.289	953	908209.	30.871	97.622
909	826281.	30.150	95.341	954	910116.	30.887	97.673
910	828100.	30.166	95.394	955	912025.	30.903	97.724
911	829921.	30.183	95.446	956	913936.	30.919	97.775
912	831744.	30.199	95.499	957	915849.	30.935	97.826
913	833569.	30.216	95.551	958	917764.	30.952	97.877
914	835396.	30.232	95.603	959	919681.	30.968	97.929
915	837225.	30.249	95.656	960	921600.	30.984	97.980
916	839056.	30.265	95.708	961	923521.	31.000	98.031
917	840889.	30.282	95.760	962	925444.	31.016	98.082
918	842724.	30.299	95.812	963	927369.	31.032	98.133
919	844561.	30.315	95.864	964	929296.	31.048	98.184
920	846400.	30.332	95.917	965	931225.	31.064	98.234
921	848241.	30.348	95.969	966	933156.	31.081	98.285
922	850084.	30.364	96.021	967	935089.	31.097	98.336
923	851929.	30.381	96.073	968	937024.	31.113	98.387
924	853776.	30.397	96.125	969	938961.	31.129	98.438
925	855625.	30.414	96.177	970	940900.	31.145	98.489
926	857476.	30.430	96.229	971	942841.	31.161	98.539
927	859329.	30.447	96.281	972	944784.	31.177	98.590
928	861184.	30.463	96.333	973	946729.	31.193	98.641
929	863041.	30.480	96.385	974	948676.	31.209	98.691
930	864900.	30.496	96.437	975	950625.	31.225	98.742
931	866761.	30.512	96.488	976	952576.	31.241	98.793
932	868624.	30.529	96.540	977	954529.	31.257	98.843
933	870489.	30.545	96.592	978	956484.	31.273	98.894
934	872356.	30.561	96.644	979	958441.	31.289	98.944
935	874225.	30.578	96.695	980	960400.	31.305	98.995
936	876096.	30.594	96.747	981	962361.	31.321	99.045
937	877969.	30.610	96.799	982	964324.	31.337	99.096
938	879844.	30.627	96.850	983	966289.	31.353	99.146
939	881721.	30.643	96.902	984	968256.	31.369	99.197
940	883600.	30.659	96.954	985	970225.	31.385	99.247
941	885481.	30.676	97.005	986	972196.	31.401	99.298
942	887364.	30.692	97.057	987	974169.	31.417	99.348
943	889249.	30.708	97.108	988	976144.	31.432	99.398

(*continued*)

TABLE A (continued)

N	N^2	$\sqrt{N}$	$\sqrt{10N}$	N	N^2	$\sqrt{N}$	$\sqrt{10N}$
989	978121.	31.448	99.448	995	990025.	31.544	99.750
990	980100.	31.464	99.499	996	992016.	31.559	99.800
991	982081.	31.480	99.549	997	994009.	31.575	99.850
992	984064.	31.496	99.599	998	996004.	31.591	99.900
993	986049.	31.512	99.649	999	998001.	31.607	99.950
994	988036.	31.528	99.700	1000	1000000.	31.623	100.000

TABLE B. **Areas Under the Normal Curve Between the Mean and z**

z	Area from Mean to z	z	Area from Mean to z	z	Area from Mean to z
0.00	.0000	0.35	.1368	0.70	.2580
0.01	.0040	0.36	.1406	0.71	.2611
0.02	.0080	0.37	.1443	0.72	.2642
0.03	.0120	0.38	.1480	0.73	.2673
0.04	.0160	0.39	.1517	0.74	.2704
0.05	.0199	0.40	.1554	0.75	.2734
0.06	.0239	0.41	.1591	0.76	.2764
0.07	.0279	0.42	.1628	0.77	.2794
0.08	.0319	0.43	.1664	0.78	.2823
0.09	.0359	0.44	.1700	0.79	.2852
0.10	.0398	0.45	.1736	0.80	.2881
0.11	.0438	0.46	.1772	0.81	.2910
0.12	.0478	0.47	.1808	0.82	.2939
0.13	.0517	0.48	.1844	0.83	.2967
0.14	.0557	0.49	.1879	0.84	.2995
0.15	.0596	0.50	.1915	0.85	.3023
0.16	.0636	0.51	.1950	0.86	.3051
0.17	.0675	0.52	.1985	0.87	.3078
0.18	.0714	0.53	.2019	0.88	.3106
0.19	.0753	0.54	.2054	0.89	.3133
0.20	.0793	0.55	.2088	0.90	.3159
0.21	.0832	0.56	.2123	0.91	.3186
0.22	.0871	0.57	.2157	0.92	.3212
0.23	.0910	0.58	.2190	0.93	.3238
0.24	.0948	0.59	.2224	0.94	.3264
0.25	.0987	0.60	.2257	0.95	.3289
0.26	.1026	0.61	.2291	0.96	.3315
0.27	.1064	0.62	.2324	0.97	.3340
0.28	.1103	0.63	.2357	0.98	.3365
0.29	.1141	0.64	.2389	0.99	.3389
0.30	.1179	0.65	.2422	1.00	.3413
0.31	.1217	0.66	.2454	1.01	.3438
0.32	.1255	0.67	.2486	1.02	.3461
0.33	.1293	0.68	.2517	1.03	.3485
0.34	.1331	0.69	.2549	1.04	.3508

(continued)

TABLE B. (continued)

z	Area from Mean to z	z	Area from Mean to z	z	Area from Mean to z
1.05	.3531	1.40	.4192	1.75	.4599
1.06	.3554	1.41	.4207	1.76	.4608
1.07	.3577	1.42	.4222	1.77	.4616
1.08	.3599	1.43	.4236	1.78	.4625
1.09	.3621	1.44	.4251	1.79	.4633
1.10	.3643	1.45	.4265	1.80	.4641
1.11	.3665	1.46	.4279	1.81	.4649
1.12	.3686	1.47	.4292	1.82	.4656
1.13	.3708	1.48	.4306	1.83	.4664
1.14	.3729	1.49	.4319	1.84	.4671
1.15	.3749	1.50	.4332	1.85	.4678
1.16	.3770	1.51	.4345	1.86	.4686
1.17	.3790	1.52	.4357	1.87	.4693
1.18	.3810	1.53	.4370	1.88	.4699
1.19	.3830	1.54	.4382	1.89	.4706
1.20	.3849	1.55	.4394	1.90	.4713
1.21	.3869	1.56	.4406	1.91	.4719
1.22	.3888	1.57	.4418	1.92	.4726
1.23	.3907	1.58	.4429	1.93	.4732
1.24	.3925	1.59	.4441	1.94	.4738
1.25	.3944	1.60	.4452	1.95	.4744
1.26	.3962	1.61	.4463	1.96	.4750
1.27	.3980	1.62	.4474	1.97	.4756
1.28	.3997	1.63	.4484	1.98	.4761
1.29	.4015	1.64	.4495	1.99	.4767
1.30	.4032	1.65	.4505	2.00	.4772
1.31	.4049	1.66	.4515	2.01	.4778
1.32	.4066	1.67	.4525	2.02	.4783
1.33	.4082	1.68	.4535	2.03	.4788
1.34	.4099	1.69	.4545	2.04	.4793
1.35	.4115	1.70	.4554	2.05	.4798
1.36	.4131	1.71	.4564	2.06	.4803
1.37	.4147	1.72	.4573	2.07	.4808
1.38	.4162	1.73	.4582	2.08	.4812
1.39	.4177	1.74	.4591	2.09	.4817

z	Area from Mean to z	z	Area from Mean to z	z	Area from Mean to z
2.10	.4821	2.45	.4929	2.80	.4974
2.11	.4826	2.46	.4931	2.81	.4975
2.12	.4830	2.47	.4932	2.82	.4976
2.13	.4834	2.48	.4934	2.83	.4977
2.14	.4838	2.49	.4936	2.84	.4977
2.15	.4842	2.50	.4938	2.85	.4978
2.16	.4846	2.51	.4940	2.86	.4979
2.17	.4850	2.52	.4941	2.87	.4979
2.18	.4854	2.53	.4943	2.88	.4980
2.19	.4857	2.54	.4945	2.89	.4981
2.20	.4861	2.55	.4946	2.90	.4981
2.21	.4864	2.56	.4948	2.91	.4982
2.22	.4868	2.57	.4949	2.92	.4982
2.23	.4871	2.58	.4951	2.93	.4983
2.24	.4875	2.59	.4952	2.94	.4984
2.25	.4878	2.60	.4953	2.95	.4984
2.26	.4881	2.61	.4955	2.96	.4985
2.27	.4884	2.62	.4956	2.97	.4985
2.28	.4887	2.63	.4957	2.98	.4986
2.29	.4890	2.64	.4959	2.99	.4986
2.30	.4893	2.65	.4960	3.00	.4987
2.31	.4896	2.66	.4961	3.01	.4987
2.32	.4898	2.67	.4962	3.02	.4987
2.33	.4901	2.68	.4963	3.03	.4988
2.34	.4904	2.69	.4964	3.04	.4988
2.35	.4906	2.70	.4965	3.05	.4989
2.36	.4909	2.71	.4966	3.06	.4989
2.37	.4911	2.72	.4967	3.07	.4989
2.38	.4913	2.73	.4968	3.08	.4990
2.39	.4916	2.74	.4969	3.09	.4990
2.40	.4918	2.75	.4970	3.10	.4990
2.41	.4920	2.76	.4971	3.11	.4991
2.42	.4922	2.77	.4972	3.12	.4991
2.43	.4925	2.78	.4973	3.13	.4991
2.44	.4927	2.79	.4974	3.14	.4992

(continued)

TABLE B. (continued)

z	Area from Mean to z	z	Area from Mean to z
3.15	.4992	3.25	.4994
3.16	.4992	3.26	.4994
3.17	.4992	3.27	.4995
3.18	.4993	3.28	.4995
3.19	.4993	3.29	.4995
3.20	.4993	3.30	.4995
3.21	.4993	3.40	.4997
3.22	.4994	3.50	.4998
3.23	.4994	3.60	.4998
3.24	.4994	3.70	.4999

TABLE C. **Values of r for the .05 and .01 Levels of Significance**

df(*N* − 2)	.05	.01	*df*(*N* − 2)	.05	.01
1	.997	1.000	31	.344	.442
2	.950	.990	32	.339	.436
3	.878	.959	33	.334	.430
4	.812	.917	34	.329	.424
5	.755	.875	35	.325	.418
6	.707	.834	36	.320	.413
7	.666	.798	37	.316	.408
8	.632	.765	38	.312	.403
9	.602	.735	39	.308	.398
10	.576	.708	40	.304	.393
11	.553	.684	41	.301	.389
12	.533	.661	42	.297	.384
13	.514	.641	43	.294	.380
14	.497	.623	44	.291	.376
15	.482	.606	45	.288	.372
16	.468	.590	46	.285	.368
17	.456	.575	47	.282	.365
18	.444	.562	48	.279	.361
19	.433	.549	49	.276	.358
20	.423	.537	50	.273	.354
21	.413	.526	60	.250	.325
22	.404	.515	70	.232	.302
23	.396	.505	80	.217	.283
24	.388	.496	90	.205	.267
25	.381	.487	100	.195	.254
26	.374	.479	200	.138	.181
27	.367	.471	300	.113	.148
28	.361	.463	400	.098	.128
29	.355	.456	500	.088	.115
30	.349	.449	1000	.062	.081

Adapted from A. L. Sockloff and J. N. Edney, Some extension of Student's *t* and Pearson's *r* central distributions, Technical Report (May 1972), Measurement and Research Center, Temple University, Philadelphia.

TABLE D. **Values of Spearman r_s for the .05 and .01 Levels of Significance**

N	.05	.01	*N*	.05	.01
6	.886	—	19	.462	.608
7	.786	—	20	.450	.591
8	.738	.881	21	.438	.576
9	.683	.833	22	.428	.562
10	.648	.818	23	.418	.549
11	.623	.794	24	.409	.537
12	.591	.780	25	.400	.526
13	.566	.745	26	.392	.515
14	.545	.716	27	.385	.505
15	.525	.689	28	.377	.496
16	.507	.666	29	.370	.487
17	.490	.645	30	.364	.478
18	.476	.625			

From E. G. Olds, Distribution of sums of squares of rank differences for small numbers of individuals, *Annals of Mathematical Statistics* 9: 133–48 (1938), and E. G. Olds, The 5% significance levels for sums of squares of rank differences and a correction, *Annals of Mathematical Statistics* 20: 117–18 (1949).

TABLE E. **Values of *t* at the .05, .01, and .001 Levels of Significance**

	Two-tailed				One-tailed		
df	*.05*	*.01*	*.001*	*df*	*.05*	*.01*	*.001*
1	12.706	63.657	636.619	1	6.314	31.821	318.309
2	4.303	9.925	31.599	2	2.920	6.965	22.327
3	3.183	5.841	12.924	3	2.353	4.541	10.215
4	2.777	4.604	8.610	4	2.132	3.747	7.173
5	2.571	4.032	6.869	5	2.015	3.365	5.893
6	2.447	3.707	5.959	6	1.943	3.143	5.208
7	2.365	3.500	5.408	7	1.895	2.998	4.785
8	2.306	3.355	5.041	8	1.860	2.897	4.501
9	2.262	3.250	4.781	9	1.833	2.821	4.297
10	2.228	3.169	4.587	10	1.813	2.764	4.144

	Two-tailed				One-tailed		
df	*.05*	*.01*	*.001*	*df*	*.05*	*.01*	*.001*
11	2.201	3.106	4.437	11	1.796	2.718	4.025
12	2.179	3.055	4.318	12	1.782	2.681	3.930
13	2.160	3.012	4.221	13	1.771	2.650	3.852
14	2.145	2.977	4.141	14	1.761	2.625	3.787
15	2.132	2.947	4.073	15	1.753	2.603	3.733
16	2.120	2.921	4.015	16	1.746	2.584	3.686
17	2.110	2.898	3.965	17	1.740	2.567	3.646
18	2.101	2.879	3.922	18	1.734	2.552	3.611
19	2.093	2.861	3.883	19	1.729	2.540	3.579
20	2.086	2.845	3.850	20	1.725	2.528	3.552
21	2.080	2.831	3.819	21	1.721	2.518	3.527
22	2.074	2.819	3.792	22	1.717	2.508	3.505
23	2.069	2.807	3.768	23	1.714	2.500	3.485
24	2.064	2.797	3.745	24	1.711	2.492	3.467
25	2.060	2.787	3.725	25	1.708	2.485	3.450
26	2.056	2.779	3.707	26	1.706	2.479	3.435
27	2.052	2.771	3.690	27	1.703	2.473	3.421
28	2.048	2.763	3.674	28	1.701	2.467	3.408
29	2.045	2.756	3.659	29	1.699	2.462	3.396
30	2.042	2.750	3.646	30	1.697	2.457	3.385
40	2.021	2.705	3.551	40	1.684	2.423	3.307
50	2.009	2.678	3.496	50	1.676	2.403	3.261
60	2.000	2.660	3.460	60	1.671	2.390	3.232
70	1.994	2.648	3.435	70	1.667	2.381	3.211
80	1.990	2.639	3.416	80	1.664	2.374	3.195
90	1.987	2.632	3.402	90	1.662	2.369	3.183
100	1.984	2.626	3.391	100	1.660	2.364	3.174
∞	1.960	2.576	3.292	∞	1.645	2.327	3.091

Adapted from A. L. Sockloff and J. N. Edney, Some extension of Student's *t* and Pearson's *r* central distributions, Technical Report (May 1972), Measurement and Research Center, Temple University, Philadelphia.

TABLE F. **Values of *F* at the 5% and 1% Levels of Significance**

df Associated with the Denominator		*df* Associated with the Numerator								
		1	2	3	4	5	6	7	8	9
1	5%	161	200	216	225	230	234	237	239	241
	1%	4052	5000	5403	5625	5764	5859	5928	5982	6022
2	5%	18.5	19.0	19.2	19.2	19.3	19.3	19.4	19.4	19.4
	1%	98.5	99.0	99.2	99.2	99.3	99.3	99.4	99.4	99.4
3	5%	10.1	9.55	9.28	9.12	9.01	8.94	8.89	8.85	8.81
	1%	34.1	30.8	29.5	28.7	28.2	27.9	27.7	27.5	27.3
4	5%	7.71	6.94	6.59	6.39	6.26	6.16	6.09	6.04	6.00
	1%	21.2	18.0	16.7	16.0	15.5	15.2	15.0	14.8	14.7
5	5%	6.61	5.79	5.41	5.19	5.05	4.95	4.88	4.82	4.77
	1%	16.3	13.3	12.1	11.4	11.0	10.7	10.5	10.3	10.2
6	5%	5.99	5.14	4.76	4.53	4.39	4.28	4.21	4.15	4.10
	1%	13.7	10.9	9.78	9.15	8.75	8.47	8.26	8.10	7.98
7	5%	5.59	4.74	4.35	4.12	3.97	3.87	3.79	3.73	3.68
	1%	12.2	9.55	8.45	7.85	7.46	7.19	6.99	6.84	6.72
8	5%	5.32	4.46	4.07	3.84	3.69	3.58	3.50	3.44	3.39
	1%	11.3	8.65	7.59	7.01	6.63	6.37	6.18	6.03	5.91
9	5%	5.12	4.26	3.86	3.63	3.48	3.37	3.29	3.23	3.18
	1%	10.6	8.02	6.99	6.42	6.06	5.80	5.61	5.47	5.35
10	5%	4.96	4.10	3.71	3.48	3.33	3.22	3.14	3.07	3.02
	1%	10.0	7.56	6.55	5.99	5.64	5.39	5.20	5.06	4.94
11	5%	4.84	3.98	3.59	3.36	3.20	3.09	3.01	2.95	2.90
	1%	9.65	7.21	6.22	5.67	5.32	5.07	4.89	4.74	4.63
12	5%	4.75	3.89	3.49	3.26	3.11	3.00	2.91	2.85	2.80
	1%	9.33	6.93	5.95	5.41	5.06	4.82	4.64	4.50	4.39
13	5%	4.67	3.81	3.41	3.18	3.03	2.92	2.83	2.77	2.71
	1%	9.07	6.70	5.74	5.21	4.86	4.62	4.44	4.30	4.19
14	5%	4.60	3.74	3.34	3.11	2.96	2.85	2.76	2.70	2.65
	1%	8.86	6.51	5.56	5.04	4.70	4.46	4.28	4.14	4.03

df Associated with the Denominator		*df* Associated with the Numerator								
		1	2	3	4	5	6	7	8	9
15	5%	4.54	3.68	3.29	3.06	2.90	2.79	2.71	2.64	2.59
	1%	8.68	6.36	5.42	4.89	4.56	4.32	4.14	4.00	3.89
16	5%	4.49	3.63	3.24	3.01	2.85	2.74	2.66	2.59	2.54
	1%	8.53	6.23	5.29	4.77	4.44	4.20	4.03	3.89	3.78
17	5%	4.45	3.59	3.20	2.96	2.81	2.70	2.61	2.55	2.49
	1%	8.40	6.11	5.18	4.67	4.34	4.10	3.93	3.79	3.68
18	5%	4.41	3.55	3.16	2.93	2.77	2.66	2.58	2.51	2.46
	1%	8.29	6.01	5.09	4.58	4.25	4.01	3.84	3.71	3.60
19	5%	4.38	3.52	3.13	2.90	2.74	2.63	2.54	2.48	2.42
	1%	8.18	5.93	5.01	4.50	4.17	3.94	3.77	3.63	3.52
20	5%	4.35	3.49	3.10	2.87	2.71	2.60	2.51	2.45	2.39
	1%	8.10	5.85	4.94	4.43	4.10	3.87	3.70	3.56	3.46
21	5%	4.32	3.47	3.07	2.84	2.68	2.57	2.49	2.42	2.37
	1%	8.02	5.78	4.87	4.37	4.04	3.81	3.64	3.51	3.40
22	5%	4.30	3.44	3.05	2.82	2.66	2.55	2.46	2.40	2.34
	1%	7.95	5.72	4.82	4.31	3.99	3.76	3.59	3.45	3.35
23	5%	4.28	3.42	3.03	2.80	2.64	2.53	2.44	2.37	2.32
	1%	7.88	5.66	4.76	4.26	3.94	3.71	3.54	3.41	3.30
24	5%	4.26	3.40	3.01	2.78	2.62	2.51	2.42	2.36	2.30
	1%	7.82	5.61	4.72	4.22	3.90	3.67	3.50	3.36	3.26
25	5%	4.24	3.39	2.99	2.76	2.60	2.49	2.40	2.34	2.28
	1%	7.77	5.57	4.68	4.18	3.86	3.63	3.46	3.32	3.22
26	5%	4.23	3.37	2.98	2.74	2.59	2.47	2.39	2.32	2.27
	1%	7.72	5.53	4.64	4.14	3.82	3.59	3.42	3.29	3.18
27	5%	4.21	3.35	2.96	2.73	2.57	2.46	2.37	2.31	2.25
	1%	7.68	5.49	4.60	4.11	3.78	3.56	3.39	3.26	3.15
28	5%	4.20	3.34	2.95	2.71	2.56	2.45	2.36	2.29	2.24
	1%	7.64	5.45	4.57	4.07	3.75	3.53	3.36	3.23	3.12
29	5%	4.18	3.33	2.93	2.70	2.55	2.43	2.35	2.28	2.22
	1%	7.60	5.42	4.54	4.04	3.73	3.50	3.33	3.20	3.09
30	5%	4.17	3.32	2.92	2.69	2.53	2.42	2.33	2.27	2.21
	1%	7.56	5.39	4.51	4.02	3.70	3.47	3.30	3.17	3.07

(continued)

TABLE F. (continued)

df Associated with the Denominator		*df* Associated with the Numerator								
		1	2	3	4	5	6	7	8	9
40	5%	4.08	3.23	2.84	2.61	2.45	2.34	2.25	2.18	2.12
	1%	7.31	5.18	4.31	3.83	3.51	3.29	3.12	2.99	2.89
60	5%	4.00	3.15	2.76	2.53	2.37	2.25	2.17	2.10	2.04
	1%	7.08	4.98	4.13	3.65	3.34	3.12	2.95	2.82	2.72
120	5%	3.92	3.07	2.68	2.45	2.29	2.18	2.09	2.02	1.96
	1%	6.85	4.79	3.95	3.48	3.17	2.96	2.79	2.66	2.56

TABLE G. **Values of Chi Square at the .05, .01, and .001 Levels of Significance**

df	.05	.01	.001	*df*	.05	.01	.001
1	3.84	6.64	10.83	16	26.30	32.00	39.29
2	5.99	9.21	13.82	17	27.59	33.41	40.75
3	7.82	11.34	16.27	18	28.87	34.80	42.31
4	9.49	13.28	18.46	19	30.14	36.19	43.82
5	11.07	15.09	20.52	20	31.41	37.57	45.32
6	12.59	16.81	22.46	21	32.67	38.93	46.80
7	14.07	18.48	24.32	22	33.92	40.29	48.27
8	15.51	20.09	26.12	23	35.17	41.64	49.73
9	16.92	21.67	27.88	24	36.42	42.98	51.18
10	18.31	23.21	29.59	25	37.65	44.31	52.62
11	19.68	24.72	31.26	26	38.88	45.64	54.05
12	21.03	26.22	32.91	27	40.11	46.96	55.48
13	22.36	27.69	34.53	28	41.34	48.28	56.89
14	23.68	29.14	36.12	29	42.56	49.59	58.30
15	25.00	30.58	37.70	30	43.77	50.89	59.70

From R. A. Fisher, *Statistical Methods for Research Workers*, 14th ed. (New York, 1970), pp. 112–13.

TABLE H. **Probabilities Associated with Values as Small as Observed Values of U in the Mann-Whitney U Test**

$N_2 = 3$

U \ N_1	1	2	3
0	.250	.100	.050
1	.500	.200	.100
2	.750	.400	.200
3		.600	.350
4			.500
5			.650

$N_2 = 4$

U \ N_1	1	2	3	4
0	.200	.067	.028	.014
1	.400	.133	.057	.029
2	.600	.267	.114	.057
3		.400	.200	.100
4		.600	.314	.171
5			.429	.243
6			.571	.343
7				.443
8				.557

$N_2 = 5$

U \ N_1	1	2	3	4	5
0	.167	.047	.018	.008	.004
1	.333	.095	.036	.016	.008
2	.500	.190	.071	.032	.016
3	.667	.286	.125	.056	.028
4		.429	.196	.095	.048
5		.571	.286	.143	.075
6			.393	.206	.111
7			.500	.278	.155
8			.607	.365	.210
9				.452	.274
10				.548	.345
11					.421
12					.500
13					.579

$N_2 = 6$

U \ N_1	1	2	3	4	5	6
0	.143	.036	.012	.005	.002	.001
1	.286	.071	.024	.010	.004	.002
2	.428	.143	.048	.019	.009	.004
3	.571	.214	.083	.033	.015	.008
4		.321	.131	.057	.026	.013
5		.429	.190	.086	.041	.021
6		.571	.274	.129	.063	.032
7			.357	.176	.089	.047
8			.452	.238	.123	.066
9			.548	.305	.165	.090
10				.381	.214	.120
11				.457	.268	.155
12				.545	.331	.197
13					.396	.242
14					.465	.294
15					.535	.350
16						.409
17						.469
18						.531

From H. B. Mann and D. R. Whitney, On a test whether one of two random variables is stochastically larger than the other, *Annals of Mathematical Statistics* 18: 50–60 (1947).

(*continued*)

TABLE H. (continued)

$N_2 = 7$

U \ N_1	1	2	3	4	5	6	7
0	.125	.028	.008	.003	.001	.001	.000
1	.250	.056	.017	.006	.003	.001	.001
2	.375	.111	.033	.012	.005	.002	.001
3	.500	.167	.058	.021	.009	.004	.002
4	.625	.250	.092	.036	.015	.007	.003
5		.333	.133	.055	.024	.011	.006
6		.444	.192	.082	.037	.017	.009
7		.556	.258	.115	.053	.026	.013
8			.333	.158	.074	.037	.019
9			.417	.206	.101	.051	.027
10			.500	.264	.134	.069	.036
11			.583	.324	.172	.090	.049
12				.394	.216	.117	.064
13				.464	.265	.147	.082
14				.538	.319	.183	.104
15					.378	.223	.130
16					.438	.267	.159
17					.500	.314	.191
18					.562	.365	.228
19						.418	.267
20						.473	.310
21						.527	.355
22							.402
23							.451
24							.500
25							.549

$N_2 = 8$

U \ N_1	1	2	3	4	5	6	7	8	t	Normal
0	.111	.022	.006	.002	.001	.000	.000	.000	3.308	.001
1	.222	.044	.012	.004	.002	.001	.000	.000	3.203	.001
2	.333	.089	.024	.008	.003	.001	.001	.000	3.098	.001
3	.444	.133	.042	.014	.005	.002	.001	.001	2.993	.001
4	.556	.200	.067	.024	.009	.004	.002	.001	2.888	.002
5		.267	.097	.036	.015	.006	.003	.001	2.783	.003
6		.356	.139	.055	.023	.010	.005	.002	2.678	.004
7		.444	.188	.077	.033	.015	.007	.003	2.573	.005
8		.556	.248	.107	.047	.021	.010	.005	2.468	.007
9			.315	.141	.064	.030	.014	.007	2.363	.009
10			.387	.184	.085	.041	.020	.010	2.258	.012
11			.461	.230	.111	.054	.027	.014	2.153	.016
12			.539	.285	.142	.071	.036	.019	2.048	.020
13				.341	.177	.091	.047	.025	1.943	.026
14				.404	.217	.114	.060	.032	1.838	.033
15				.467	.262	.141	.076	.041	1.733	.041
16				.533	.311	.172	.095	.052	1.628	.052
17					.362	.207	.116	.065	1.523	.064
18					.416	.245	.140	.080	1.418	.078
19					.472	.286	.168	.097	1.313	.094
20					.528	.331	.198	.117	1.208	.113
21						.377	.232	.139	1.102	.135
22						.426	.268	.164	.998	.159
23						.475	.306	.191	.893	.185
24						.525	.347	.221	.788	.215
25							.389	.253	.683	.247
26							.433	.287	.578	.282
27							.478	.323	.473	.318
28							.522	.360	.368	.356
29								.399	.263	.396
30								.439	.158	.437
31								.480	.052	.481
32								.520		

Critical Values of *U* for a One-Tailed Test at $\alpha = .001$ or for a Two-Tailed Test at $\alpha = .002$

N_1 \ N_2	9	10	11	12	13	14	15	16	17	18	19	20
1												
2												
3									0	0	0	0
4		0	0	0	1	1	1	2	2	3	3	3
5	1	1	2	2	3	3	4	5	5	6	7	7
6	2	3	4	4	5	6	7	8	9	10	11	12
7	3	5	6	7	8	9	10	11	13	14	15	16
8	5	6	8	9	11	12	14	15	17	18	20	21
9	7	8	10	12	14	15	17	19	21	23	25	26
10	8	10	12	14	17	19	21	23	25	27	29	32
11	10	12	15	17	20	22	24	27	29	32	34	37
12	12	14	17	20	23	25	28	31	34	37	40	42
13	14	17	20	23	26	29	32	35	38	42	45	48
14	15	19	22	25	29	32	36	39	43	46	50	54
15	17	21	24	28	32	36	40	43	47	51	55	59
16	19	23	27	31	35	39	43	48	52	56	60	65
17	21	25	29	34	38	43	47	52	57	61	66	70
18	23	27	32	37	42	46	51	56	61	66	71	76
19	25	29	34	40	45	50	55	60	66	71	77	82
20	26	32	37	42	48	54	59	65	70	76	82	88

Critical Values of U for a One-Tailed Test at $\alpha = .01$ or for a Two-Tailed Test at $\alpha = .02$

N_1 \ N_2	9	10	11	12	13	14	15	16	17	18	19	20
1												
2					0	0	0	0	0	0	1	1
3	1	1	1	2	2	2	3	3	4	4	4	5
4	3	3	4	5	5	6	7	7	8	9	9	10
5	5	6	7	8	9	10	11	12	13	14	15	16
6	7	8	9	11	12	13	15	16	18	19	20	22
7	9	11	12	14	16	17	19	21	23	24	26	28
8	11	13	15	17	20	22	24	26	28	30	32	34
9	14	16	18	21	23	26	28	31	33	36	38	40
10	16	19	22	24	27	30	33	36	38	41	44	47
11	18	22	25	28	31	34	37	41	44	47	50	53
12	21	24	28	31	35	38	42	46	49	53	56	60
13	23	27	31	35	39	43	47	51	55	59	63	67
14	26	30	34	38	43	47	51	56	60	65	69	73
15	28	33	37	42	47	51	56	61	66	70	75	80
16	31	36	41	46	51	56	61	66	71	76	82	87
17	33	38	44	49	55	60	66	71	77	82	88	93
18	36	41	47	53	59	65	70	76	82	88	94	100
19	38	44	50	56	63	69	75	82	88	94	101	107
20	40	47	53	60	67	73	80	87	93	100	107	114

(continued)

TABLE I. (continued)

Critical Values of U for a One-Tailed Test at $\alpha = .025$ or for a Two-Tailed Test at $\alpha = .05$

N_1 \ N_2	9	10	11	12	13	14	15	16	17	18	19	20
1												
2	0	0	0	1	1	1	1	1	2	2	2	2
3	2	3	3	4	4	5	5	6	6	7	7	8
4	4	5	6	7	8	9	10	11	11	12	13	13
5	7	8	9	11	12	13	14	15	17	18	19	20
6	10	11	13	14	16	17	19	21	22	24	25	27
7	12	14	16	18	20	22	24	26	28	30	32	34
8	15	17	19	22	24	26	29	31	34	36	38	41
9	17	20	23	26	28	31	34	37	39	42	45	48
10	20	23	26	29	33	36	39	42	45	48	52	55
11	23	26	30	33	37	40	44	47	51	55	58	62
12	26	29	33	37	41	45	49	53	57	61	65	69
13	28	33	37	41	45	50	54	59	63	67	72	76
14	31	36	40	45	50	55	59	64	67	74	78	83
15	34	39	44	49	54	59	64	70	75	80	85	90
16	37	42	47	53	59	64	70	75	81	86	92	98
17	39	45	51	57	63	67	75	81	87	93	99	105
18	42	48	55	61	67	74	80	86	93	99	106	112
19	45	52	58	65	72	78	85	92	99	106	113	119
20	48	55	62	69	76	83	90	98	105	112	119	127

Critical Values of U for a One-Tailed Test at $\alpha = .05$ or for a Two-Tailed Test at $\alpha = .10$

N_1 \ N_2	9	10	11	12	13	14	15	16	17	18	19	20
1											0	0
2	1	1	1	2	2	2	3	3	3	4	4	4
3	3	4	5	5	6	7	7	8	9	9	10	11
4	6	7	8	9	10	11	12	14	15	16	17	18
5	9	11	12	13	15	16	18	19	20	22	23	25
6	12	14	16	17	19	21	23	25	26	28	30	32
7	15	17	19	21	24	26	28	30	33	35	37	39
8	18	20	23	26	28	31	33	36	39	41	44	47
9	21	24	27	30	33	36	39	42	45	48	51	54
10	24	27	31	34	37	41	44	48	51	55	58	62
11	27	31	34	38	42	46	50	54	57	61	65	69
12	30	34	38	42	47	51	55	60	64	68	72	77
13	33	37	42	47	51	56	61	65	70	75	80	84
14	36	41	46	51	56	61	66	71	77	82	87	92
15	39	44	50	55	61	66	72	77	83	88	94	100
16	42	48	54	60	65	71	77	83	89	95	101	107
17	45	51	57	64	70	77	83	89	96	102	109	115
18	48	55	61	68	75	82	88	95	102	109	116	123
19	51	58	65	72	80	87	94	101	109	116	123	130
20	54	62	69	77	84	92	100	107	115	123	130	138

From D. Auble, Extended tables for the Mann–Whitney statistic, *Bulletin of the Institute of Educational Research at Indiana University*, Vol. 1, No. 2(1953).

TABLE J. **Probabilities Associated with Values as Small as Observed Values of x in the Binomial Test**

N \ x	0	1	2	3	4	5	6	7	8	9	10	11	12	13	14	15
5	031	188	500	812	969	†										
6	016	109	344	656	891	984	†									
7	008	062	227	500	773	938	992	†								
8	004	035	145	363	637	855	965	996	†							
9	002	020	090	254	500	746	910	980	998	†						
10	001	011	055	172	377	623	828	945	989	999	†					
11		006	033	113	274	500	726	887	967	994	†	†				
12		003	019	073	194	387	613	806	927	981	997	†	†			
13		002	011	046	133	291	500	709	867	954	989	998	†	†		
14		001	006	029	090	212	395	605	788	910	971	994	999	†	†	
15			004	018	059	151	304	500	696	849	941	982	996	†	†	†
16			002	011	038	105	227	402	598	773	895	962	989	998	†	†
17			001	006	025	072	166	315	500	685	834	928	975	994	999	†
18			001	004	015	048	119	240	407	593	760	881	952	985	996	999
19				002	010	032	084	180	324	500	676	820	916	968	990	998
20				001	006	021	058	132	252	412	588	748	868	942	979	994
21				001	004	013	039	095	192	332	500	668	808	905	961	987
22					002	008	026	067	143	262	416	584	738	857	933	974
23					001	005	017	047	105	202	339	500	661	798	895	953
24					001	003	011	032	076	154	271	419	581	729	846	924
25						002	007	022	054	115	212	345	500	655	788	885

†p = 1.0 or approximately 1.0.

TABLE K. **Critical Values of *T* in the Wilcoxon Matched-Pairs Signed-Ranks Test**

	Level of Significance for one-tailed test		
	.025	.01	.005
	Level of Significance for two-tailed test		
N	**.05**	**.02**	**.01**
6	1	—	—
7	2	0	—
8	4	2	0
9	6	3	2
10	8	5	3
11	11	7	5
12	14	10	7
13	17	13	10
14	21	16	13
15	25	20	16
16	30	24	19
17	35	28	23
18	40	33	28
19	46	38	32
20	52	43	37
21	59	49	43
22	66	56	49
23	73	62	55
24	81	69	61
25	90	77	68

From F. Wilcoxon and R. Wilcox, *Some Rapid Approximate Statistical Procedures* (New York, 1964), p. 28. Copyright 1964 by the American Cyanamid Co. Reproduced with the permission of the American Cyanamid Company.

TABLE L. **Random Numbers**

Col. 1	Col. 2	Col. 3	Col. 4	Col. 5	Col. 6	Col. 7	Col. 8
3831	7167	1540	1532	6617	1845	3162	0210
6019	4242	1818	4978	8200	7326	5442	7766
6653	7210	0718	2183	0737	4603	2094	1964
8861	5020	6590	5990	3425	9208	5973	9614
9221	6305	6091	8875	6693	8017	8953	5477
2809	9700	8832	0248	3593	4686	9645	3899
1207	0100	3553	8260	7332	7402	9152	5419
6012	3752	2074	7321	5964	7095	2855	6123
0300	0773	5128	0694	3572	5517	3689	7220
1382	2179	5685	9705	9919	1739	0356	7173
0678	7668	4425	6205	4158	6769	7253	8106
8966	0561	9341	8986	8866	2168	7951	9721
6293	3420	9752	9956	7191	1127	7783	2596
9097	7558	1814	0782	0310	7310	5951	8147
3362	3045	6361	4024	1875	4124	7396	3985
5594	1248	2685	1039	0129	5047	6267	0440
6495	8204	9251	1947	9485	3027	9946	7792
9378	0804	7233	2355	1278	8667	5810	8869
2932	4490	0680	8024	4378	9543	4594	8392
2868	7746	1213	0396	9902	4953	2261	8117
3047	6737	5434	9719	8026	9283	6952	1883
3673	2265	5271	4542	2646	1744	2684	4956
0731	8278	9597	0745	9682	8007	7836	2771
2666	3174	0706	6224	4595	2273	0802	9402
5879	3349	9239	2808	8626	8569	6660	9683
7228	8029	3633	6194	9030	1279	2611	3805
4367	2881	3996	8336	7933	6385	5902	1664
1014	9964	1346	4850	1524	1919	7355	4737
6316	4356	7927	6709	1375	0375	8855	3632
2302	6392	5023	8515	1197	9182	4952	1897
7439	5567	1156	9241	0438	0607	1962	0717
1930	7128	6098	6033	5132	5350	1216	0518
4598	6415	1523	4012	8179	9934	8863	8375
2835	5888	8616	7542	5875	2859	6805	4079
4377	5153	9930	0902	8208	6501	9593	1397
3725	7202	6551	7458	4740	8234	4914	0878

Col. 1	Col. 2	Col. 3	Col. 4	Col. 5	Col. 6	Col. 7	Col. 8
7868	7546	5714	9450	6603	3709	7328	2835
2168	2879	8000	8755	5496	3532	5173	4289
1366	5878	6631	3799	2607	0769	8119	7064
7840	6116	6088	5362	7583	6246	9297	9178
1208	7567	2984	1555	5633	2676	8668	9281
5492	1044	2380	1283	4244	2667	5864	5325
1049	9457	3807	8877	6857	6915	6852	2399
7834	8324	6028	6356	2771	1686	1840	3035
5907	6128	9673	4251	0986	3668	1215	2385
3405	6830	2171	9447	4347	6948	2083	0697
1785	4670	1154	2567	8965	3903	4669	4275
6180	3600	8393	5019	1457	2970	9582	1658
4614	8527	8738	5658	4017	0815	0851	7215
6465	6832	7586	3595	9421	9498	8576	4256
0573	7976	3362	1807	2929	0540	8721	3133
7672	3912	8047	0966	6692	4444	7690	8525
9182	1221	2215	0590	4784	5374	7429	5422
2118	5264	7144	8413	4137	6178	8670	4120
0478	5077	0991	3657	9242	5710	2758	0574
3386	1570	5143	4332	2599	4330	4999	8978
2053	4196	1585	4340	1955	6312	7903	8253
0483	3044	4609	4046	4614	4566	7906	0892
3825	9228	2706	8574	0959	6456	7232	5838
3426	9307	7283	9370	5441	9659	6478	1734
8365	9252	5198	2453	7514	5498	7105	0549
7915	3351	8381	2137	9695	0358	5163	1556
7521	7744	2379	2325	3585	9370	4879	6545
1262	0960	5816	3485	8498	5860	5188	3178
9110	8181	0097	3823	6955	1123	6794	5076
9979	5039	0025	8060	2668	0157	5578	0243
2312	2169	5977	8067	2782	7690	4146	6110
3960	1408	3399	4940	3088	7546	1170	6054
5227	6451	4868	0977	5735	0359	7805	8250
2599	3800	9245	6545	6181	7300	2348	4378
9583	3746	4175	0143	3279	0809	7367	2923
8740	4326	1105	0498	3910	2074	3623	9890
6541	2753	2423	4282	2195	1471	0852	6604
1237	2419	4572	3829	1274	9378	2393	4028
7397	4135	8132	3143	3638	0515	1133	9975

(*continued*)

TABLE L. (continued)

Col. 1	Col. 2	Col. 3	Col. 4	Col. 5	Col. 6	Col. 7	Col. 8
9105	3396	9469	0966	6128	3808	7073	7779
3348	5436	1171	5853	2392	7643	2011	0538
7792	4714	5799	1211	0409	5036	7879	6173
7523	0348	5237	2533	0635	2382	5092	3497
2674	2435	5979	7697	3260	2939	2511	7318
6825	3660	2688	9560	1329	4268	2532	5024
0639	6884	8337	5308	2054	3454	8745	1877
2467	2505	4916	1683	0034	7758	4458	9918
9513	2949	9337	7234	8458	3329	9691	4278
9116	6846	0205	1158	6112	9916	0723	3769
4012	3863	4817	6294	7865	1672	0137	6557
7698	0651	9756	1816	1154	6708	2522	8296
7158	8463	6406	0779	1185	7660	3065	8941
8412	5905	5612	7028	2545	2392	8434	1551
3134	3962	3147	9631	2881	3091	4678	4465
5840	1940	0754	0457	9533	0108	4523	8441
3237	4236	5504	3282	2838	5002	6614	2463
1990	9392	4943	9505	4925	8313	3108	7681
6724	8147	1557	1342	3352	4421	3707	2445
6521	8766	0654	2300	1696	0145	3257	3496
1888	6629	5385	8725	7185	6826	2279	5200
5567	1138	7139	8157	4906	2872	8842	0890
4511	3021	7370	0264	2690	6187	9110	0941
2188	3642	8905	8172	3930	0152	6931	4340
4086	8745	0988	4815	6192	9608	8686	7459
6817	9456	9157	3036	4769	9362	0074	0837
2914	8776	4833	3214	7643	4345	3304	6137
9122	4766	1599	5271	2257	8502	9560	2833
3558	1472	7664	7256	7181	0088	2257	2503
1928	8097	3520	2187	5124	7295	2525	1891
8032	1390	6606	7195	2724	7239	3888	5582
1846	9648	8699	9716	7752	9886	6299	9129
8691	5849	1005	6629	1632	1463	9288	8600
1884	3228	6397	1733	9543	9868	3611	4828
8211	8273	3941	1484	2627	8257	8493	6354
4070	3899	3121	6736	0668	0782	1398	7729
4463	5758	3905	1545	4699	4338	1235	9547
9961	4716	1687	2448	0815	3022	1220	4055
0420	8921	1593	4599	3401	7209	7877	6001

Col. 1	Col. 2	Col. 3	Col. 4	Col. 5	Col. 6	Col. 7	Col. 8
7927	6608	5190	9268	8431	0324	6619	6159
4007	1367	5975	8972	6629	1259	7204	6556
9515	5611	3025	2016	9209	0290	6236	7360
6670	0458	2062	7235	6818	7619	8698	0110
7485	8847	7234	9278	9453	4900	9119	9216
9177	4212	3238	2358	1109	9441	7591	3901

From B. G. Andreas, *Experimental Psychology* (New York, 1972), pp. 579–82. Copyright 1972 by John Wiley and Sons, Inc. Reprinted by permission of the publisher.

TABLE M. Critical Values of the Studentized Range Statistic, q

df for denominator	k (Number of Means)									
	α	2	3	4	5	6	7	8	9	10
1	.05	18.0	27.0	32.8	37.1	40.4	43.1	45.4	47.4	49.1
	.01	90.0	135	164	186	202	216	227	237	246
2	.05	6.09	8.3	9.8	10.9	11.7	12.4	13.0	13.5	14.0
	.01	14.0	19.0	22.3	24.7	26.6	28.2	29.5	30.7	31.7
3	.05	4.50	5.91	6.82	7.50	8.04	8.48	8.85	9.18	9.46
	.01	8.26	10.6	12.2	13.3	14.2	15.0	15.6	16.2	16.7
4	.05	3.93	5.04	5.76	6.29	6.71	7.05	7.35	7.60	7.83
	.01	6.51	8.12	9.17	9.96	10.6	11.1	11.5	11.9	12.3
5	.05	3.64	4.60	5.22	5.67	6.03	6.33	6.58	6.80	6.99
	.01	5.70	6.97	7.80	8.42	8.91	9.32	9.67	9.97	10.2
6	.05	3.46	4.34	4.90	5.31	5.63	5.89	6.12	6.32	6.49
	.01	5.24	6.33	7.03	7.56	7.97	8.32	8.61	8.87	9.10
7	.05	3.34	4.16	4.69	5.06	5.36	5.61	5.82	6.00	6.16
	.01	4.95	5.92	6.54	7.01	7.37	7.68	7.94	8.17	8.37
8	.05	3.26	4.04	4.53	4.89	5.17	5.40	5.60	5.77	5.92
	.01	4.74	5.63	6.20	6.63	6.96	7.24	7.47	7.68	7.78
9	.05	3.20	3.95	4.42	4.76	5.02	5.24	5.43	5.60	5.74
	.01	4.60	5.43	5.96	6.35	6.66	6.91	7.13	7.32	7.49
10	.05	3.15	3.88	4.33	4.65	4.91	5.12	5.30	5.46	5.60
	.01	4.48	5.27	5.77	6.14	6.43	6.67	6.87	7.05	7.21

(continued)

TABLE M. (continued)

df for denominator	k (Number of Means)									
	α	2	3	4	5	6	7	8	9	10
11	.05	3.11	3.82	4.26	4.57	4.82	5.03	5.20	5.35	5.49
	.01	4.39	5.14	5.62	5.97	6.25	6.48	6.67	6.84	6.99
12	.05	3.08	3.77	4.20	4.51	4.75	4.95	5.12	5.27	5.40
	.01	4.32	5.04	5.50	5.84	6.10	6.32	6.51	6.67	6.81
13	.05	3.06	3.73	4.15	4.45	4.69	4.88	5.05	5.19	5.32
	.01	4.26	4.96	5.40	5.73	5.98	6.19	6.37	6.53	6.67
14	.05	3.03	3.70	4.11	4.41	4.64	4.83	4.99	5.13	5.25
	.01	4.21	4.89	5.32	5.63	5.88	6.08	6.26	6.41	6.54
16	.05	3.00	3.65	4.05	4.33	4.56	4.74	4.90	5.03	5.15
	.01	4.13	4.78	5.19	5.49	5.72	5.92	6.08	6.22	6.35
18	.05	2.97	3.61	4.00	4.28	4.49	4.67	4.82	4.96	5.07
	.01	4.07	4.70	5.09	5.38	5.60	5.79	5.94	6.08	6.20
20	.05	2.95	3.58	3.96	4.23	4.45	4.62	4.77	4.90	5.01
	.01	4.02	4.64	5.02	5.29	5.51	5.69	5.84	5.97	6.09
24	.05	2.92	3.53	3.90	4.17	4.37	4.54	4.68	4.81	4.92
	.01	3.96	4.54	4.91	5.17	5.37	5.54	5.69	5.81	5.92
30	.05	2.89	3.49	3.84	4.10	4.30	4.46	4.60	4.72	4.83
	.01	3.89	4.45	4.80	5.05	5.24	5.40	5.54	5.56	5.76
40	.05	2.86	3.44	3.79	4.04	4.23	4.39	4.52	4.63	4.74
	.01	3.82	4.37	4.70	4.93	5.11	5.27	5.39	5.50	5.60
60	.05	2.83	3.40	3.74	3.98	4.16	4.31	4.44	4.55	4.65
	.01	3.76	4.28	4.60	4.82	4.99	5.13	5.25	5.36	5.45
120	.05	2.80	3.36	3.69	3.92	4.10	4.24	4.36	4.48	4.56
	.01	3.70	4.20	4.50	4.71	4.87	5.01	5.12	5.21	5.30
∞	.05	2.77	3.31	3.63	3.86	4.03	4.17	4.29	4.39	4.47
	.01	3.64	4.12	4.40	4.60	4.76	4.88	4.99	5.08	5.16

Adapted from H. Leon Harter, Tables of range and studentized range, *Annals of Mathematical Statistics 31*: 1122–47 (1960).

Answers to Selected Exercises

Chapter 1

1. (a) R; (b) N; (c) R; (d) I; (e) O; (f) I; (g) O; (h) N
2. (a) N; (b) I; (c) R; (d) N; (e) I; (f) O; (g) R
3. (a) $\sum_{i=1}^{4} X_i$ (b) $\sum_{i=1}^{3} X_i^2$ (c) $\sum_{i=3}^{6} X_i$ (d) $\sum_{i=1}^{N} X_i$ or ΣX
4. (a) $\sum_{i=7}^{15} X_i$ (b) $\sum_{i=1}^{8} X_i$ (c) $\sum_{i=1}^{N} X_i$ or ΣX (d) $\sum_{i=1}^{N} X_i^2$ or ΣX^2
5. (a) Add all scores from the second through the twentieth.
   (b) Add the fifth, sixth, and seventh scores.
   (c) Square each value of Y and add all the squared values.
   (d) Add all the scores in the X column.

Chapter 2

1. (a) 49.5–99.5, 74.5, 50
   (b) 0.5–2.5, 1.5, 2
   (c) 1.95–2.45, 2.2, 0.5
   (d) 60.5–63.5, 62, 3
   (e) 19.45–19.95; 19.7, 0.5
   (f) 79.5–89.5, 84.5, 10
   (g) 99.5–124.5, 112, 25
   (h) 49.5–79.5, 64.5, 30
   (i) 24.5–29.5, 27, 5
   (j) .015–.045, 0.03, 0.03

2. (a) 119.5–129.5, 124.5, 10
(b) 48.5–51.5, 50, 3
(c) 2.45–2.95, 2.7, 0.5
(d) 24.5–49.5, 37, 25
(e) .055–.085, 0.07, 0.03
(f) 99.5–149.5, 124.5, 50
(g) 64.5–69.5, 67, 5
(h) 69.5–99.5, 84.5, 30
(i) 16.95–17.45, 17.2, 0.5
(j) 2.5–4.5, 3.5, 2

3. (a) D; (b) C; (c) C; (d) D; (e) D; (f) C; (g) C; (h) D; (i) D; (j) C

4. (a) D; (b) C; (c) D; (d) C; (e) C; (f) D; (g) C; (h) C; (i) D; (j) D

15. (a) Positive; (b) normal; (c) negative; (d) negative; (e) positive; (f) positive

Chapter 3

1. $\overline{X} = 25.5$
2. $\overline{X} = 61.3$
3. $\overline{X} = 18.0$, *Med* = 18.0
4. $\overline{X} = 10.5$, *Med* = 10.3
5. $\overline{X} = 34{,}761.72$
6. $\overline{X} = 39{,}227.00$
7. $\overline{X} = 45.97$, *Med* = 47.13
8. $\overline{X} = 5.43$, *Med* = 5.75
9. $\overline{X} = 5.97$, *Med* = 6.68, negative skewness
10. $\overline{X} = 4.70$, *Med* = 5.0
11. $\overline{X} = 3.2553$
12. $\overline{X} = 2.8975$
13. $\overline{X} = 12.2$, *Med* = 13.0
14. $\overline{X} = 266$

Chapter 4

1. 11 years, 2 months; 8th month of 5th grade
2. Third; 27

3. $PR = 44$; $PR = 23$
5. 20
6. 15
9. (a) 69.5; (b) 62; (c) 60
10. (a) 47; (b) 93; (c) 85
11. (a) 17.6; (b) 15; (c) 6
12. (a) 7.7; (b) 12; (c) 9
13. (a) 41 and 46.5; (b) 54%; (c) 69%
14. (a) 34.5 and 39.5; (c) $GMS = 77$, $NP = 61$

Chapter 5

1. $S = 2.39$ for all three methods
2. $S = 2.36$ for all three methods
3. Range = 401; $S = 124.45$
4. Range = 1.11; $S = 0.2792$
5. $\overline{X} = 18$; $S = 5.62$
6. $\overline{X} = 10.5$; $S = 3.48$
7. $\overline{X} = 25.5$; $S = 2.36$
8. $\overline{X} = 61.3$; $S = 6.67$
11. $z = .95, .72, .92, .88$, respectively
12. $z = 1.13, .74, .78, .56$, respectively
13. $z = 1.48, -1.20, .59$, respectively
14. $z = -.53, -1.47, 1.50$, respectively
15. Approximately 5.2 to 6.5 for Exercise 5 and 3.0 to 3.75 for Exercise 6.
16. Approximately 2.0 to 2.5 for Exercise 7 and 4.4 to 5.5. for Exercise 8.

Chapter 6

1. (a) 41.15; (b) 27.34; (c) 91.32; (d) 95.00; (e) 2.50; (f) 0.49; (g) 92.65; (h) 97.50
2. (a) 45.82; (b) 47.50; (c) 78.15; (d) 99.02; (e) 96.41; (f) 2.50; (g) 84.13; (h) 1.22
3. (a) 1.58; (b) 9.69; (c) 203.86 pounds; (d) 24.22; (e) 112.06 to 195.94 pounds; (f) 181.39 pounds

4. (a) 5.37; (b) 0.78; (c) 60.83 inches; (d) 20.90;
(e) 57.60 to 70.40 inches; (f) 69.08 inches

5. (a) 2.28; (b) 93.32; (c) 68.8; (d) 41.6

6. (a) 4.75 or 5%; (b) 18.14; (c) 199.77 or 200 mg/dl; (d) 182.28.

7. (a) .20; (b) .32; (c) 32 to 68, 68 to 32

8. (a) .18; (b) .18; (d) 18 to 82, 82 to 18

9. (a) .0158; 2 to 98; 98 to 2
(b) .0969; 10 to 90; 90 to 10

10. (a) .0537; 5 to 95; 95 to 5
(b) .0078; 1 to 99; 99 to 1

11. (a) .0228; (b) .9332

12. (a) .0475; (b) .1814

Chapter 7

3. 5%

4. .05

5. 20.46 to 22.74

6. 16.30 to 18.30

7. 193.77 to 208.23

9. 99.51 to 108.49

11. $S^2 = 31.53$; $s^2 = 33.5$

12. $S^2 = 5.58$; $s^2 = 6.09$

Chapter 8

1. (a) +; (b) 0; (c) +; (d) −; (e) +; (f) +; (g) −

2. (a) −; (b) +, then −; (c) +; (d) 0; (e) −; (f) +; (g) +

3. $r = .43$

4. (a) $r = .17$; (b) $r = .15$

5. $r = .53$

6. $r = .74$

7. $r = .71$

8. $r = .93$

9. $r = .87$

10. $r = .76$

13. $r_s = .94$

14. $r_s = -.73$

15. $r_s = .49$

16. $r_s = .78$

Chapter 9

1. 3; 5

2. −2; 10

3. (a) $Y' = 1.13X - 7.25$; (b) 60.55 or 60,550 plants

4. (a) $Y' = .0198X + 2.7217$; (b) 3.12

7. (b) $Y' = .91X + 7.11$; (c) 79.91; (d) $s_E = 3.55$, 76.23 to 83.59

8. (b) $Y' = .32X + 4.11$; (c) 7.31; (d) $s_E = 2.15$, 5.00 to 9.62

9. (a) $\overline{X} = 39.225$, $S_X = 4.76$, $\overline{Y} = 8.675$, $S_Y = 2.21$; (c) $Y' = .34X - 4.80$; (d) 5.4; (e) $s_E = 1.49$, 3.82 to 6.98

10. (a) $\overline{X} = 10.7$, $S_X = 4.03$, $\overline{Y} = 7.8$, $S_Y = 2.16$; (b) $r = .78$; (d) $Y' = .42X + 3.33$; (e) 9.63; (f) $s_E = 1.35$, 8.21 to 11.05

Chapter 10

3. $t = 5.75$, $p < .001$

4. $t = 1.92$, $p > .05$

5. $t = 3.16$, $p < .05$

6. $t = 3.25$, $p < .01$

7. $t = 2.49$, $p < .05$

8. $t = 1.29$, $p > .05$

9. $t = 2.87$, $p < .05$

10. $t = -2.92$, $p < .05$

Chapter 11

1. $SS_T = 92$; $SS_{BG} = 66$; $SS_{WG} = 26$

2. $F = 8.89$, $p < .05$

4. $SS_T = 126$; $SS_{BG} = 70$; $SS_{WG} = 56$

5. $F = 7.49$, $p < .01$

7. 3.55

8. 2.96

9. 3.61

10. 3.90

11. $F = 9.78$, $p < .01$; $\overline{X}_1$ and $\overline{X}_3$: $q = 5.78$, $p < .01$; $\overline{X}_2$ and $\overline{X}_3$: $q = 4.91$, $p < .01$; $\overline{X}_1$ and $\overline{X}_2$: $q = 0.87$, $p > .05$

12. $F = 4.76$, $p < .05$. Although F is significant at the .05 level, the largest q in Tukey's procedure did not quite reach significance at the .05 level.

13. $F = 4.22$, $p < .05$; $\overline{X}_1$ and $\overline{X}_2$: $q = 4.21$, $p < .05$; $\overline{X}_1$ and $\overline{X}_3$: $q = 2.64$, $p > .05$; $\overline{X}_2$ and $\overline{X}_3$: $q = 1.44$, $p > .05$.

14. $F = 6.00$, $p < .01$; $\overline{X}_1$ and $\overline{X}_2$: $q = 4.38$, $p < .05$; $\overline{X}_1$ and $\overline{X}_3$: $q = 4.11$, $p < .05$; $\overline{X}_2$ and $\overline{X}_3$: $q = 0.27$, $p > .05$.

15. $F = 11.11$, $p < .01$; $\overline{X}_1$ and $\overline{X}_2$: $q = 6.62$, $p < .01$; $\overline{X}_1$ and $\overline{X}_3$: $q = 2.34$, $p > .05$; $\overline{X}_2$ and $\overline{X}_3$: $q = 3.77$, $p < .05$.

Chapter 12

3.

Source	*df*	*SS*	*MS*	*F*
Image quality	1	90	90	25.86, $p < .01$
Stress	1	78.4	78.4	22.53, $p < .01$
Interaction	1	10	10	2.87, $p > .05$
Within groups	36	125.2	3.48	

4.

Source	*df*	*SS*	*MS*	*F*
Sex of stationary person	1	10	10	5.49, $p < .05$
Sex of approaching person	1	0.4	0.4	<1, $p > .05$
Interaction	1	14.4	14.4	7.93, $p < .01$
Within groups	36	65.6	1.82	

5.

Source	*df*	*SS*	*MS*	*F*
Sex typing	1	173.4	173.4	39.50, $p < .01$
Sex	1	9.6	9.6	2.19, $p > .05$
Interaction	1	38.4	38.4	8.75, $p < .01$
Within groups	56	245.6	4.39	

6. Source	df	SS	MS	F
Grade	1	722.5	722.5	16.94, $p < .01$
Code	1	202.5	202.5	4.75, $p < .05$
Interaction	1	10	10	<1, $p > .05$
Within groups	36	1,535	42.64	

7. Source	df	SS	MS	F
Dieting	1	2,048	2,048	3.96, $p > .05$
Preloading	1	5,832	5,832	11.29, $p < .01$
Interaction	1	9,248	9,248	17.90, $p < .01$
Within groups	28	14,468	516.71	

8. Source	df	SS	MS	F
Locus of control	1	15.63	15.63	5.88, $p < .01$
Information	1	1.23	1.23	<1, $p > .05$
Interaction	1	38.02	38.02	14.29, $p < .01$
Within groups	36	95.9	2.66	

Chapter 13

1. $\chi^2 = 8.58, p < .01$
2. $\chi^2 = 10.72, p < .01$
3. $\chi^2 = 97.60, p < .001$
4. $\chi^2 = 10.01, p < .01$
5. $\chi^2 = 5.58, p < .05$
6. $\chi^2 = 3.38, p > .05$
7. $U = 34, p > .05$
8. $U = 1, p < .001$, two-tailed
9. $H = 8.43, p < .05$
10. $H = 7.08, p < .05$
11. $x = 2, p = .033$
12. $x = 4, p = .05$, two-tailed
13. $T = 46, p > .05$

14. $T = 9$, $p < .01$, two-tailed

15. $\chi_r^2 = 2.11$, $p > .05$

16. $\chi_r^2 = 11.1$, $p < .01$

Chapter 14

1. $r_{oe} = .85$; $r_{tt} = .92$

2. $S_{EM} = 5.05$; 23.95 to 34.05

3. $r_{oe} = .76$; $r_{tt} = .86$

4. $S_{EM} = 2.94$; 38.06 to 43.94

5. $r_{oe} = .84$; $r_{tt} = .91$

6. $S_{EM} = 1.65$; 29.35 to 32.65

7. $r = .15$

8. $r = .44$

9. $r_{oe} = .73$; $r_{tt} = .84$

10. $S_{EM} = 2.81$; 63.19 to 68.81

11. $r = .11$

Index